Principles of Plant and Animal Taxonomy

Principles of Plant and Animal Taxonomy

Edited by **Austin Balfour and Dominic Fasso**

SYRAWOOD
PUBLISHING HOUSE
New York

Published by Syrawood Publishing House,
750 Third Avenue, 9th Floor,
New York, NY 10017, USA
www.syrawoodpublishinghouse.com

Principles of Plant and Animal Taxonomy
Edited by Austin Balfour and Dominic Fasso

International Standard Book Number: 978-1-68286-265-0 (Hardback)

Printed in China.

Contents

Preface

Taxonomy is a science that specifically pertains to grouping and nomenclature of organisms. It plays a crucial role in understanding biodiversity, its management and conservation. This book provides significant information to the reader about latest developments of the field as well as important concepts of taxonomy, like alpha taxonomy, beta taxonomy, microtaxonomy, macrotaxonomy, etc. It is an important tool for the students of taxonomy, biology, zoology and associated fields of study. This text will also provide various innovative topics for research which interested readers can take up further.

This book unites the global concepts and researches in an organized manner for a comprehensive understanding of the subject. It is a ripe text for all researchers, students, scientists or anyone else who is interested in acquiring a better knowledge of this dynamic field.

I extend my sincere thanks to the contributors for such eloquent research chapters. Finally, I thank my family for being a source of support and help.

Editor

1

New Baltic amber leafhoppers representing the oldest Aphrodinae and Megophthalminae (Hemiptera, Cicadellidae)

Christopher H. DIETRICH & Ana Clara GONÇALVES

Illinois Natural History Survey, Prairie Research Institute, University of Illinois,
1816 S. Oak St., Champaign, IL 61820

Abstract. Three fossil leafhopper inclusions from Eocene Baltic amber, representing three new extinct genera and species, are described and illustrated. *Eomegophthalmus lithuaniensis* gen. et sp. nov. is tentatively placed in Megophthalminae, although it may represent the stem group from which Megophthalminae, Ulopinae, and Membracidae arose. *Xestocephalites balticus* gen. et sp. nov. and *Brevaphrodella nigra* gen. et sp. nov. are placed in Aphrodinae: Xestocephalini based on the structure of the head, leg chaetotaxy, and male genital capsule. These new genera and species represent the oldest known representatives of their respective subfamilies and the latter is the oldest known brachypterous adult leafhopper.

Keywords. Morphology, fossil, evolution, phylogeny, Paleogene.

Introduction

The fossil record of leafhoppers (Cicadellidae), as reported in the literature, is very sparse, consisting of a few reports of rock fossils from the lower Cretaceous of Brazil, Eurasia and Australia (Shcherbakov 1986; Hamilton 1990, 1992), a single wing impression from the upper Cretaceous of western North America (Oman 1937), several genera and species from Eocene Baltic amber (Keilbach 1982; Spahr 1988, Szwedo 2002) and several genera and species from Oligo-Miocene Dominican amber (Dietrich & Vega 1995). Several fossil leafhoppers from Miocene deposits in Europe and North America have also been described (Metcalf & Wade 1966). Known Cretaceous leafhopper fossils are mostly either too poorly preserved to facilitate placing them to subfamily or, with the exception of two Brazilian fossils placed in Ledrinae (Shcherbakov 1992), have generalized morphology without distinctive features that would allow them to be placed with certainty in modern subfamilies. In contrast, representatives of several extant subfamilies have been reported from Baltic and Dominican amber. The cicadellid subfamilies (*sensu* Dietrich 2005) reported thus far from Baltic amber (reviewed by Szwedo 2002) are Bathysmatophorinae, Macropsinae, Mileewinae, Ledrinae, and Typhlocybinae (*sensu* Dietrich 2005, 2011; Wei *et al.* 2010). An extinct subfamily, Nastlopiinae, description based on a first instar nymph (Szwedo & Gebicki 2002), needs more study to determine its status and relationship to other groups. Oligo-Miocene Dominican amber has yielded representatives of Aphrodinae (Xestocephalini), Cicadellinae (Cicadellini), Deltocephalinae (Athysanini), Evacanthinae (Nirvanini), Iassinae (Krisnini), Megophthalminae (Agalliini), Neocoelidiinae (Krocodonini) and Typhlocybinae (Dikraneurini; Dietrich & Vega 1995 and unpublished observations). While Dominican amber fossils are mostly referable to modern genera, most of the fossil leafhoppers present in the older (Eocene) Baltic amber appear to belong to extinct genera and, in some cases, cannot be assigned with confidence to modern tribes. This suggests that much tribe- and genus-level diversification in leafhoppers occurred during the Paleogene. Thus, Eocene Baltic amber documents an important stage in the evolution of the major lineages of Cicadellidae.

Germar & Berendt (1856) were the first authors to report on Cicadellidae from Baltic amber, describing seven species and placing two of these in *Typhlocyba* Germar, 1833, two in *Jassus* Fabricius, 1803, two in *Tettigonia* Fabricius, 1775, and one in *Bythoscopus* Germar, 1833. Unfortunately, the type material for these species was apparently destroyed during World War II and the original descriptions and illustrations are not sufficiently detailed to allow for their proper placement although, as noted by Szwedo (2002), the two species of *Typhlocyba* are probably correctly placed in subfamily Typhlocybinae, and *Bythoscopus* may be placed with reasonable confidence in Macropsini, but they cannot be placed to genus based on the information available at present. The original illustrations of *Tettigonia proavia* Germar-Berendt, 1856 and *T. terebrans* Germar-Berendt, 1856 (Germar & Berendt 1856) indicate that they belong to Bathysmatophorinae. *Tettigonia terebrans* appears to be similar to *Ambericarda skalskii* Szwedo & Gebicki, 1998, so far the only other bathysmatophorine formally described from Baltic amber. Germar & Berendt mentioned that their two "*Jassus*" species were somewhat similar to modern European species of Deltocephalinae (*Speudotettix* Ribaut, 1942 and *Thamnotettix* Zetterstedt, 1840), but they did not mention or illustrate characters that would unequivocally place these species in that subfamily. Their drawings and descriptions of these two taxa are consistent with Deltocephalinae, but they could also apply to other subfamilies; thus these two "*Jassus*" species should be treated as *species incertae sedis*.

Based on the original descriptions and illustrations, additional cicadellid species described from Baltic amber by Bervoets (1910) may be tentatively placed as follows: *Bythoscopus punctatus* Bervoets, 1910 and *Pediopsis minuta* Bervoets, 1910 belong in tribe Macropsini; *Eupteryx minuta* Bervoets, 1910 belongs in subfamily Typhlocybinae (tribal placement uncertain); *Acocephalus resinosus* Bervoets, 1910 may belong in Bathysmatophorini, but the original description and illustration are not sufficient to confirm this placement.

More recently described cicadellids from Baltic amber (Szwedo & Gebicki 1999; Gebicki & Szwedo 2001) were placed in Ledrinae and Mileewinae, respectively. Although the subfamily placements of these fossils appear to be correct, the placement of the two ledrine nymphs in the modern genus *Camptelasmus* Spinola, 1850 is questionable, given that the type material of the type species of this genus is lost, and nymphs of modern representatives have not been previously described (Jones & Deitz 2009). Jones & Deitz (2009) excluded *Camptelasmus* from Ledrinae and considered the genus to be of uncertain position within Cicadellidae. The nymphs described and illustrated by Szwedo and Gebicki (1999) do not resemble those of other known members of Ledrini (*sensu* Jones & Deitz 2009, = Petalocephalini), which are much more strongly flattened with lamelliform extensions of the tergites but, nevertheless, are consistent with Ledrinae in having the frontoclypeus narrower than the distance between the frontoclypeus and the eye, and in having setal row AD of the hind tibia with enlarged, spiniform bases. These fossils should be considered unplaced within Ledrinae until the nymphs of modern ledrines are better known.

The fossil taxa from Eocene Baltic amber described herein are remarkable in that they represent the oldest known representatives of Aphrodinae and Megophthalminae and exhibit morphological character combinations not found in modern representatives of their respective groups. The geologic age of Baltic amber is generally considered to be Late Eocene (37.7±3 Ma; Perkovsky *et al.* 2007) although some evidence suggests that a Middle Eocene (44.1±1 Ma) or even older origin is possible (reviewed by Szwedo & Sontag 2009; Weitschat & Wichard 2010).

Material and Methods

All specimens studied were obtained by the first author from amber dealers in Palanga, Lithuania, where the material was originally collected. The specimens are deposited in the Insect Collection of the Illinois Natural History Survey (INHS). Digital photographs were taken using a Q-Imaging digital camera attached to an Olympus SZX-12 stereomicroscope.

Abbreviations

AD	=	anterodorsal
AM1	=	apical anteromedial
AV	=	anteroventral
CuA	=	anterior Cubitus
M	=	Media
PD	=	posterodorsal
PV	=	posteroventral
R	=	Radius

Results

Order Hemiptera Linnaeus, 1758
Suborder Auchenorrhyncha Dumeril, 1806
Infraorder Cicadomorpha Evans, 1946
Superfamily Membracoidea Rafinesque, 1815
Family Cicadellidae Latreille, 1825
Subfamily Megophthalminae Kirkaldy, 1906

Eomegophthalmus gen. nov.
urn:lsid:zoobank.org:act:8F89B113-8148-4FA5-9B5F-275F46E02416

Type-species

Eomegophthalmus lithuaniensis sp. nov.

Diagnosis

Same as for *Eomegophthalmus lithuaniensis* sp. nov., due to monotypy of this genus.

Etymology

The genus name, a feminine Latinized Greek noun, combines *eos* ("dawn") with *Megophthalmus*, the name of a related modern genus, and refers to the apparent position of the genus as the earliest known representative of its lineage.

Eomegophthalmus lithuaniensis sp. nov.
urn:lsid:zoobank.org:act:2A1548C9-BC26-4F05-BB99-E0EE5FB55AB9
Figs 1A-B, 2A-F

Diagnosis

This species resembles extant members of the tribe Megophthalmini in having the crown of the head short and broad, the eyes relatively large and extended well laterad of the pronotum, the ocelli on the face distant from the eyes, and lateral frontal sutures of the head distinctly carinate. It is however easily distinguished by its relatively large and more dorsally placed ocelli, elongate wings, and short, broad second valvulae.

Etymology

The species name refers to the country in which the holotype was collected.

Material examined

Holotype female, Eocene Baltic amber, Palanga, Lithuania (INHS).

Description

Measurements. (in mm). Body length including forewings at rest 9.0; head width 2.9; pronotum width 2.5; front femur length 1.1, tibia 1.4; middle femur length 1.4, tibia 2.0; hind femur length 2.0, tibia 3.5, tarsus 1.1; forewing length 7.1; ovipositor length 2.0.

Structure. Body elongate, somewhat depressed, uniformly dark brown. Head broad and short, much wider than pronotum; eyes bulbous; crown very short and poorly developed, coronal suture not visible, posterodorsal margin of head elevated and forming vertical rim above pronotum; ocelli large, slightly closer to midline than to eyes, in depressions near dorsal margin of face; antennal ledges oblique, flattened, slightly extended over antennal pits; antenna nearly as long as width of head; frontoclypeus narrow, rugulose, elevated and shelflike ventrolaterally in relation to gena, concave dorsomedially, extended to dorsal margin of face in anteroventral view, evenly tapered from antennal pit to anteclypeus; lateral frontal sutures complete, weakly carinate, extended from antennal pit to midline; ocellocular area broad; clypeal suture complete; anteclypeus tapered, slightly convex, apex rounded, extended slightly beyond lower margin of gena; lorum flat, well separated from genal margin, ventral 2/3 bordering anteclypeus; gena very narrow; maxillary cleft absent. Proepisternum small, flat, largely exposed. Pronotum depressed, transversely rugulose, anterior margin roundly produced but not extended anterad of eyes, lateral margin short, carinate. Exposed part of mesonotum and scutellum depressed, scutellum acuminate. Episternum divided by suture into anepisternum and katepisternum, without processes. Front femur with numerous scattered, poorly undifferentiated setae on dorsal and anterior surface, ventral rows poorly differentiated, AM1 slightly enlarged; tibia slender, dorsal surface flat, longitudinal rows present but poorly differentiated and surfaces of tibia between rows with numerous scattered fine setae, ventral setae scattered and short. Middle leg similar to front leg in shape and chaetotaxy. Hind femur apical macrosetae 2+0; hind tibia compressed, with setal rows PD, AD, AV and PV with 15, 10, 11, and ~53 setae, respectively; with PD and AD closer to each other than distance from AD to AV; PD setae slightly smaller than those of AD; AD macrosetae with bases enlarged but not spinelike, intercalary setae absent; AV extended from basal third to apex; PV setae fine and subequal in length with macrosetae of similar lengths; pecten with single row of macrosetae with spinelike bases, 2 lateral setae longer than 4 medial setae; tarsus ca. 1/3 length of tibia; tarsomere I without dorsoapical pair of setae, ventral setae scattered, pecten with 3 platellae. Forewings elongate, macropterous, with venation poorly delimited, texture glabrous except rugulose near base of clavus; vein R branched in basal 1/3, with 6 branches, 4 extended to costal margin with basal branch arising near midlength, crossvein s present; M with 2 branches; only distal r-m crossvein visible; inner apical cell long, tapered in distal 3/4, not extended to wing apex; CuA connected to submarginal vein slightly distad of apex of clavus; clavus occupying ~3/4 total wing length; appendix absent. Hind wing macropterous but not well preserved in holotype.

Female genitalia. Ovipositor with first valvulae sculpturing imbricate; second valvulae abruptly broadened near base and tapered distally, with 2 large teeth near base and several smaller serrations more distad.

Male. Unknown.

Age and occurrence

Baltic region. Baltic amber, Middle Eocene, ca. 44 Ma.

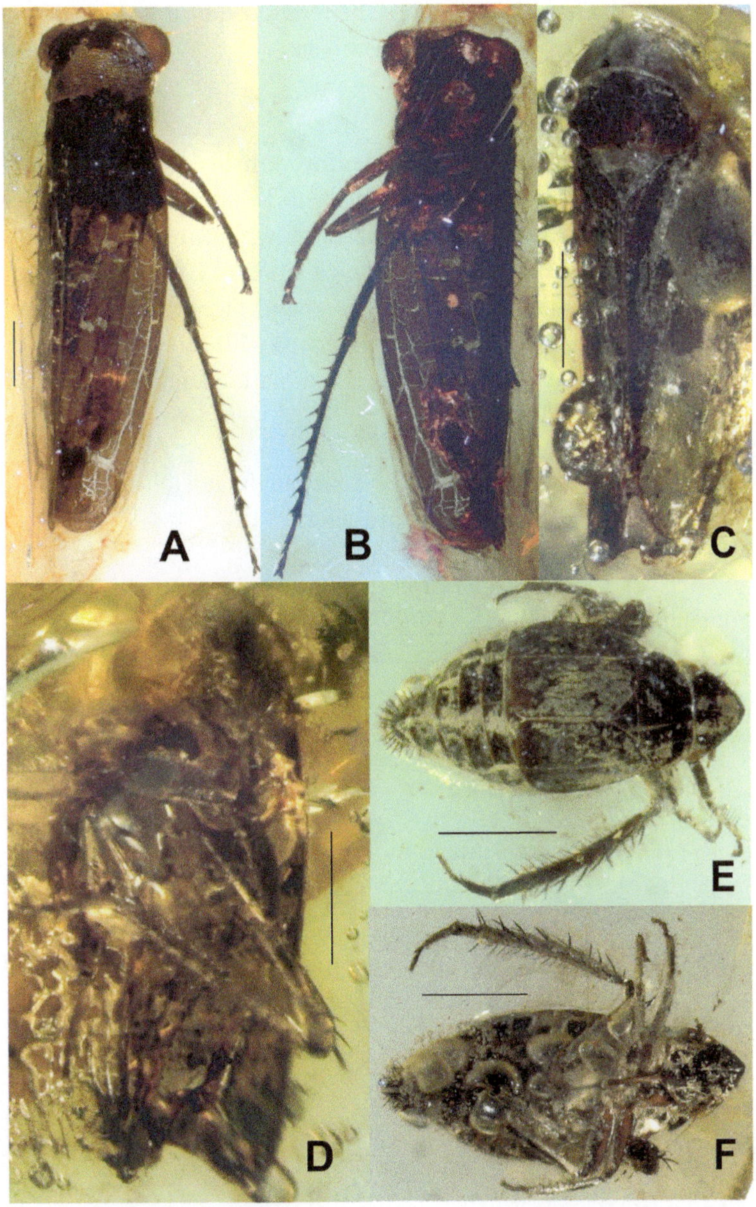

Fig. 1. Habitus in dorsal and ventral view. **A-B**. *Eomegopthalmus lithuaniensis* sp. nov. ♀. **C-D**. *Xestocephalites balticus* sp. nov. ♂. **E-F**. *Brevaphrodella nigra* sp. nov. ♂. Scale bars = 1 mm.

Remarks

The structure of the head (crown absent, ocelli on face distant from eyes) places this leafhopper within the lineage comprising Eurymelinae, Megophthalminae, Ulopinae, and treehoppers. The carinate facial sutures extended ventrad of the ocelli suggest that it is closest to Megophthalmini but, unlike *Eomegophthalmus* gen. nov., modern members of that tribe have the gena expanded and concealing the proepisternum and the wings reduced in size. The presence of macropterous forewings with supernumerary crossveins suggests a relationship to Adelungiini, but modern members of the latter tribe have the head smoothly rounded and shagreen in texture (except in *Adelungia* Melichar, 1902, which has a compressed, bladelike median dorsal process), and lack conspicuous transverse rugae on the pronotum. The broad second valvulae are more membracid-like than those of modern Megophthaminae, which have the second valvulae relatively narrow with serrations restricted to the distal half. The relatively dorsal position of the ocelli also suggests a relationship with Ulopinae and Membracidae. Thus, the subfamily placement must be considered tentative until the relationship of *Eomegophthalmus* gen. nov. to other leafhoppers can be elucidated by phylogenetic analysis.

Specimen notes

The type specimen, embedded in a clear piece of orange-yellow polished amber with most aspects of the body visible, apparently underwent considerable decay prior to fossilization. The abdomen is missing except for most of the first and second valvulae of the ovipositor, which are exposed and situated in their original resting position near the apex, beneath the wings. The rest of the exoskeleton is remarkably well preserved, although there is some distortion of the head and thorax due to compression, and the forewing integument has numerous small fractures that appear as white areas and give the specimen the appearance of having reticulate forewing venation. Close examination suggests that only a few of these fractures correspond to veins.

Subfamily Aphrodinae Haupt, 1927 (=Acocephalinae Dohrn, 1859)
Tribe Xestocephalini Baker, 1915

Xestocephalites gen. nov.
urn:lsid:zoobank.org:act:3D2E3FA1-905A-4892-AF4F-DCD7CFC9F04F

Type-species

Xestocephalites balticus sp. nov.

Diagnosis

Same as for *Xestocephalites balticus* sp. nov., due to monotypy of this genus.

Etymology

The genus name, a masculine Latin noun, was formed by replacing the suffix of *Xestocephalus*, a related modern genus, with *-ites* ("stone"), a common suffix used to designate fossil taxa.

Xestocephalites balticus sp. nov.
urn:lsid:zoobank.org:act:FC8D9A1F-FF96-4DD0-8A1D-5C587489CC76
Figs 1C-D, 2G-I

Diagnosis

This species resembles the modern genus *Xestocephalus* Van Duzee, 1892 in the structure of the head and in the leg chaetotaxy but differs in having the head wider than the pronotum, the crown more strongly depressed, the forewing elongate, and the front femur lacking an enlarged ventral seta near the midlength.

Etymology

The species name refers to the Baltic region, where the holotype originated.

Material examined

Holotype male, Eocene Baltic amber, Palanga, Lithuania (INHS).

Description

MEASUREMENTS. (in mm). Body length including forewing at rest 5.3; head width 1.5; pronotum width 1.4; front femur length 0.8, tibia 0.9; middle femur length 0.8, tibia 1.1; hind femur length 1.5, tibia 2.1, tarsus 0.7; forewing length 4.2.

STRUCTURE. Body elongate, weakly depressed, dark brown colored, without conspicuous pattern. Head slightly wider than pronotum; crown well developed, roundly produced, longer medially than next to eye, uniformly shagreen; ocelli well developed but small, on crown just posterad of anterior margin, slightly closer to midline than to eye; transition from crown to face rounded; antennal ledges oblique, weakly developed; antennal pits deep; antenna shorter than half width of head; frontoclypeus weakly convex; lateral margins evenly tapered ventrad from antennal pits to just dorsad of clypeal suture, thence abruptly narrowed; clypeal suture weakly delimited; lateral frontal sutures extended dorsomesad from antennal pits to near dorsal margin of face; anteclypeus parallel-sided, weakly convex, apex rounded, extended slightly beyond genal margin; lorum flat, narrowly separated from genal margin ventrally, bordering anteclypeus for half length; gena broadly rounded; rostrum extended to just beyond base of hind legs. Pronotum depressed smooth, with sparse fine punctures and very inconspicuous transverse striations lateral margin less than half-length of eye, carinate. Exposed part of mesonotum and scutellum together slightly shorter than broad. Front femur chaetotaxy poorly visible in holotype, without conspicuous large ventral setae, tibia without conspicuous dorsal preapical macrosetae, ventral rows with several conspicuous macrosetae. Hind femur macrosetae 2+1+1; hind tibia flattened and bowed, with setal rows PD, AD, AV with 12, 11, and 17 setae, respectively (PV not visible on specimen); with macrosetae of dorsal rows approximately equal in length, AD and PD with approximately equal numbers of macrosetae extended from near base to apex, without short intercalary setae between successive large macrosetae, AV with 3 stout preapical macrosetae in addition to numerous close set, slender setae of approximately equal length; tarsomere I elongate, with conspicuous dorsoapical pair of macrosetae and well differentiated ventral longitudinal row of short, stout setae. Forewings macropterous, venation not visible in holotype; clavus occupying ¾ length of wing, appendix absent. Male subgenital plates short and broad, somewhat compressed, fused along midline, with numerous conspicuous long, fine setae along dorsal and apical margins and few stout setae ventromedially, apices obliquely truncate.

FEMALE. Unknown.

Age and occurrence

Baltic region. Baltic amber, Middle Eocene, ca. 44 Ma.

Remarks

This genus is referable to Aphrodinae based on the position of the ocelli on the crown slightly posterad of the anterior margin and distant from the eye. Among modern aphrodines, it most closely resembles *Xestocephalus* (Xestocephalini) in the structure of the head. *Xestocephalus* differs in having the head less strongly depressed and narrower than the pronotum with the gena broader and the anteclypeus strongly flattened. *Xestocephalites* gen. nov. also differs from most modern aphrodines in having the wings fully macropterous. Species of the related tribe Portanini also have fully developed wings, but differ in having the head narrower, the ocelli on the margin of the crown, and the lorum greatly enlarged.

Specimen notes

The type specimen is embedded in a clear orange-yellow piece of amber but it is partly degraded with the left front leg (femur and segments distad) detached, and parts of the dorsum, including the forewings, largely concealed by air bubbles such that the wing venation is not visible. The anterior margin of the head is obscured by a horizontal fracture plane.

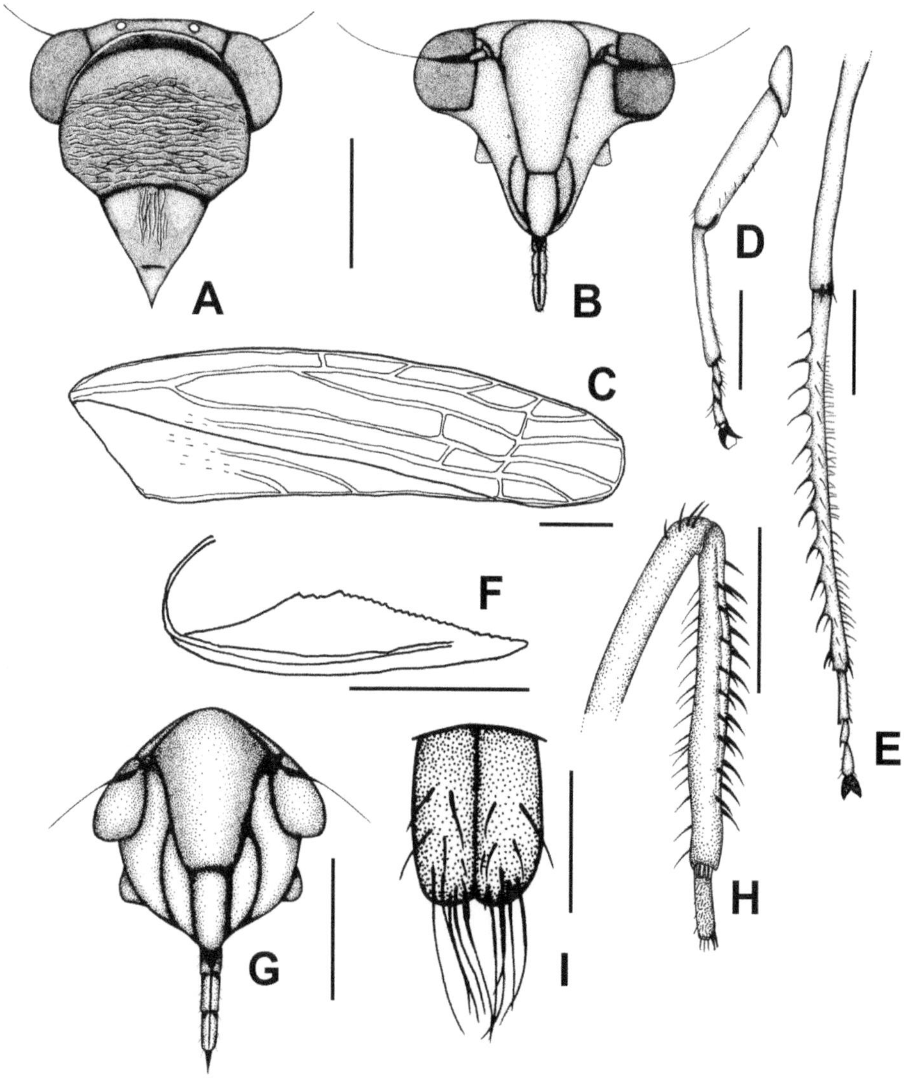

Fig. 2. A-F. *Eomegopthalmus lithuaniensis* sp. nov. ♀. **A**. Crown, pronotum, and mesonotum, dorsal view. **B**. Face, anteroventral view. **C**. Right forewing, lateral view. **D**. Prothoracic leg, lateral view. **E**. Metathoracic leg, dorsal view. **F**. Second valvulae of ovipositor, general lateral view. **G-I**. *Xestocephalites balticus* sp. nov. ♂. **G**. Face. **H**. Apical portion of femur, tibia, and first tarsomere of metathoracic leg, lateral view. **I**. Subgenital plates, ventral view. Scale bars: A-H = 1 mm; I = 0.5 mm.

Brevaphrodella gen. nov.

urn:lsid:zoobank.org:act:898D14D3-6580-4EE5-8C6D-D9FBDE5EB92E

Type-species

Brevaphrodella nigra sp. nov.

Diagnosis

Same as for *Brevaphrodella nigra* sp. nov., due to monotypy of this genus.

Etymology

The genus name, a feminine Latin noun, combines *brevis* ("short"), *Aphrodes* (the name of a related extant leafhopper genus), and *-ella* (a diminutive suffix), refering to the small size and short forewings of the genus.

Brevaphrodella nigra sp. nov.

urn:lsid:zoobank.org:act:0E415A40-3645-4C63-9B59-6F15E2512BB2

Figs 1E-F, 3

Diagnosis

Although the placement in Aphrodinae is somewhat tentative because the ocelli are not visible on the specimen, the overall structure of the head (flattened, produced crown; broad, flattened face with wide, angulate gena); leg chaetotaxy (2+2+1 hind femoral macrosetal formula); and male genital capsule (with numerous, scattered macrosetae and ligulate subgenital plates) are consistent with Aphrodinae. The presence of a single enlarged ventral seta near the midlength of the front femur suggests that the new genus is related to *Xestocephalus* and supports its placement in tribe Xestocephalini.

Etymology

The species name refers to the black overall coloration of the holotype.

Material examined

Holotype male, Eocene Baltic amber, Palanga, Lithuania (INHS).

Description

Measurements. (in mm). Body length including forewings at rest 3.7; head width 0.9; pronotum width 1.0; front femur length 0.7, tibia 0.7; middle femur length 0.7, tibia 0.8; hind femur length 1.1, tibia 1.9, tarsus 0.7; forewing length 1.2.

Structure. Body small, depressed, ovoid, entirely black. Head triangularly produced; crown weakly convex, shagreen; coronal suture not visible; eyes small, anterior margin slightly emarginate adjacent to antennal base; ocelli not visible. Antennal pits deep, pedicel enlarged, base of flagellum divided into 5 subsegments. Lorum large, flat, narrowly separated from gena ventrally; gena angulate below eye, concealing proepisternum. Anteclypeus flat, tapered distally, apex rounded, extended slightly beyond gena. Rostrum long, extended to base of hind coxae. Pronotum with lateral margins divergent posteriorly, long, carinate. Mesonotum reduced, almost completely concealed by pronotum, scutellum small. Forewing short, coriaceous, elytralike, apex truncate, extended nearly to posterior margin of abdominal tergite IV, venation not delimited. Front femur with AM1 small, near midheight of femur, intercalary row well differentiated with 5-6 setae; AV with single long stout seta near midlength; hind tibia with setal rows PD, AD, AV and PV with 15, 10, 11, and ~53 setae, respectively; with dorsal preapical macrosetae 1+2, row AV with 13 setae becoming longer from base to apex, basal 5 abruptly shorter than others. Middle trochanter with stout ventroapical seta; femur with 4 AV and 1 PV setae, tibia with pair of dorsal setae near base and another near apex, AV with several setae. Hind femur macrosetae 2+2+1, penultimate pair close-set with posterior seta much smaller than anterior; tibia rows AD and PD with approximately equal numbers of macrosetae; AD macrosetae each with one intercalary seta, AD1 slightly offset toward middle of dorsal surface and distad of PD1, pecten with two pairs of shorter setae between two longer lateral and on long medial seta; tarsomere I with dorsoapical pair of macrosetae and two well differentiated rows of short, stout ventral setae; pecten with 5 tapered setae, medial seta longest. Male pygofer with numerous large macrosetae scattered over distal half. Valve concealed by sternite VIII. Subgenital plates ligulate, boatlike, with numerous large macrosetae scattered over most of length, mesal margins straight and closely appressed, apex evenly tapered.

FEMALE. Unknown.

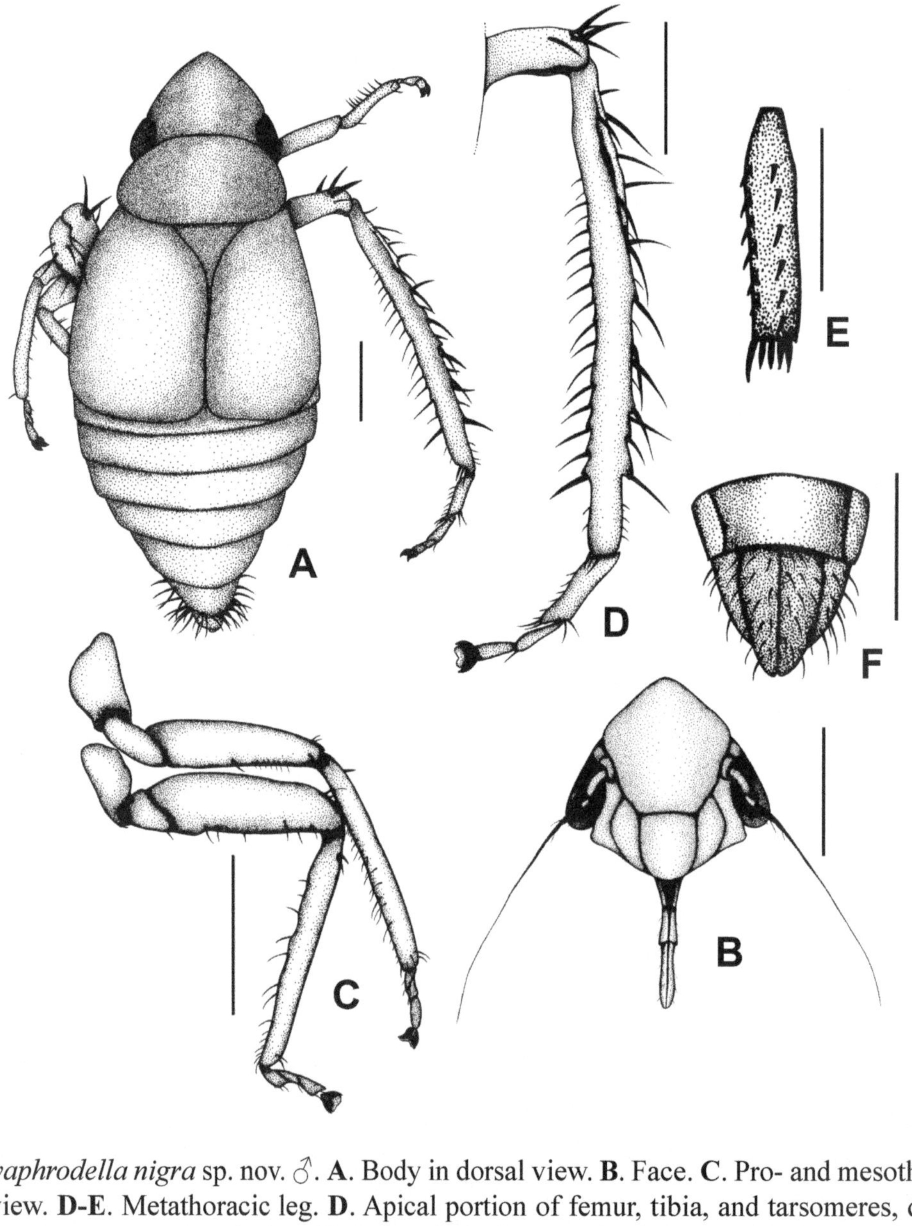

Fig. 3. *Brevaphrodella nigra* sp. nov. ♂. **A**. Body in dorsal view. **B**. Face. **C**. Pro- and mesothoracic legs in ventral view. **D-E**. Metathoracic leg. **D**. Apical portion of femur, tibia, and tarsomeres, dorsolateral view. **E**. First tarsomere, ventral view. **F**. Subgenital plates, and valve, ventral view. Scale bars: A-D, F = 0.5 mm; E = 0.25 mm.

Age and occurrence

Baltic region. Baltic amber, Middle Eocene, ca. 44 Ma.

Remarks

Placement of *Brevaphrodella* gen. nov. in Aphrodinae is somewhat tentative because the ocelli are not visible on the specimen. Either they are absent or greatly reduced (as in some modern brachypterous leafhoppers), or they are on the anterior margin of the head and not visible due to the presence of a fracture plane concealing this part of the head in the only available specimen. Nevertheless, the overall structure of the head (flattened, produced crown; broad, flattened face with wide, angulate gena); leg chaetotaxy (2+2+1 hind femoral macrosetal formula); and male genital capsule (with numerous, scattered macrosetae and ligulate subgenital plates) are consistent with Aphrodinae. The presence of a single enlarged ventral seta near the midlength of the front femur suggests that the new genus is related to *Xestocephalus* and supports its placement in tribe Xestocephalini. This appears to be the oldest known example of a brachypterous adult leafhopper, along with the specimen of an undescribed species and genus from Baltic amber illustrated by Szwedo (2002: fig. 24).

Specimen notes

The holotype is a specimen in excellent condition with apparently very little decomposition or loss of original color, embedded in a clear, light yellow piece of amber with dorsum and venter clearly visible but with numerous air bubbles concealing parts of the venter and a fracture plane obscuring the view of the anterodorsal margin of the head. The abdomen appears to have longitudinal pale stripes but these are asymmetrical and appear to be caused by pockets of air between the integument and the matrix.

Discussion

Although, to date, very few fossil Cicadellidae from Eocene Baltic amber have been described and illustrated adequately, leafhoppers are quite common in the Baltic amber fauna and the fossil taxa described so far from this source appear to be of great importance to the study of leafhopper phylogeny. The species represented by these fossils lived during a time when modern subfamilies and tribes were first beginning to appear; thus the fossils document evolutionary changes in morphology that were associated with the transition from the more morphologically generalized leafhoppers recorded from the Cretaceous period (Hamilton 1990, 1992) and the thoroughly modern leafhoppers recorded from the more recent Dominican amber (Dietrich & Vega 1995). Because the preservation of Baltic amber fossils is often very good, detailed study of their morphology is beginning to yield insights into the evolution of characters important for inferring the ecology and behavior of leafhoppers as well as for diagnosing higher taxa.

Eomegophthalmus gen. nov. exhibits a unique combination of features shared with some modern leafhopper subfamilies, as well as the treehopper families Aetalionidae, Melizoderidae and Membracidae. This provides further support for phylogenetic results suggesting that treehoppers arose from within a lineage comprising Megophthalminae and Ulopinae (Dietrich *et al.* 2001).

Xestocephalites gen. nov. and *Brevaphrodella* gen. nov. reveal that diversification of the lineage, that eventually gave rise to the large modern leafhopper subfamily Deltocephalinae, involved parallel acquisition of characters such as brachyptery and specialized leg chaetotaxy that now occur in several independent lineages of leafhoppers.

These recent discoveries, in addition to those recently published by Szwedo and colleagues, suggest that Baltic amber may soon yield representatives of other modern leafhopper subfamilies and tribes, most of which remain undocumented in the fossil record.

Acknowledgements

Fellowships from Conselho Nacional de Desenvolvimento Científico e Tecnológico (CNPq) to A. C. G. through Ciência sem Fronteiras (CsF) program, process 237427/2012-5, and from Lemann Foundation, are acknowledged. S.W. Heads and J. Thomas provided helpful advice on photographing specimens. We also thank the anonymous reviewers for constructive comments that greatly improved the manuscript.

References

Bervoets R. 1910. Diagnoses de quelques nouvelles espèces de cicadines de l'ambre de la Baltique. *Annales Historico-Naturales Musei Hungarici* 8: 125-128.

Dietrich C.H. 2005. Keys to the families of Cicadomorpha and subfamilies and tribes of Cicadellidae (Hemiptera: Auchenorrhyncha). *Florida Entomologist* 88 (4): 502-517.

Dietrich C.H. 2011. Tungurahualini, a new tribe of Neotropical leafhoppers, with notes on the subfamily Mileewinae (Hemiptera: Cicadellidae). *ZooKeys* 124: 19-39.

Dietrich C.H., Rakitov R.A., Holmes J.L. & Black W.C. IV. 2001. Phylogeny of the major lineages of Membracoidea (Insecta: Hemiptera: Cicadomorpha) based on 28S rDNA sequences. *Molecular Phylogenetics and Evolution* 18 (2): 293-305.

Dietrich C.H. & Vega F.E. 1995. Leafhoppers (Homoptera: Cicadellidae) from Dominican amber. *Annals of the Entomological Society of America* 88 (3): 236-270.

Gebicki C. & Szwedo J. 2001. The first record of fossil Mileewinae from Eocene Baltic amber (Hemiptera: Membracoidea: Cicadellidae). *Annales Zoologici* 51 (4): 417-422.

Germar E.F. & Berendt G.C. 1856. Die im Bernstein befindlichen Hemipteren und Orthopteren der Vorwelt. *In*: Berendt G.C. (ed.) *Die im Bernstein befindlichen organischen Reste der Vorwelt gesammelt in Verbindung mit Mehreren, bearbeitet und herausgeben von G. C. Berendt* 2 (1): 1-40. Nicolaische Buchhandlung, Berlin.

Hamilton K.G.A. 1990. Homoptera. *In*: Grimaldi D.A. (ed.). *Insects from the Santana Formation. Lower Cretaceous of Brazil*: 82-122. Bulletin of the American Museum of Natural History 195, American Museum of Natural History, New York.

Hamilton K.G.A. 1992. Lower Cretaceous Homoptera from the Koonwarra Fossil Bed in Australia, with a New Superfamily and Synopsis of Mesozoic Homoptera. *Annals of the Entomological Society of America* 85 (4): 423-430.

Jones J.R. & Deitz L.L. 2009. Phylogeny and systematics of the leafhopper subfamily Ledrinae (Hemiptera: Cicadellidae). *Zootaxa* 2186: 1-120.

Keilbach R. 1982. Bibliographie und Liste der Arten tierischer Einschlüsse in fossilen Harzen sowie ihrer Aufbewahrungsorte. *Deutsche Entomologische Zeitschrift, Neue Folge* 29 (1-3): 129-286.

Metcalf Z.P. & Wade V. 1966. *A Catalogue of the Fossil Homoptera (Homoptera: Auchenorhyncha).* General Catalogue of the Homoptera: a supplement to Fascicle I, North Carolina State University, Raleigh.

Oman P.W. 1937. Fossil Hemiptera from the Fox Hills sandstone (Cretaceous) of Colorado. *Journal of Paleontology* 11 (1): 38.

Perkovsky E.E., Rasnitsyn A.P., Vlaskin A.P. & Taraschuk M.V. 2007. A comparative analysis of the Baltic and Rovno amber arthropod faunas: representative samples. *African Invertebrates* 48 (1): 229-245.

Shcherbakov D.E. 1986. Cicadina (= Auchenorrhyncha). *In*: Rasnitsyn A.P. (ed.) *Insects in the Early Cretaceous ecosystems of West Mongolia*: 47-50. Transactions of the Joint Soviet-Mongolian Palaeontological Expedition 28, Nauka Moscow (in Russian).

Shcherbakov D.E. 1992. The earliest leafhoppers (Hemiptera: Karajassidae n. fam.) from the Jurassic of Karatau. *Neues Jahrbuch für Geologie und Paläontologie Monatshefte* 1: 39-51.

Spahr U. 1988. Ergänzungen und Berichtigungen zu R. Keilbachs Bibliographie und Liste der Bernsteinfossilien - Überordnung Hemipteroidea. *Stuttgarter Beiträge zur Naturkunde Serie B (Geologie und Paläontologie)* 144: 1-60.

Szwedo J. 2002. Amber and amber inclusions of planthoppers, leafhoppers and their relatives (Hemiptera, Archaeorrhyncha et Clypaeorrhyncha). *In*: Holzinger W. (ed.). *Zikaden – Leafhoppers, Planthoppers and Cicadas (Insecta: Hemiptera: Auchenorrhyncha)*: 37-56. Denisia 4, Biologiezentrum/ Oberösterreichisches Landesmuseum, Linz.

Szwedo J. & Gebicki C. 1998. *Ambericarda skalskii* gen. et sp. n. from Baltic amber (Homoptera: Cicadellidae). *Polskie Pismo Entomologiczne* 67 (3-4): 179-184.

Szwedo J. & Gebicki C. 1999. Representatives of Petalocephalini (Homoptera, Cicadellidae: Ledrinae). *Estudios del Museo de Ciencias Naturales de Alava* 14, Numero especial 2: 201-206.

Szwedo J. & Gebicki C. 2002. Nastlopiinae nov. subfam. to comprise *Nastlopia nigra* gen. and sp. n. from Eocene Baltic amber (Hemiptera: Cicadomorpha: Cicadellidae). *Mitteilungen aus dem Geologisch-Paläontologischen Institut der Universität Hamburg* 86: 211-220.

Szwedo J. & Sontag E. 2009. The traps of the "amber trap". How inclusions could trap scientists with enigmas. *In*: *Amber - Archive of Deep Time*: 155-169. Denisia 26, Biologiezentrum/Oberösterreichisches Landesmuseum, Linz.

Wei C., Zhang Y. & Dietrich C H. 2010. First record of the tribe Malmaemichungiini Kwon from China with description of a new species of the genus *Malmaemichungia* Kwon (Hemiptera: Cicadellidae). *Zootaxa* 2689: 48-56.

Weitschat W. & Wichard W. 2010. Baltic amber. *In*: Penney D. (ed.). *Biodiversity of fossils in amber from the major world deposits*: 80-115. Siri Scientific Press, Manchester, UK.

2

Redescription of *Tapinesthis inermis* (Araneae, Oonopidae), with detailed information on its ultrastructure

Arnaud HENRARD[1,2], Rudy JOCQUÉ[1] & Barbara C. BAEHR[3,4]

[1] Section Invertebrates non-insects, Royal Museum for Central Africa, Leuvense Steenweg 13, B-3080 Tervuren, Belgium
[2] Earth and Life Instititute, Biodiversity Research Center, Université Catholique de Louvain, Place Croix du Sud 1–4, B-1348 Louvain la Neuve, Belgium
[3] Queensland Museum, PO Box 3300, South Brisbane, QLD 4101, Australia
[4] CSER, School of Environmental and Life Sciences, University of Newcastle, Callaghan, NSW 2308, Australia

Abstract. *Tapinesthis inermis* Simon, 1882, the only species in the genus, is widely distributed in western Europe. This redescription provides the first information on the ultrastructure of the species using SEM. The morphology of the spinnerets, tarsal claws and tarsal organs, and the internal structure of the female genitalia and the male palp are described and illustrated in detail. The combination of these structures is very similar to those encountered in some dysderoid spiders and supports the basal placement of *Tapinesthis* among Oonopinae. The phylogenetic relationships of the species are discussed. The only female among the three syntypes is designated as the lectotype.

Key words. Female genitalia, male palp, organ receptors, spinnerets, systematics

Introduction

The spider genus *Tapinesthis* Simon, 1914 is monospecific, containing the single species *T. inermis* (Simon, 1882), widely distributed in Europe (Le Peru 2011; Platnick 2013). This rare species is often found in houses or in the neighbourhood of habitations (Kraus 1967; Heimer & Nentwig 1991; Lazarov *et al.* 2001; Korenko *et al.* 2007; Van Keer *et al.* 2006). *Tapinesthis inermis* thus appears to be strongly synanthropic, as it is often found in urban areas. Saaristo & Marusik (2009) even employed the term "hemisynanthropic" to describe its habitat preference. However, it has also been collected in natural habitats such as forests (Dalmas 1920; Le Peru 2007), garrigues and dry grasslands (Le Peru 2011) or near cliffs with a dense cover of ivy (e.g., Rochers de Marches les Dames in Belgium, this study). In Belgium, *T. inermis* appears to be abundant in litter and among ivy (Van Keer *et al.* 2006; pers. obs.).

The species is often found together with *Oonops domesticus* and *O. pulcher* (Van Keer *et al.* 2006, 2010), and, at least in Belgium, also with *Harpactea hombergi* (Scopoli, 1763) (pers. obs.).

Tapinesthis inermis is easy to identify and has been reviewed by many authors (see Platnick 2013). Originally, the species was described as *Oonops inermis* and only the female was known (Simon 1882). Later, Simon (1914) created the genus *Tapinesthis* for this species and described both sexes.

A remarkable redescription of both males and females was provided by Kraus (1967), with detailed drawings of the genitalia and other body parts. Other significant illustrations and photos have been published by Korenko *et al.* (2007) (female prosoma, vulva) and by Saaristo & Marusik (2009) (male and female habitus, male palp, female genitalia). However, this is the first detailed morphological study of *T. inermis* using SEM images, providing detailed information about the ultrastructure of the female genitalia, the spinnerets, mouthparts and legs. We also compare the genitalia with those of related taxa and discuss the relationships of this species with other oonopids.

The results of this work have previously been used for the polarization of characters in a phylogenetic study of Afrotropical *Orchestina* Simon, 1882 (Henrard & Jocqué 2012). However, one particular character, the number of tarsal organ receptors, is updated here and the consequences are dealt with in the discussion.

Material and methods

Specimens were observed with a Zeiss Stemi 2000 stereo microscope. Measurements and photographs of the habitus and details of the mouthparts and detached male palps and genitalia were taken with a Leica MZ16 using the LAS automontage software (ver. 3.8). Female genitalia were dissected, digested with pancreatin and then immersed in 75% ethanol. One male palp was detached and temporarily mounted in a clearing mixture of methyl salicylate and cedukol (Merck, Darmstadt), observed with a Leitz Dialux 22 microscope and subjected to automontage with the Syncroscopy software (ver. 5.03.61). For SEM photos, specimens were dried in HMDS (36 h), gold coated, and examined and photographed with a JEOL 6480 LV scanning electron microscope. All measurements are in millimetres.

This paper follows the terminology of Burger's series of papers (Burger 2009, 2011a, 2011b, 2013; Burger & Michalik 2010) with a few adaptations.

The following abbreviations are used:

ALS = anterior lateral spinnerets
ALE = anterior lateral eyes
ARe = anterior receptaculum
AS = anterior sclerite
D = duct
Ht = holotype
FA = flattened apodeme
MNHN = Muséum national d'Histoire naturelle, Paris (C. Rollard)
MRAC = Musée Royal de l'Afrique Centrale, Tervuren (R. Jocqué)
PBI = Planetary Biodiversity Inventory project
PER = posterior eye row
PLE = posterior lateral eyes
PLS = posterior lateral spinnerets
PME = posterior median eyes
PMS = posterior median spinnerets
Pp = posterior plate

Pr = protrusion
PRe = posterior receptaculum
Pt = paratype
RBINS = Royal Belgian Institute of Natural Sciences, Brussels
V = vesicle
Te = tendon
Tw = thin-walled duct

Results

Class Arachnida Cuvier, 1812
Order Araneae Clerck, 1757
Family Oonopidae Simon, 1890

Tapinesthis Simon, 1914

Diagnosis

Soft-bodied, pale Oonopidae with strongly sloping cephalothorax provided with dark net-shaped pattern, without leg spines, with unipectinate tarsal claws; pedicel with meshed texture; male palp with short sclerified sperm duct and female genitalia with wide median T-shaped structure visible by transparency, flanked by two rounded apodemes.

Description

Male

CEPHALOTHORAX. *Carapace* in vivo pale yellowish-orange (Fig. 1A–B) to pinkish, in ethanol yellowish-white to pale orange (Figs 2A–B, 3A–D), with dark net-shaped pattern and dark margins; non-marginal pars cephalica setae present in three rows, ovoid in dorsal view, pars cephalica strongly sloping in lateral view, reaching highest point at level of second coxa (Fig. 2E), anteriorly narrowed to 0.49 times its maximum width or less, with rounded posterolateral corners (Fig. 2A–B), surface of elevated portion of pars cephalica smooth, sides smooth, with radiating patches of smooth platelets (Fig. 6A–B); thorax without depressions, fovea absent, lateral margin undulate, smooth; pars cephalica setae needle-like. *Clypeus* margin curved downwards in front view (Fig. 2G), sloping forward in lateral view (Fig. 2E), high, ALE separated from edge of carapace by their radius or more. *Eyes* (Fig. 2A, E): six, well developed, all subequal, ALE oval, PME circular, PLE circular; posterior eye row recurved from both above and front; ALE separated by their radius to diameter, ALE–PLE separated by less than ALE radius, PME touching, PLE–PME separated by less than PME radius, ALE–PME separated by less than PME radius. *Sternum* (Fig. 2C) yellowish white, with faint radiating dark stripes, longer than wide, with radial smooth furrows between coxae I–II, II–III and III–IV, anterior margin with semicircular depression on the middle half, posterior margin extending posteriorly beyond anterior edges of coxae IV as single extension, distance between coxae approximately equal; setae abundant, evenly scattered, without hair tufts. *Mouthparts*: chelicerae, endites and labium yellowish white. Chelicerae (Fig. 6F) straight; without teeth on both promargin and retromargin; fangs directed medially, paturon inner margin with scattered setae. Labium (Figs. 2C, 6C–D, F) rectangular, fused to sternum, anterior margin indented at middle, basal corners with small circular depression, sclerotization as in sternum; with 6 or more setae on anterior margin, subdistal portion with unmodified setae. Endites (Fig. 6C–G) distally not excavated, serrula present in single row, anteromedian tip unmodified, posteromedian part unmodified, sclerotization as in sternum; anterior margin with ventral row of spatulate setae and dorsal row of distally pectinate setae, median margin with finely barbed setae.

ABDOMEN (Fig. 2A–B, D–F). Ovoid, rounded posteriorly,without scuta; dorsum soft portions yellow-brown, without color pattern. Dorsum setae present, needle-like. *Epigastric area* setae uniform, needle-like. Postepigastric area setae present, needle-like. *Pedicel* with meshed texture (Fig. 6C) as in Orchestina (see Henrard & Jocqué 2012, figs. 21, 113, 114, 404), abdomen extending anteriad of pedicel. *Colulus* present (Fig. 9A), with four setae. *Spinnerets* (Fig. 9B–D): ALS with one major ampullate gland spigot emerging from cylindrical tubercle and two piriform gland spigots emerging from shallow tubercle; PMS with one minor ampullate gland spigot emerging from cylindrical tubercle; PLS with two aciniform gland spigots emerging from cylindrical tubercle.

LEGS. White, without color pattern; femur IV not thickened, same size as femora I–III, patella plus tibia I longer than carapace, tibia I unmodified, metatarsi of at least leg II in the middle and IV at basal quarter, with eye-shaped smooth gland outlet with single pore (Fig. 12C–E). *Leg spines* absent. *Tarsal claws* (Fig. 10A–F) unipectinate, tarsal proclaws and retroclaws inner face smooth; tarsus I–IV superior claws with six teeth on lateral surface of proclaw, six teeth on lateral surface of retroclaw. Tarsi claw tuft with three carpeted setae. Tarsi I to IV without inferior claw. *Trichobothria* (Fig. 12D–E) with rounded bothrium, internal texture of aperture not grate-like, hood covered by numerous low, closely spaced ridges; sometimes accompanied by eye shaped structure with unknown function (Fig. 12C–D). *Tarsal organ* (Fig. 11A–D, H–I) pear-shaped, margin raised above surroundings, with 4 receptors on legs I and II, with 3 receptors on legs III and IV. In a probably unusual case, the tarsal organ has only 1 receptor on both legs IV (Fig. 11H–I), whereas I–III of the same specimen have the usual 4-4-3 formula (not illustrated). According to observations made on other oonopids the formula appears to be 3-3-2-2, since the distalmost receptor is distally bifid, but both parts stem from a common base as in *Stenoonops* Simon, 1891 (Platnick & Dupérré 2010).

Fig. 1. *Tapinesthis inermis* photographed in vivo. **A.** ♂ (PBI 33564). **B.** Same. **C.** ♀ (PBI 33267). **D.** Same. **E.** Same. Remark color differences of cephalothorax between ♂ and ♀. For explanation, see text.

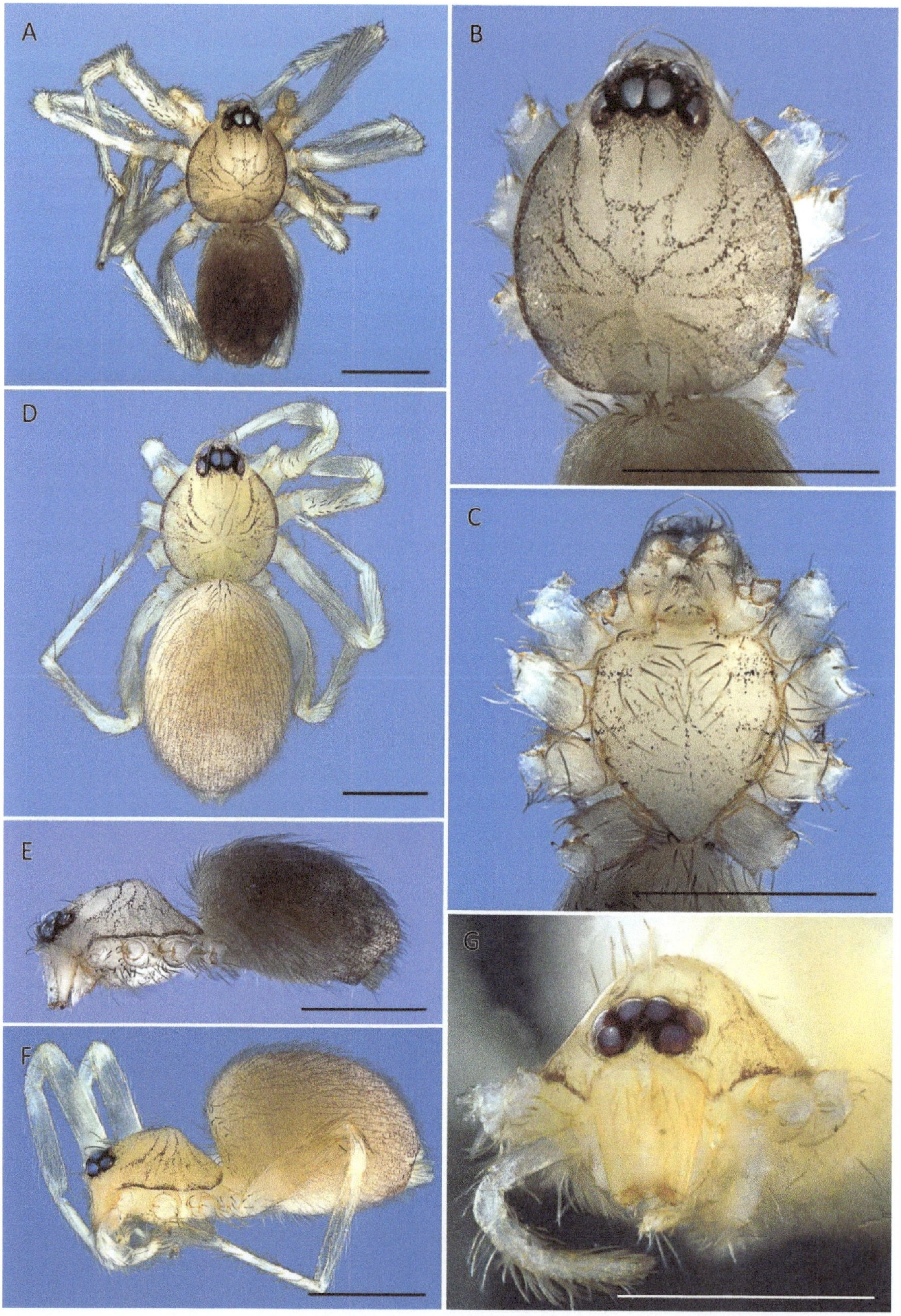

Fig. 2. *Tapinesthis inermis.* **A**. ♂ (PBI 33564), habitus, dorsal view. **B**. Same, carapace. **C**. Same, sternum, ventral view. **D**. ♀ (PBI 33267), habitus, dorsal view. **E**. ♂ (PBI 33564), habitus, lateral view. **F**. Same, ♀ (PBI 33267). **G**. ♀ (PBI 08123), carapace, anterior view. Scale bars = 0.5 mm.

GENITALIA (Figs. 4A–E, 7A–D). *Epigastric region* with sperm pore small, narrow, slit-like, situated at level of anterior spiracles. *Palp* normal size, not strongly sclerotized, right and left palps symmetrical, proximal segments white; trochanter normal size, unmodified; femur normal size, two or more times as long as trochanter, without posteriorly rounded lateral dilation, attaching to patella basally; patella shorter than femur, not enlarged, without prolateral row of ridges, setae unmodified; tibia ventral margin swollen; cymbium white, ovoid in dorsal view, not fused with bulb, not extending beyond distal tip of bulb, plumose setae absent, without stout setae, without distal patch of setae; bulb white, 1 to 1.5 times as wide as cymbium, stout, piriform with frontal margin slightly concave, tapering apically with long, medially bent embolus, distal sclerotized part of seminal duct (D) short, restricted to distal part, abruptly interrupted, internally with poorly defined vesicle (V), opening into embolus through duct with basal part thin-walled (Tw), distal part sclerotized; tendon (Te) running from base of cymbium to distal end of vesicle (V); embolus simple, dark, tube-shaped, flattened apically, tip blunt, grooved, appearing pale.

Female (Figs 1C–E, 2D, F–G)

As male except as noted.

CEPHALOTHORAX. Female palpal claw absent; spines absent. Tarsal organs (Fig. 11A–D) as in male.

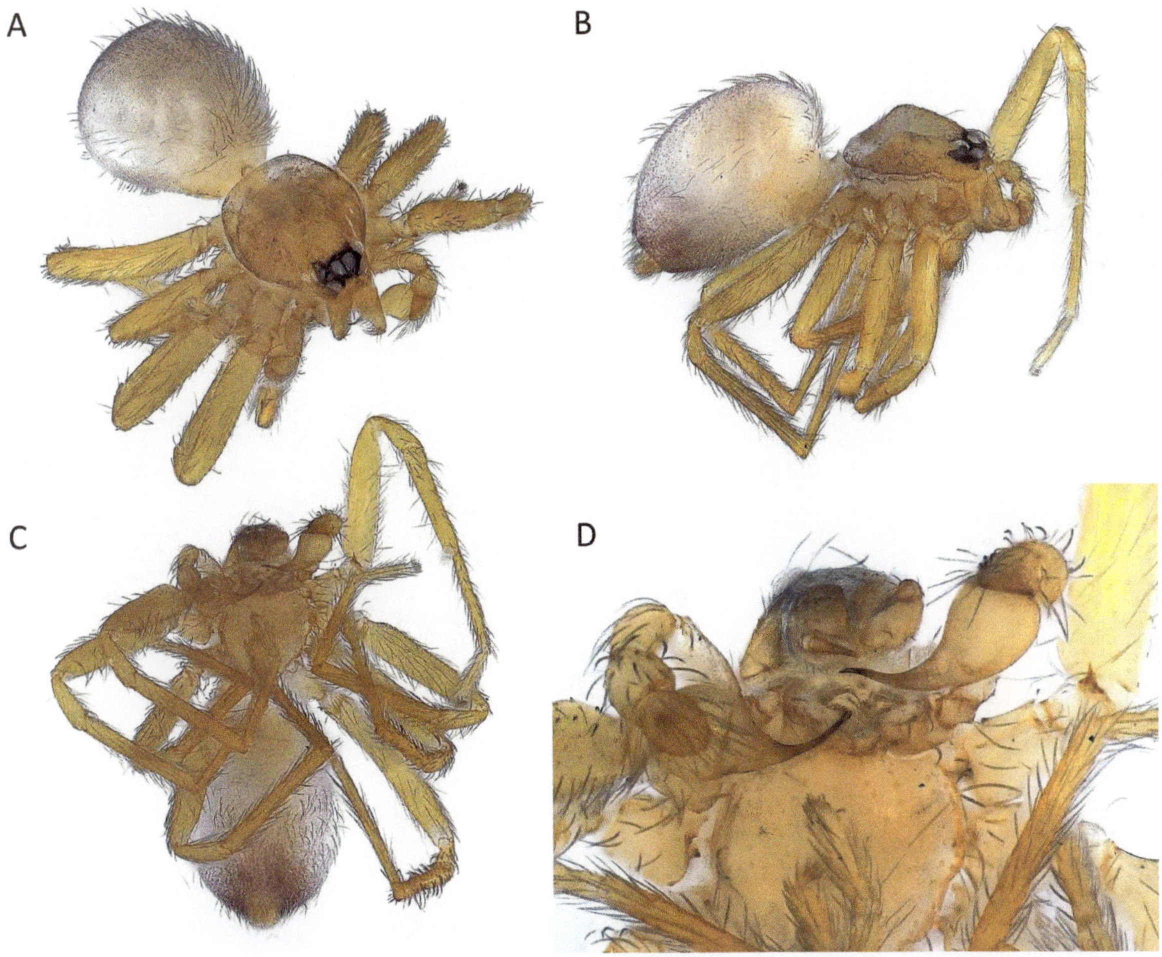

Fig. 3. *Tapinesthis inermis*, ♂ from the Czech Republic (PBI 32902). **A**. Habitus, dorsal view. **B**. Same, lateral view. **C**. Same, lateral view. **D**. Prosoma, ventral view.

Abdomen. *Spinnerets* (Fig. 9E–H): ALS with one major ampullate gland spigot emerging from cylindrical tubercle and three piriform gland spigots emerging from shallow tubercle; PMS with minor ampullate gland spigot and three aciniform gland spigots; PLS with six aciniform gland spigots.

Genitalia (Figs 5A–D, 8A–G). Shape: anterior epigastric margin strongly sclerotized, preceded by T-shaped structure visible in transparency (AS), with larger, square part (ARe) in between, the latter provided at base with fine transverse sclerite (T) ending in two faint procurved lobes (FA) extending as far as lateral protrusions (Pr) of frontal T-shape (AS). SEM view, details: anterior sclerite (AS) short, T-shaped, with lateral protrusions (Pr) well developed and provided with ramified ridges on anterior side and dorsally near extremities; AS connected with anterior receptaculum (ARe) antero-ventrally; ARe about half as wide as AS including Pr, square, provided dorsally and apically with rows of gland ducts (GD) confined in frontal groove; ARe adjacent to transverse sclerite (T) ending laterally in well developed,

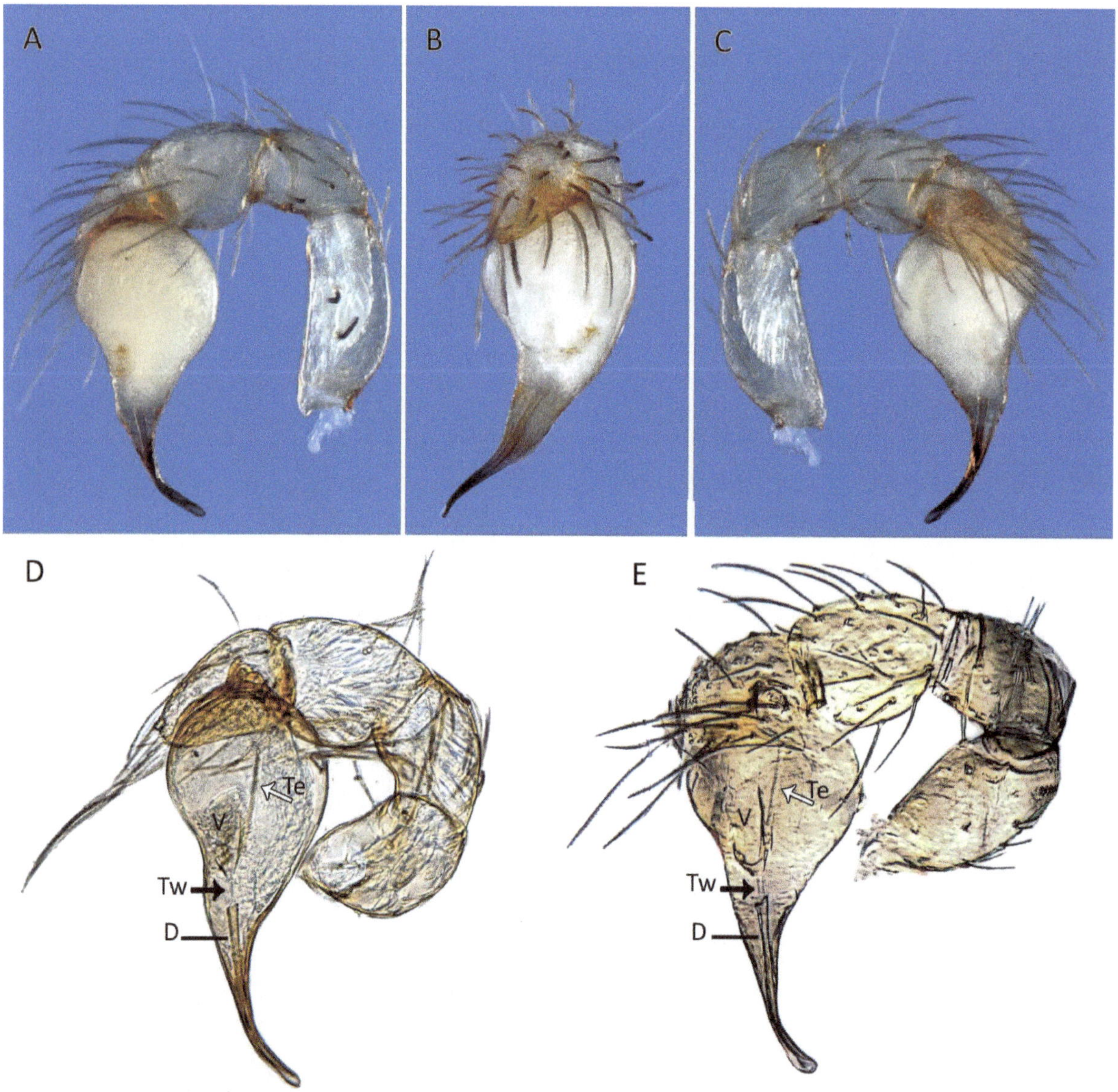

Fig. 4. *Tapinesthis inermis*. – **A–C**. ♂ (PBI 33564), palp. **A**. Retrolateral view. **B**. Same, frontal view. **C**. Same, prolateral view. – **D**. ♂ (PBI 09021), palp, prolateral view, transmitted light. – **E**. ♂ palp, lateral view after Saaristo & Marusik, 2009 (modified). – D: duct; Te: tendon; Tw: thin-walled duct; V: vesicle.

bilobed, flattened apodeme (FA), its dorsal extremity anchoring into the posterior receptaculum (PRe); PRe well developed, egg-shaped, emerging from base of ARe. Uterus externus not visible.

Tapinesthis inermis (Simon, 1882)

Oonops inermis Simon, 1882: 240.
Tapinesthis inermis. – Simon 1914: 88, 90. — Kaston 1948: 61, figs 4–6. — Kraus 1967: 382, figs 1–10. — Heimer & Nentwig 1991: 52, fig. 110. — Lazarov, Deltshev & Blagoev 2001: 8, figs 5–6. — Trotta 2005: 170, figs 369–370. — Van Keer 2006: 36, fig. 3. — Korenko, Rezác & Pekár 2007: 6, figs 1–3. — Saaristo & Marusik 2009: 64, figs 11–14, 23, 27. — Le Peru 2011: 307, fig. 537.

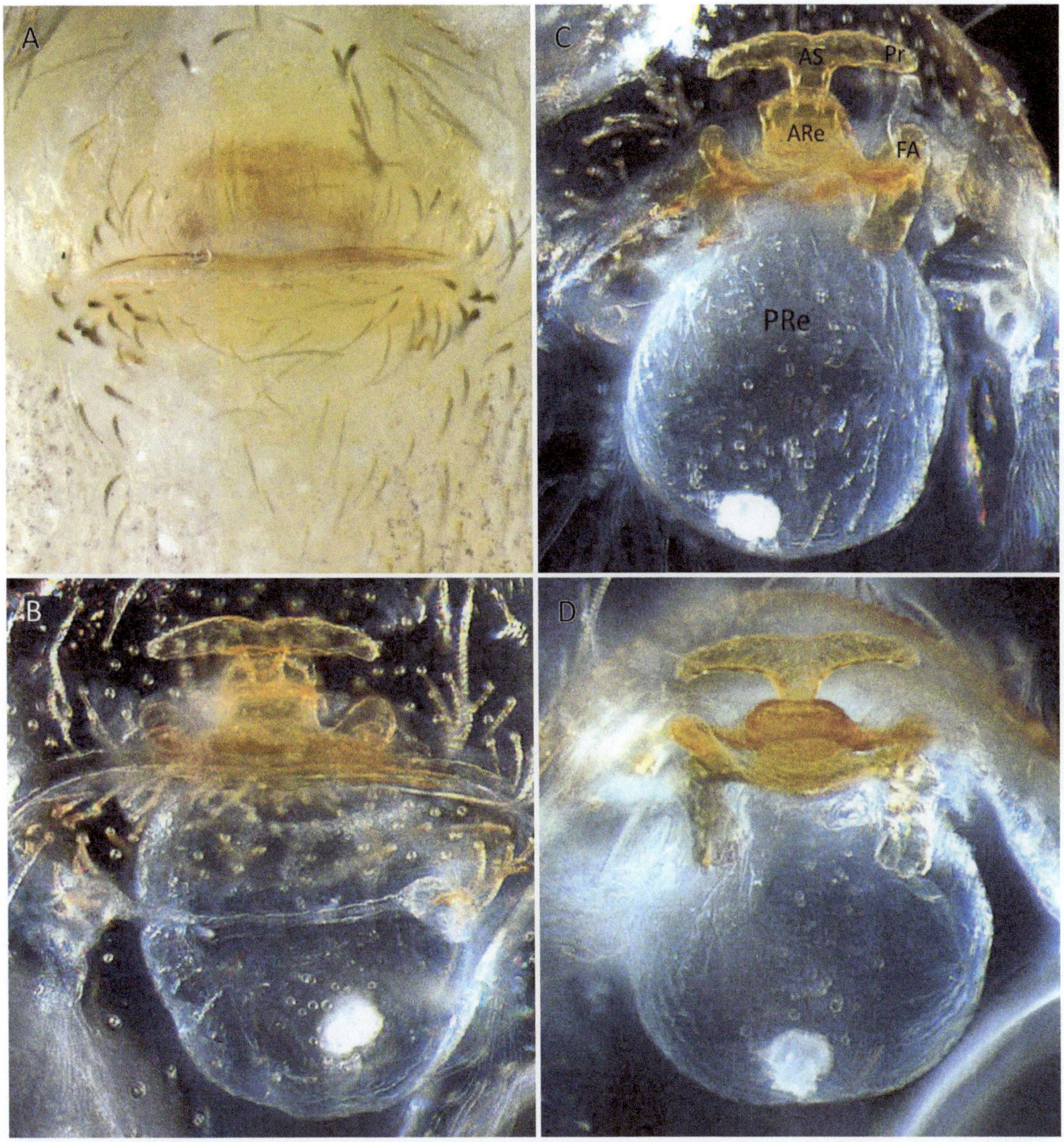

Fig. 5. *Tapinesthis inermis*, genitalia of ♀ (PBI 33267) **A**. Genital area, ventral view. **B**. Same, after digestion, ventral view. **C**. Same, dorsal view. **D**. Same, anterior view. – ARe: Anterior Receptaculum; AS: Anterior Sclerite; FA: Flattened Apodeme; Pr: lateral Protrusion; PRe: Posterior Receptaculum.

Material examined

Lectotype (here designated)

♀, *Oonops inermis* E.S. [= E. Simon], Gall. m. [= Gallia mediterranea?], MNHN AR1468 *pars* (PBI OON 6262 *pars*).

Paralectotypes

2 ♂♂, *Oonops inermis* E.S. [= E. Simon], Gall. m. [= Gallia mediterranea?], MNHN AR1468 *pars* (PBI OON 6262 *pars*, PBI OON 6263).

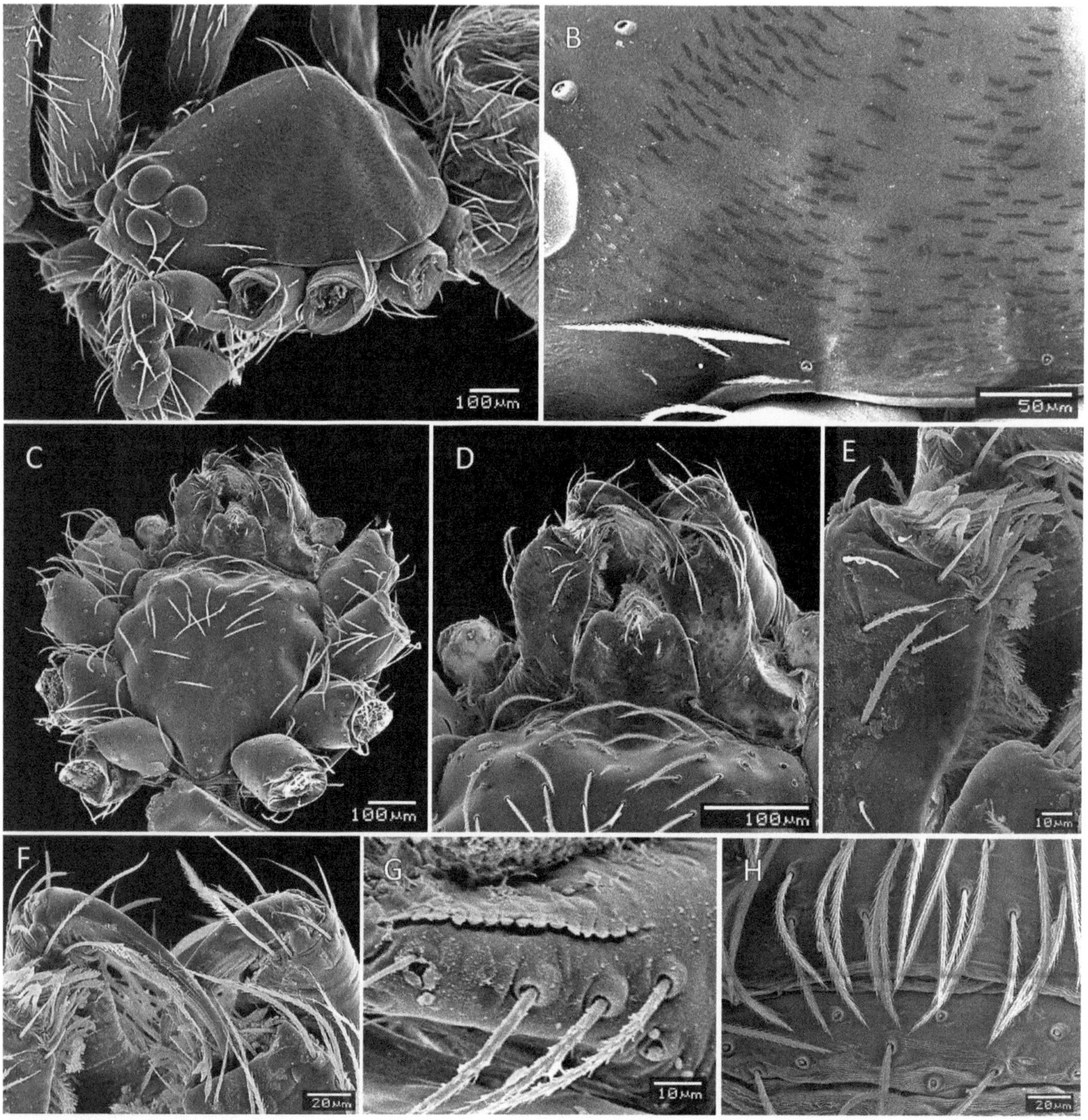

Fig. 6. *Tapinesthis inermis*, SEM views. **A.** ♂ (PBI 08123), carapace, antero-lateral view. **B.** Same, close up on the distribution of the smooth platelets. **C.** ♂ (PBI 33564), sternum, ventral view. **D.** Same, mouthparts. **E.** Same, close up of endite setae. **F.** Same, cheliceral fangs. **G.** ♂ (PBI 08123), endites, anterior view with detail of serrula. **H.** ♂ (PBI 33564), genital region, ventral view.

Other material

BELGIUM: Antwerpen, Turnhout, forest remnant, under liana, 27 m, 51.31666° N, 4.93333° E, 19 Oct. 2006, H. De Coninck, 5 ♂♂, RBINS I.G.32708 *pars* (PBI OON 8123 *pars*), 3 ♀♀, RBINS I.G.32708 *pars* (PBI OON 8123 *pars*). Namur, Marche-les-Dames, 50.48250° N, 4.94917° E, 25 May 2010, A. Henrard, 1 ♀, RBINS I.G.32708 *pars* (PBI OON 33267); Marche-les-Dames, understory, litter among ivy, 50.48250° N, 4.94917° E, 7 Aug. 2011, A. Henrard, 1 ♂, RBINS I.G.32708 *pars* (PBI OON 33564).

CZECH REPUBLIC: Zlínský Kraj, Lhota u Vsetina, Lhotské Paseky Sklep, building, 545 m, 49.324528° N, 17.936269° E, 8 Aug. 2010, coll. Andrěj Mácháč, det. Jørgen Lissner, 1 ♂ (leg. J. Lissner, collection (V), 6821) (PBI OON 32902).

Description

Male (PBI OON 33564)

As for the genus except as noted. Total length 2.02.

Female (PBI OON 6262)

As male except as noted. Total length 2.21.

Distribution

Austria, Belgium, Bulgaria, Czech Republic, France, Germany, Italy, Netherlands, Spain, Switzerland (after Le Peru 2011, extended).

Remarks

It was the count de Dalmas (1916), and then Kaston (1948), who pointed out that the specimen mentioned in Emerton's note on the New England spiders (Emerton 1909) was not *Orchestina saltitans* but *Tapinesthis inermis*. This single specimen is the only one ever recorded from the United States and was found in the cellar of the Boston Society of Natural History building, probably hidden in a package shipped from Europe to the Society (Kaston 1948). The mention 'introduced in USA' is therefore not valid as no further specimens were found, which indicates that no viable population of this species persists in the USA.

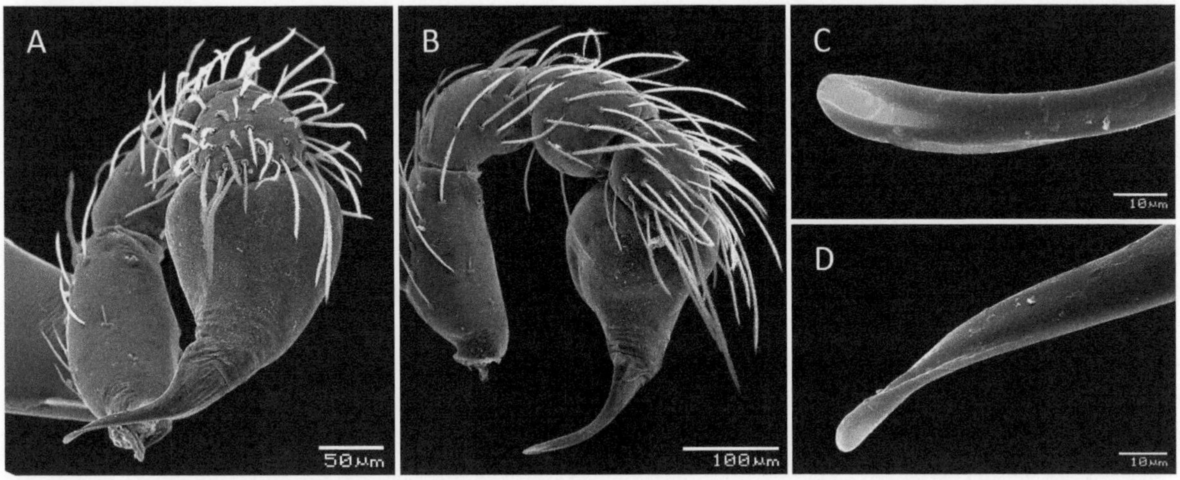

Fig. 7. *Tapinesthis inermis*, SEM views of ♂ (PBI 33564). **A**. Palp, frontal view. **B**. Same, prolateral view. **C**. Tip of embolus, ventral view. **D**. Same, retrolateral view.

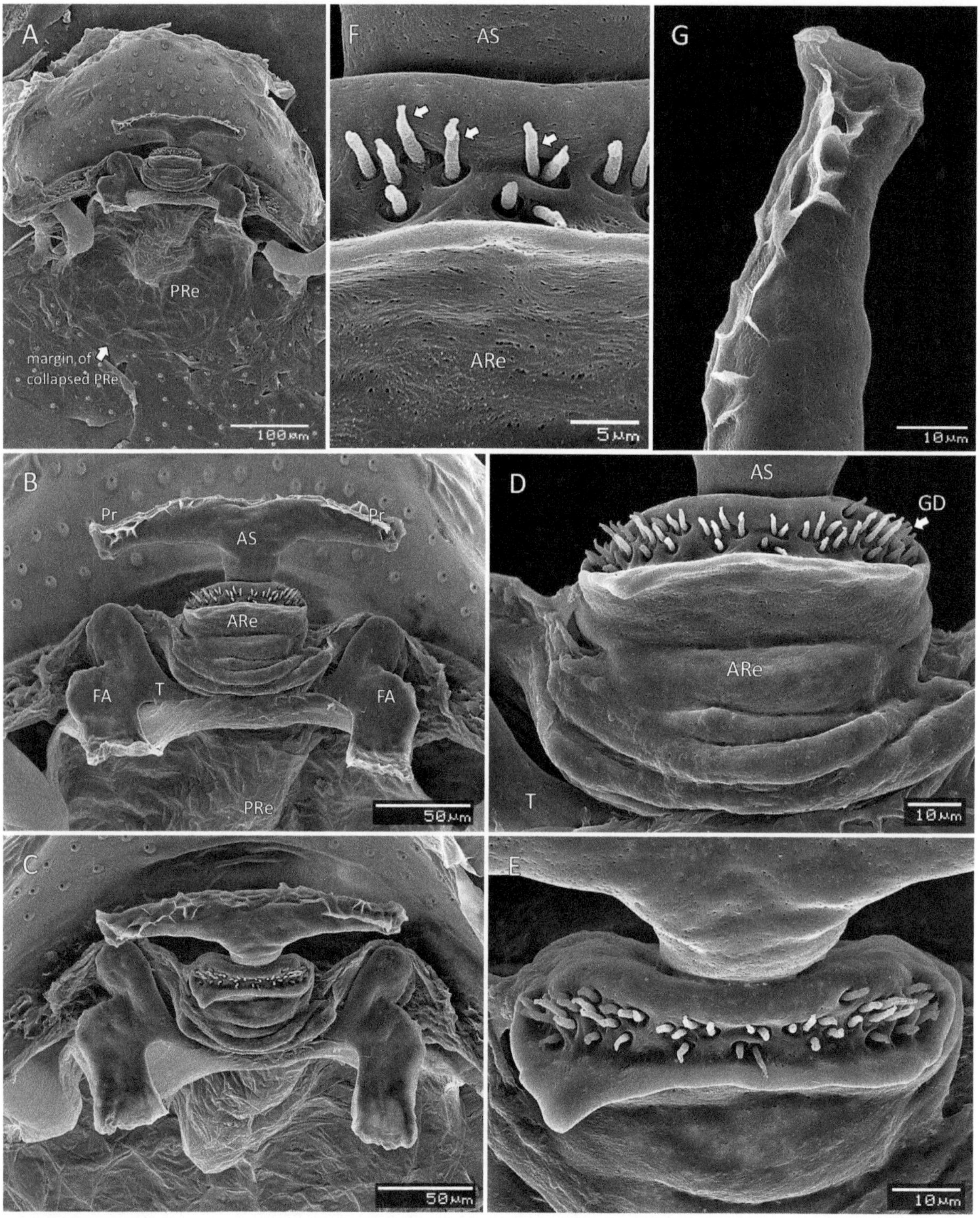

Fig. 8. *Tapinesthis inermis*, SEM views of ♀ genitalia (PBI 33267). **A**. Overview of genital area, dorsal view. **B**. Same, close up. **C**. Same, view slightly anterior. **D**. Anterior receptaculum (ARe), close up, arrow indicates gland ducts (GD). **E**. Same, anterior view. **F**. Same, detail of gland ducts (arrows). **G**. Lateral protrusion (Pr), detail, dorsal view. – ARe: Anterior Receptaculum; AS: Anterior Sclerite; FA: Flattened Apodeme; GD: Gland Duct; Pr: lateral Protrusion; PRe: Posterior Receptaculum; T: transverse sclerite.

Discussion

Tapinesthis inermis is perhaps one of the easiest oonopid species to identify. This spineless "molles" oonopid, without abdominal scuta, can easily be recognized by the absence of ventral spines on the first tibiae, the net-shaped pattern on the carapace, the clumped eyes with slightly recurved PER and by the plesiomorphic aspect of the male palp and the female genital organs. Although this species is not easily seen, it cannot be considered as "rare". However, in the Czech Republic for instance, it is rarely found and only females were caught before this study (J. Král, pers. comm.; see also Korenko *et al.* 2007).

Dalmas (1916) mentions the occurrence of two adult seasonal forms, one in spring and one in summer, differing by the colouration and the hair cover. This observation is of interest since it is the only spider

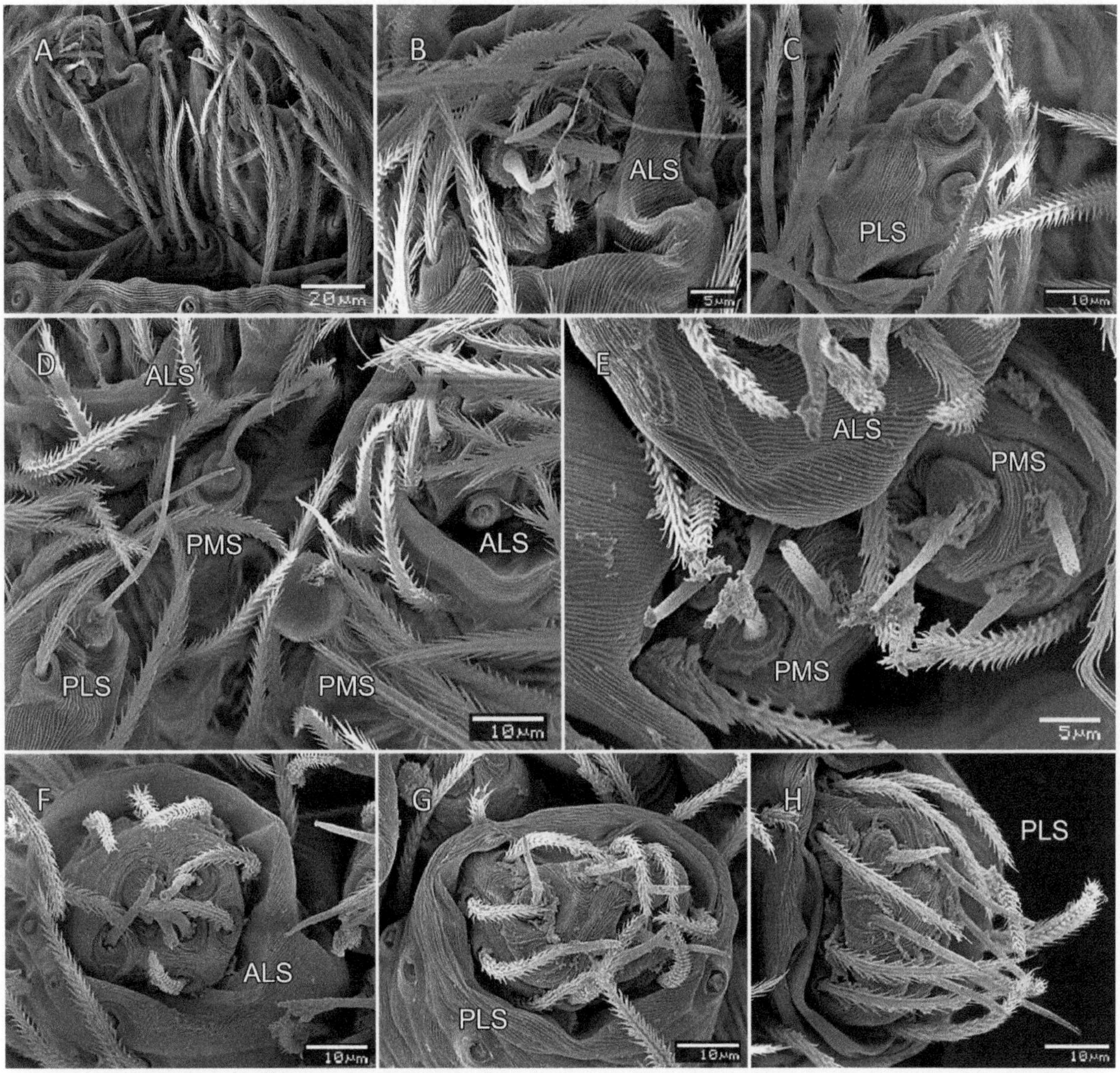

Fig. 9. *Tapinesthis inermis*, SEM views of spinneret area of ♂ (PBI 33564) and ♀ (PBI 33267). **A**. ♂, colulus, ventral view. **B**. Same, anterior lateral spinnerets, ventral view. **C**. Same, posterior lateral spinnerets. **D**. Same, focus on posterior median spinnerets. **E**. ♀, posterior median spinnerets, ventral view. **F**. Same, anterior lateral spinnerets. **G**. Same, posterior lateral spinnerets. **H**. Same, lateral view. – ALS: anterior lateral spinnerets; PLS: posterior lateral spinnerets; PMS: posterior median spinnerets.

species for which seasonal forms have been recognized. However, our observations do not point in the same direction. The illustrations of Fig. 1 clearly show the colour differences between a specimen caught in summer (Fig. 1A–B) and another one photographed in spring (Fig. 1C–E). The spring specimen (female) is pale, whereas the one from August (male) is orange. The hair cover is similar in both cases. Since we know that the differences do not depend on the sex, we assume that the two forms are simply specimens of different age: they become darker, possibly by incorporating pigment from their prey, and might in some cases loose part of the hair cover. Dalmas (1916) possibly observed dark specimens that were born in autumn and had passed the winter, which, in the Mediterranean area where he made his observations, might have been perfectly possible.

Genitalia, male palp

The male palp of *T. inermis* is an excellent example of the basic plesiomorphic structure of the haplogyne palp: a piriform bulbus with a simple tubular embolus, without appendage as found in some segestriids. The latter character is the main difference with *Oonopinus angustatus* Simon, 1882, another spineless species (Simon 1882; Saaristo & Marusik 2009; Le Peru 2011).

A number of features of the male palp can only be seen when it is cleared. The seminal duct has apparently two parts: a short, well sclerotized part leading into the embolus and a thin-walled part connecting to the faintly delimited, vesicle-like reservoir. Next to the latter, a thin linear structure runs from the base of the bulb to the distal end of the sperm reservoir. This structure is also visible in *Spinestis nikita* Saaristo & Marusik, 2009 and *Megaoonops avrona* Saaristo, 2007 (Saaristo & Marusik 2009) and some other oonopids. Saaristo and Marusik (2009) regard the linear structure adjacent to the sperm reservoir as a sclerite, but Grismado & Ramirez (2013) and Grismado *et al.* (in press) clearly showed that this is a tendon, which is in accordance with the findings of Huber (2004).

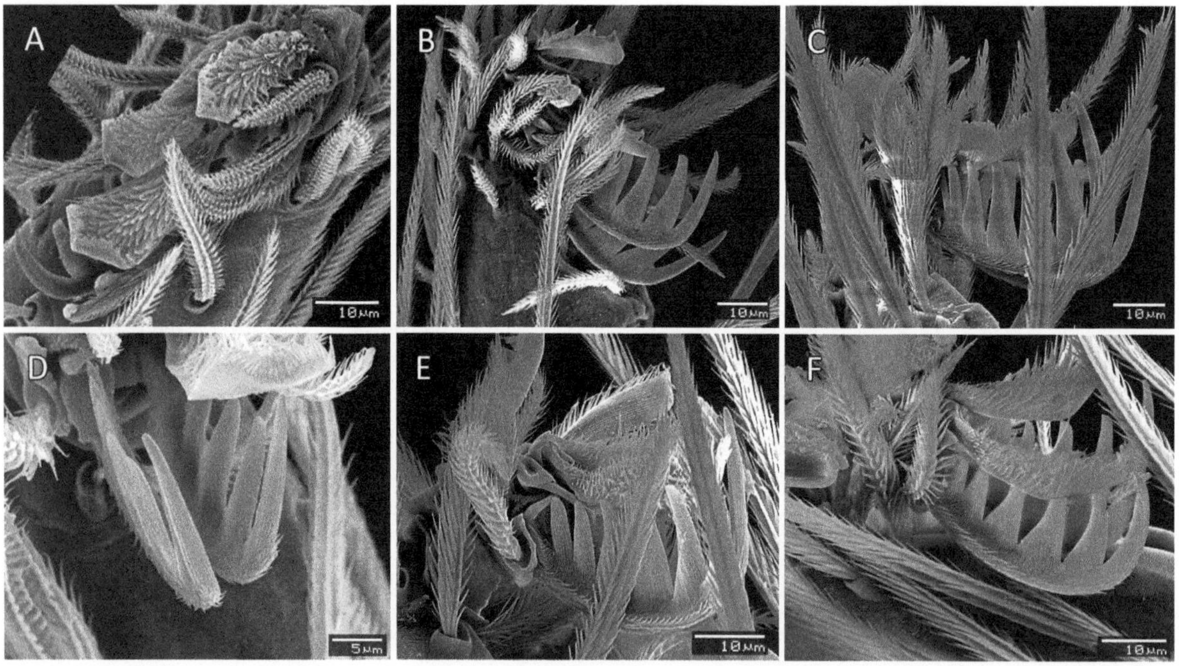

Figs. 10. *Tapinesthis inermis*, SEM views of tarsal claws of ♂ (PBI 33564). **A.** Leg I, anterior view. **B.** Same, lateral view. **C.** Leg II, lateral view. **D.** Same, anterior view. **E.** Leg III, lateral view. **F.** Leg IV, lateral view.

Female genitalia

It is now well established that the internal genital structures of some haplogynes are highly diverse and can be remarkably complex (Uhl 2000; Huber 2002; Burger *et al.* 2003, 2006; Burger 2007). According to some authors, the complexity of the female genital organ may be correlated to its function for postcopulatory female choice (Burger *et al.* 2003; Burger 2011a, 2011b, 2011c).

The female genitalia of *Tapinesthis inermis* are characterized by the presence of a large T-shaped anterior sclerite (AS) based antero-ventrally to the anterior receptaculum (ARe), associated with gland ducts dorsally and apically to the ARe, strong, flattened apodemes and a large, egg-shaped membranous posterior receptaculum.

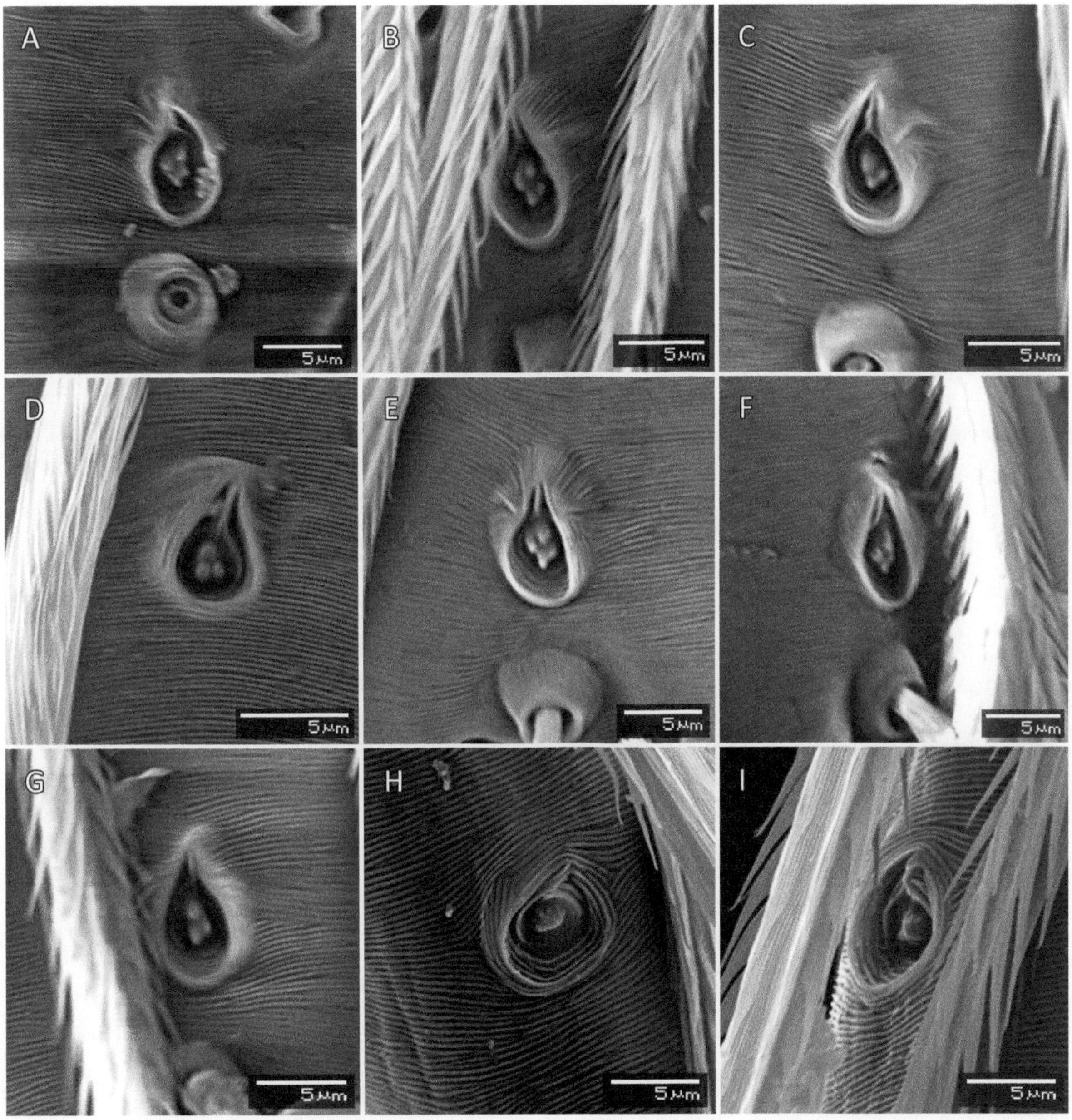

Fig. 11. *Tapinesthis inermis*, SEM, A–H dorsal views, I lateral view. – **A–D**. ♂ (PBI 33882). **A**. Leg I. **B**. Leg II. **C**. Leg III. **D**. Leg IV. – **E–G**. ♀ (PBI 33883). **E**. Leg I. **F**. Leg II. **G**. Leg III. – **H–I**. ♂ (PBI 33564). **H**. Leg IV. **I**. Tarsal organ, leg IV.

The combination of these structures is very similar to the one encountered in many dysderoid spiders such as Dysderidae (Forster & Platnick 1985; Ulh 2000; Burger & Kropf 2007), Segestriidae (Brignoli 1976), Orsolobidae (Forster & Platnick 1985; Izquierdo & Labarque 2010) and some basal Oonopidae such as *Puan chechehet* Izquierdo, 2012 (Izquierdo *et al.* 2012) and *Unicorn catleyi* Platnick & Brescovit, 1995 (Izquierdo & Rubio 2011). It thus supports the basal placement of *Tapinesthis* among Oonopinae.

Burger's terminology for receptacula is based on sections. If a sclerite anterior of the uterus didn't contain sperm in one of the different females analyzed, the use of the term "receptaculum" was considered ambiguous and the term "anterior sclerite" was preferred over "anterior receptaculum". Burger (2010, 2011a, 2013) assumes that in "higher" oonopids, the receptacula are reduced to sclerites and that sperm is stored only in the uterus in order to improve the control over fertilization processes through cryptic female choice (Burger *et al.* 2003; Burger 2011a, 2011b, 2011c).

In the present work, no sections were made to verify the presence of sperm in those sclerites. However, the use of "ARe" for this sclerite was preferred to differentiate the upper part, i.e., the T-shaped structure

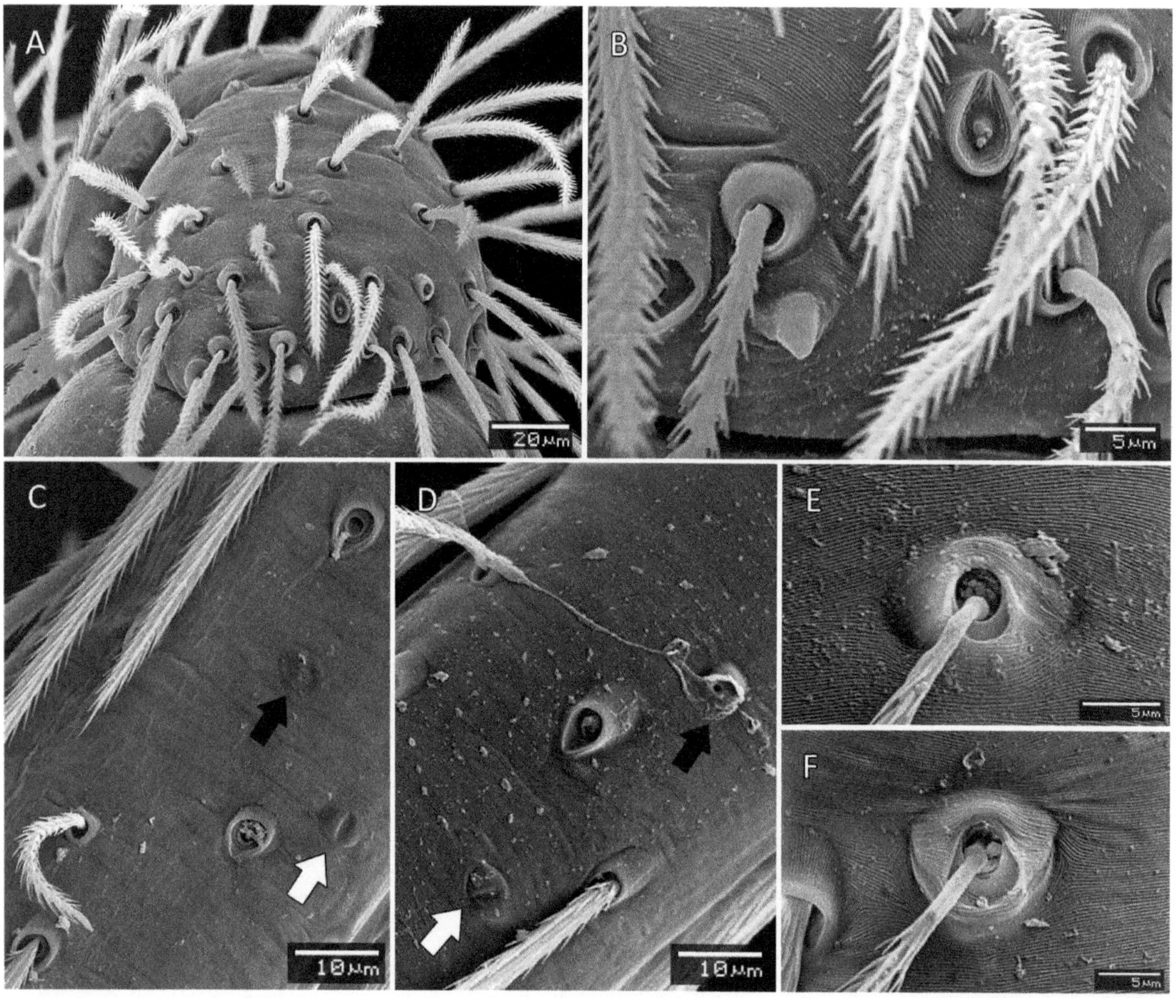

Fig. 12. *Tapinesthis inermis* ♂ (PBI 33564), SEM views. **A**. Cymbium, frontal view. **B**. Same. **C**. Metatarsus II, dorsal view, black arrow showing pore and white arrow indicating eye-shaped structure. **D**. Metatarsus IV, black arrow as in C. **E**. Trichobothrium, dorsal view of bothrium on tibia IV. **F**. Dorsal view of bothrium on metatarsus.

called here anterior sclerite (As), which has a different function (muscle attachment, see Izquierdo & Rubio 2011).

It was not possible to observe the uterus externus (UE). The tissue of that organ typically appears as a very flat, membranous structure on SEMs and the opening for the UE is supposed to be somewhere between the ARe and the PRe (M. Burger, pers. comm.). Again, cross sections are needed to certify the position of the UE. We therefore avoid the term 'Anterior Uterine Sclerite (AUS)' used by Fannes (2010) and Henrard & Jocqué (2012) for what is called here the AS (e.g.). The term 'uterine' can indeed be confusing, especially since we do not know the position of the uterus on SEMs.

In *Tapinesthis*, gland ducts are confined to a frontal groove on the apical part of the ARe. In other oonopids, GD are located in the middle of the AS (e.g., *Puan chechehet*, see Izquierdo *et al.* 2012) or at the base of the AS (e.g., *Heteroonops spinimanus*, see Burger 2011a; *Oonopinus kilikus*, see Burger 2013). Note that GD are also present on the AS in other dysderoid genera of the families Orsolobidae (Izquierdo & Labarque 2010), Segestriidae (Brignoli 1976; Forster & Platnick 1985) and Dysderidae (Uhl 2000). According to Burger (2011b), gland ducts associated with the anterior sclerites may produce part of the secretion surrounding the spermatozoa in the posterior receptacula. This secretion may either serve to store and fix the spermatozoa within the receptaculum or provide nutrition for the spermatozoa (e.g., Uhl 1994, 1996).

Relationships

Simon (1893) recognized two informal groups, 'molles' and 'loricatae', based on the presence of abdominal scuta and sclerotization of the body. Petrunkevitch (1923) and subsequent authors recognized these groups as subfamilies, Oonopinae and Gamasomorphinae, respectively. Later, a third subfamily, the monotypic Orchestininae, was added for the genus *Orchestina* (Chamberlin & Ivie 1942). In a recent paper, Platnick *et al.* (2012) profoundly changed the subfamilial structure of the Oonopidae. The main characters they used were the structure of the tarsal organs and more precisely their number of receptors, the sclerotization of the spermduct in the male palp and the position of the eyes. On the basis of seven characters, the family was subdivided into three subfamilies: the Sulsulinae, Oonopinae and Orchestininae, which remained monotypic.

The three main characters on which the recent subfamilial structure is based are not straightforward in all the oonopids and may need an additional state. In *Tapinesthis*, and a few other genera, it might indeed be questioned whether the eyes are clumped and whether the male palpal bulbus lacks a sclerotized sperm duct. The situation in *Tapinesthis* can be considered as intermediate between Orchestininae and Oonopinae: the eyes are not really clumped but also not as far apart as in *Orchestina* and they nearly have an H-shaped disposition with a slightly recurved PER. The absence of a clearly thick-walled, heavily sclerotized sperm duct within the male palp leaves no doubt about the placement of *Tapinesthis* among Oonopinae. The sperm duct in the bulbus of *Tapinesthis* has a thin-walled part, but the distal part is still clearly sclerotized. However, this structure cannot be considered identical to the sclerotized duct as found in *Orchestina* and the outgroups used in Platnick *et al.* (2012); hence an intermediate state for this set-up would be necessary. The tarsal organ receptor pattern that gives the impression to be 4433 at a first glance turns out to be 3322, since the distalmost receptor appears to be bifid. Here again, this configuration would best be considered as a separate state different from both the "simple" 3322 without a bifid receptor and the plesiomorphic 4433 receptor formulas.

In view of these modifications, a more extensive set of characters will be needed to clarify the position of *Tapinesthis* within the Oonopinae. However, the combination of the characters discussed above supports a relatively basal placement of *Tapinesthis* in the Oonopinae. In addition, recent molecular results of De

Busschere *et al.* (submitted), which place *Tapinesthis* relatively basally among the Oonopinae, appear to corroborate our findings on the basis of morphology.

Acknowledgements

The present study is part of a Planetary Biodiversity Inventory (PBI) project of the family Oonopidae, coordinated by the American Museum of Natural History (AMNH), New York. We are indebted to Jørgen Lissner for the loan of material (*Tapinesthis inermis* ♂ from the Czech Republic). We thank Y. Marusik for sharing the nice picture of the ♂ palp (Fig. 4E). M. Burger and M. Izquierdo shared their knowledge of the Oonopidae in order to solve difficulties with the interpretation of the structure of the female genitalia. B. Huber, P. Michalik and M. Ramirez commented on the structure in the cleared male palp. This paper is publication BRC 307 of the Biodiversity Research Center (Université Catholique de Louvain).

References

Brignoli P.M. 1976. Ragni d'Italia XXIV. Note sulla morfologia dei genitali interni dei Segestriidae e cenni sulle specie italiane (Araneae). *Fragmenta Entomologica* 12: 19–62.

Burger M. 2009. Female genitalia of goblin spiders (Arachnida: Araneae: Oonopidae): a morphological study with functional implications. *Invertebrate Biology* 128: 340–358.

Burger M. 2010. Goblin spiders without distinct receptacula seminis (Arachnida: Araneae: Oonopidae). *Journal of Morphology* 271: 1110–1118.

Burger M. 2011a. Structure and function of the female reproductive system in three species of goblin spiders (Arachnida: Araneae: Oonopidae). *Invertebrate Biology* 130: 148–160.

Burger M. 2011b. Functional morphology of female goblin spider genitalia (Arachnida: Araneae: Oonopidae) with notes on fertilization in spiders. *Zoologischer Anzeiger* 250: 123–133.

Burger M. 2011c. Structure and function of the female reproductive system in three species of goblin spiders (Arachnida: Araneae: Oonopidae). *Invertebrate Biology* 130: 148–160.

Burger M. 2013. Genital morphology of female goblin spiders (Arachnida: Araneae: Oonopidae) with functional implications. *Acta Zoologica* (Stockholm) 92: 280–290.

Burger M., Graber W., Michalik P. & Kropf C. 2006. *Silhouettella loricatula* (Arachnida, Araneae, Oonopidae) a haplogyne spider with complex female genitalia. *Journal of Morphology* 258: 663–677.

Burger M., Izquierdo M. & Carrera P. 2010. Female genital morphology and mating behaviour of *Orchestina* (Arachnida: Araneae: Oonopidae). *Zoology* 113: 100–109.

Burger M. & Kropf C. 2007. Genital morphology of the haplogyne spider *Harpactea lepida* (Arachnida, Araneae, Dysderidae). *Zoomorphology* 126**:** 45–52.

Burger M. & Michalik P. 2010. The male genital system of goblin spiders: evidence for the monophyly of Oonopidae (Arachnida: Araneae). *American Museum Novitates* 3675: 1–13.

Burger M., Nentwig W. & Kropf C. 2003. Complex genital structures indicate cryptic female choice in a haplogyne spider (Arachnida, Araneae, Oonopidae, Gamasomorphinae). *Journal of Morphology* 255: 80–93.

Dalmas R. de 1916. Révision du genre *Orchestina* E.S., suive de la description de nouvelles espèces du genre *Oonops* et d'une étude sur les Dictynidae du genre *Scotolathys*. *Annales de la Société entomologique de France* 85: 203–258.

Dalmas R. de. 1920. Captures d'araignées rares dans la région parisienne. *Bulletin de la Société entomologique de France* 1920: 59–60.

Emerton J.H. 1909. Supplement to the New England spiders. *Transactions of the Connecticut Academy of Arts and Sciences* 14: 171–236.

Forster R.R. & Platnick N.I. 1985. A review of the austral spider family Orsolobidae (Arachnida, Araneae), with notes on the superfamily Dysderoidea. *Bulletin of the American Museum of Natural History* 181: 1–230.

Grismado C.J. 2010. Description of *Birabenella*, a new genus of goblin spiders from Argentina and Chile (Araneae: Oonopidae). *American Museum Novitates* 3693: 1–21.

Grismado C.J., Deeleman C., Piacentini L., Izquierdo M.A. & Ramírez M.J. In press. A taxonomic review of the goblin spiders of the genus *Dysderoides* Fage and their Himalayan relatives of the genera *Trilacuna* Tong & Li, and *Himalayana* new genus (Araneae: Oonopidae). *Bulletin of the American Museum of Natural History.*

Grismado C.J. & Ramírez M.J. 2013. The New World goblin spiders of the new genus *Neotrops* (Araneae, Oonopidae). Part 1. *Bulletin of the American Museum of Natural History* 383: 1–150.

Henrard A. & Jocqué R. 2012. An overview of Afrotropical canopy-dwelling *Orchestina* (Araneae, Oonopidae), with a wealth of remarkable sexual dimorphic characters. *Zootaxa* 3284: 1–104.

Heimer S. & Nentwig W. 1991. *Spinnen Mitteleuropas: Ein Bestimmungsbuch.* Verlag Paul Parey, Berlin.

Huber B.A. 2004. Evolutionary transformation from muscular to hydraulic movements in spider (Arachnida, Araneae) genitalia: a study based on histological serial sections. *Journal of Morphology* 261: 364–376.

Izquierdo M.A. & Labarque F.M. 2010. Description of the female of *Orsolobus pucara* Forster & Platnick 1985, with comments on the functional morphology of the female genitalia in Dysderoidea (Araneae: Dysderoidea: Orsolobidae). *Journal of Arachnology* 38: 511–520.

Izquierdo M.A., Ferretti N. & Pompozzi G. 2012. On *Puan*, a new genus of goblin spiders from Argentina (Araneae, Dysderoidea, Oonopidae). *American Museum Novitates* 3757: 1–22.

Izquierdo M.A. & Rubio G.D. 2011. Male genital mutilation in the high-mountain goblin spider, *Unicorn catleyi*. *Journal of Insect Science* 118 (11): 1–8.

Kaston B.J. 1948. Spiders of Connecticut. *Bulletin – State Geological and Natural History Survey of Connecticut* 70: 1–874.

Knoflach B., Pfaller K. & Stauder F. 2010. *Cortestina thaleri* – a new dwarf six-eyed spider from Austria and Italy (Araneae: Oonopidae: Oonopinae). *Contributions to Natural History* 12: 743–771.

Korenko S., Rezác M. & Pekár S. 2007. Spiders (Araneae) of the family Oonopidae in the Czech Republic. *Arachnologische. Mitteilungen* 34: 6–8.

Kraus O. 1967. Zur Spinnenfauna Deutschlands, I. *Tapinesthis inermis*, eine für Deutschland neue Oonopide (Arachnida: Araneae: Oonopidae). *Senckenbergiana Biologica* 48: 381–385.

Le Peru B. 2007. Catalogue et répartition des araignées de France. *Revue Arachnologique* 16: 1–468.

Le Peru B. 2011. The spiders of Europe. Synthesis of data – Volume 1. Atypidae to Theridiidae. *Mémoires de la Société Linnéenne de Lyon* 2: 1–522.

Lazarov S., Deltshev C. & Blagoev G. 2001. The spiders (Araneae) of Sashtinska Sredna Gora Mountain (Bulgaria). Faunistic and zoogeographical analysis. *Acta Zoologica Bulgarica* 53: 3–28.

Petrunkevitch A.I. 1923. On families of spiders. *Annals of the New York Academy of Sciences* 29: 145–180.

Platnick N.I., Abrahim N., Alvarez-Padilla F., Andriamalala D., Baehr B., Baert L., Bonaldo A.B., Brescovit A.D., Chousou-Polydouri N., Dupérré N., Eichenberger B., Fannes W., Gaublomme E., Gillespie R.G., Grismado C.J., Griswold C.E., Harvey M.S., Henrard A., Hormiga G., Izquierdo M.A., Jocqué R., Kranz-Baltensperger Y., Kropf C., Ott R., Ramírez M.J., Raven R.J., Rheims C.A., Ruiz G.R.S., Santos A.J., Saucedo A., Sierwald P., Szuts T., Ubick D. & Wang X.P. 2012. Tarsal organ morphology and the phylogeny of goblin spiders (Araneae, Oonopidae), with notes on basal genera. *American Museum Novitates* 3736: 1–52.

Platnick N.I. 2013. The World Spider Catalog, version 14.5. American Museum of Natural History.

Platnick N.I. & Dupérré N. 2010. The goblin spider genera *Stenoonops* and *Australoonops* (Araneae, Oonopidae), with notes on related taxa. *Bulletin of the American Museum of Natural History* 340: 1–111.

Saaristo M.I. & Marusik Y.M. 2009. A new genus and species of oonopid spider (Araneae, Oonopidae) from Ukraine. *ZooKeys* 24: 63–74.

Simon E. 1882. Etudes arachnologiques. 13e mémoire. XX. Descriptions d'espèces et de genres nouveaux de la famille des Dysderidae. *Annales de le Société entomologique de France*, Série 6, 2: 201–240.

Simon E. 1893. *Histoire Naturelle des Araignées*. Vol. 1, part 2. Paris, pp. 257–488.

Simon E. 1914. Les arachnides de France. *Synopsis Générale et Catalogue des Espèces Françaises de l'Ordre des Araneae*, Paris, 6 (1): 1–308.

Trotta A. 2005. Introduzione al ragni italiani (Arachnida Araneae). *Memorie della Società Entomologica Italiana* 83: 3–178.

Uhl G. 1994. Genital morphology and sperm storage in *Pholcus phalangioides* (Fuesslin, 1775) (Pholcidae; Araneae). *Acta Zoologica* (Stockholm) 75: 1–12.

Uhl G. 1996. Sperm storage secretion of female cellar spiders (*Pholcus phalangioides*; Araneae): A gel-electrophoretic analysis. *Journal of Zoology* 240: 153–161.

Uhl G. 2000. Two distinctly different sperm storage organs in female *Dysdera erythrina* (Araneae: Dysderidae). *Arthropod Structure & Development* 29: 163–169.

Van Keer K., De Koninck H., Vanuytven H. & Van Keer J. 2006. Some – mostly southern European – spider species (Araneae), new or rare to the Belgian fauna, found in the city of Antwerp. *Nieuwsbrief van de Belgische Arachnologische Vereniging* 21: 33–40.

Van Keer K., Vanuytven H., De Koninck H. & Van Keer J. 2010. More than one third of the Belgian spider fauna (Araneae) found within the city of Antwerp: faunistics and some reflections on urban ecology. *Nieuwsbrief van de Belgische Arachnologische Vereniging* 25: 160–180.

3

The Lanternfly genus *Polydictya* (Hemiptera: Fulgoromorpha: Fulgoridae) from Sulawesi and neighbouring islands, with the description of three new species

Jérôme CONSTANT

Royal Belgian Institute of Natural Sciences, O.D. Phylogeny and Taxonomy, Entomology, Vautier street 29, B-1000 Brussels, Belgium.

Abstract. The species of the genus *Polydictya* Guérin-Méneville, 1844 from Sulawesi and adjacent islands are reviewed and three new species are described: *P. bantimurung* sp. nov. (Sulawesi), *P. basirubra* sp. nov. (Sulawesi, Wowoni and Buton) and *P. pelengana* sp. nov. (Peleng). An identification key for the 6 species recorded in the area is given. *Habitus*, detailed illustrations and a distribution map are provided for all species. The male genitalia of the three new species are illustrated and described, and the male genitalia of *P. thanatos* Chew Kea Foo, Porion & Audibert, 2010 are figured and described for the first time. The genus *Polydictya* now contains 24 species.

Keywords. Indonesia, lanternbug, Fulgoroidea, Homoptera, planthopper.

Introduction

The genus *Polydictya* was erected by Guérin-Méneville (1844) to accommodate one species, *Eurybarchis* (sic!) *basalis* Hope, 1843, described from Silhet (Myanmar) and mentioned from Penang Island (Malaysia) by Guérin-Méneville (1844). Species were progressively added to the genus (Bourgoin 2014).

The genus *Thaumastodictya*, described by Kirkaldy (1902) based on one species, *T. krisna* Kirkaldy, 1902 from Sri Lanka, was synonymized under *Polydictya* by Distant (1906a). Lallemand (1959, 1963), Nagai & Porion (1996, 2004), Constant & Pham (2008), Constant (2009, 2010) and Chew Kea Foo *et al.* (2010) more recently added new species and proposed nomenclatural changes within the genus.

The genus presently contains 21 species (Bourgoin 2014). It is distributed in the Oriental region: from Sri Lanka it extends over northern India, Thailand, Vietnam and southwards to Sulawesi and its adjacent islands.

One species, *P. crassa*, was described from "South Celebes" (Distant 1906b). It was the only species known from Sulawesi until the recent descriptions of *P. katsurakoae* Nagai & Porion, 1996 and *P. thanatos* Chew Kea Foo, Porion & Audibert, 2010.

The study of recent material in the collections of BMNH, MNHL and RBINS led to the discovery of three new species which are here described, illustrated and keyed together with the three previously decribed ones.

Materials and methods

Type specimens of all species have been examined. The male genitalia were dissected as follows: the pygofer was cut from the abdomen of the softened specimen with a needle blade, then boiled for about one hour in a 10% solution of potassium hydroxide (KOH). The phallic complex was dissected with a needle blade and all pieces examined in ethanol, the whole placed in glycerine for preservation. Observations were done with a Leica MZ8 stereo microscope. Pictures were taken with a Canon EOS 300 D camera with Sigma DG Macro lens and optimized with Adobe Photoshop CS3. The inflation of the phallus was not done due to the difficulty to obtain good and replicable results, and because it is not indispensable to separate the species in the genus *Polydictya.*

For the transcription of the labels of the types, the wording on each single label is limited by square brackets.

The measurements were taken as in Constant (2004) and the following abbreviations are used:

BF = breadth of the frons
BTg = breadth of the tegmen
BV = breadth of the vertex
LF = length of the frons
LT = total length
LTg = length of the tegmen
LV = length of the vertex

Acronyms used for the collections (name of the curator in parentheses):

BMNH = Natural History Museum, London, United Kingdom (M. Webb)
EUM = Ehime University Museum, Matsuyama, Japan (H. Yoshitomi)
MHNL = Muséum d'Histoire naturelle de Lyon, France (H. Labrique, J. Clary and C. Audibert)
MNHN = Museum national d'Histoire naturelle, Paris, France (T. Bourgoin and A. Soulier)
RBINS = Royal Belgian Institute of Natural Sciences, Brussels, Belgium (W. Dekoninck)

Results

Class Hexapoda Blainville, 1816
Order Hemiptera Linnaeus, 1758
Suborder Auchenorrhyncha Duméril, 1806
Infraorder Fulgoromorpha Evans, 1946
Superfamily Fulgoroidea Latreille, 1807
Family Fulgoridae Latreille, 1807

Genus ***Polydictya*** Guérin-Méneville, 1844

Polydictya Guérin-Méneville, 1844: 358. Type species: *Polydictya basalis* (Hope, 1843) by monotypy. *Thaumastodictya* Kirkaldy, 1902: 307. Type species: *Polydictya krisna* Kirkaldy, 1902 by original designation (junior synonym of *Polydictya pantherina* Gerstaecker, 1895, see Nagai & Porion 1996).

Polydictya – Walker 1851: 289 (list of species). — Stål 1866: 135 (key). — Atkinson 1885: 155 (described). — Distant 1888: 487 (compared to *Myrilla* Distant, 1888). — Karsch 1890: 63 (compared to *Anecphora* Karsch, 1890). — Gerstaecker 1895: 10 (compared to *Holodictya* Gerstaecker, 1895). — Kirkaldy 1902: 307 (compared to *Thaumastodictya* Kirkaldy, 1902). — Melichar 1903: 71 (erroneous synonymy with *Chalia* Walker, 1858). — Distant 1906a: 199 (key to Oriental genera), 215 (described, senior synonym of *Thaumastodictya*). — Kirkaldy 1907: 59 (note on publication date). — Schmidt 1907: 113 (compared to *Myrilla*, copied from Distant 1888). — Jacobi 1910: 101 (compared to *Coelodictya* Jacobi, 1910). — Schmidt 1912: 71 (compared to *Coelodictya* and *Holodictya*, comments on species). — Lallemand 1963: 7 (key to genera), 14 (key to species). — Nagai & Porion 1996: 13 (list of species, synonymies). — Constant & Pham 2008 (notes). — Constant 2009 (notes, key to species with hind wings red basally). — Constant 2010 (notes).
Thaumastodictya – Melichar 1903: 71 (description). — Distant 1906a: 215 (junior synonym of *Polydictya*).

The characters defining the genus were given by Lallemand (1963), i.e., head broad, slightly narrower than the pronotum; frons as long as broad, not carinate, largely rounded dorsally, broadened above clypeus and strongly broader than the latter; vertex at least 4 times broader than long, excavate; tegmina at least 2.5 times longer than broad, densely reticulate; clavus closed with vein A1+A2 extending up to sutural margin; hind wings almost entirely reticulate; anterior tibiae as long as femora, sometimes slightly foliaceous; hind tibiae with 4–6 lateral spines.

Identification key to the species of *Polydictya* from Sulawesi and neighbouring islands

1. Dorsum of abdomen black (Figs 5A, 6A) 2
– Dorsum of abdomen red or brown (Figs 3A, 4A) 3

2. Tegmina dark olivaceous brown with small whitish spots along costal margin, claval suture and disc (Fig. 5A); frons and vertex yellow brown (Fig. 5C–D); anterior tibiae slightly laminate (Fig. 5A–B) .... ***Polydictya pelengana*** sp. nov.
– Tegmina black with irregular white tranverse bands (Fig. 6A); frons and vertex red brown (Fig. 6C–D); anterior tibiae slender, not laminate (Fig. 6A–B) .... ***Polydictya thanatos*** Chew Kea Foo, Porion & Audibert, 2010

3. Frons and vertex unicolorous (Figs 3C–D, 4C–D) 4
– Frons black-brown with baso-lateral angles yellow brown (Fig. 1C–D) .... ***Polydictya bantimurung*** sp. nov.

4 Hind wings basally without large bright red patch (Figs 3A, 4A) 5
– Hind wings with a large basal bright red patch (Fig. 2 A) .... ***Polydictya basirubra*** sp. nov.

5. Tegmina nearly unicolorous, brown with a small whitish spot along claval suture and 3 small spots on disc of membrane, no spot along costal margin (Fig. 3A) and not tinged with red ventrally (Fig. 3B); tegmina slightly broader: LTg/BTg = 2.72 .... ***Polydictya crassa*** Distant, 1906
– Tegmina brown, tinged with reddish ventrally (Fig. 4B), and with whitish spots: large one at apex of clavus, 3 along claval suture, 3–4 along costal margin and 3 on disc of membrane (Fig. 4A); tegmina slightly narrower: LTg/BTg = 2.86 .... ***Polydictya katsurakoae*** Nagai & Porion, 1996

Figs 1–2. **1**. *Polydictya bantimurung* sp. nov. **2**. *Polydictya basirubra* sp. nov. **A**. Habitus, dorsal view. **B**. Habitus, ventral view. **C**. Vertex, pro and mesonotum, dorsal view. **D**. Frons, normal view. **E**. Habitus, lateral view. **F**. Head and prothorax, lateral view. Scale bars = 10 mm.

Figs 3–4. 3. *Polydictya crassa* Distant, 1906. **4**. *Polydictya katsurakoae* Nagai & Porion, 1996. **A**. Habitus, dorsal view. **B**. Habitus, ventral view. **C**. Vertex, pro and mesonotum, dorsal view. **D**. Frons, normal view. **E**. Labels. Scale bars = 10 mm.

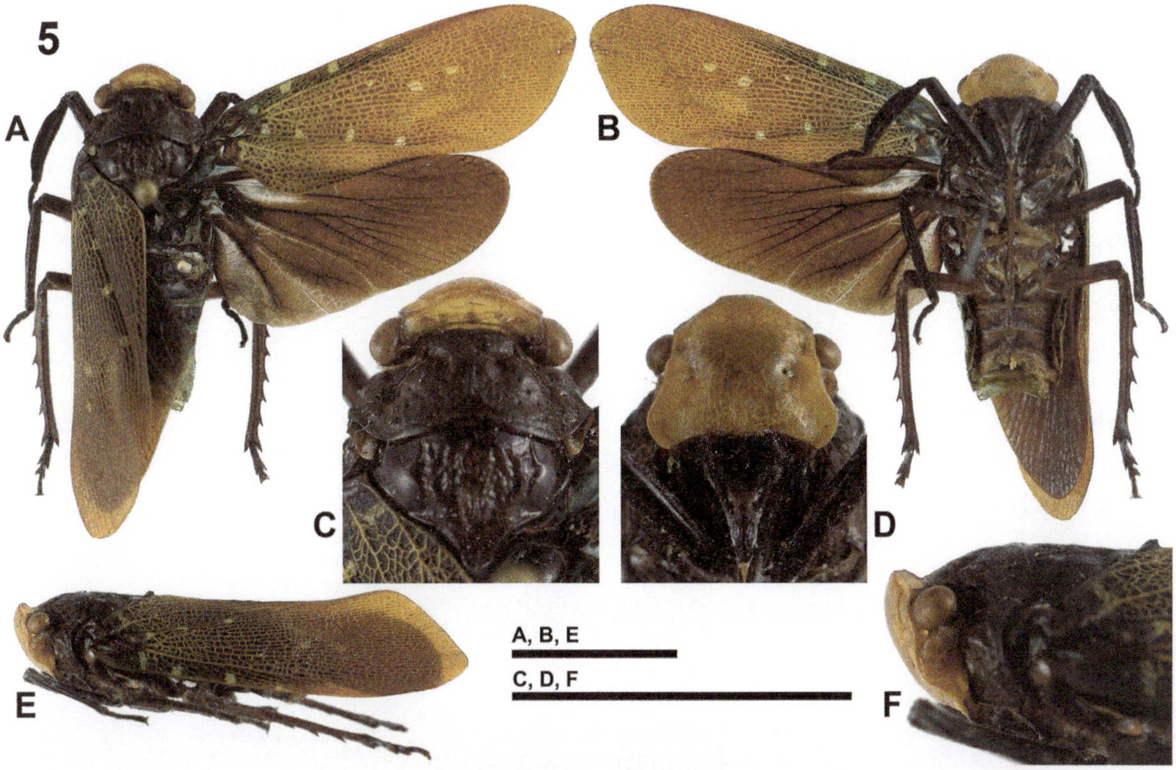

Figs 5–6. 5. *Polydictya pelengana* sp. nov. **6**. *Polydictya thanatos* Chew Kea Foo, Porion & Audibert, 2010. **A**. Habitus, dorsal view. **B**. Habitus, ventral view. **C**. Vertex, pro and mesonotum, dorsal view. **D**. Frons, normal view. **E**. Habitus, lateral view. **F**. Head and prothorax, lateral view. Scale bars = 10 mm.

Polydictya bantimurung sp. nov.
urn:lsid:zoobank.org:act:CB4E8FA9-FDC8-4197-AA6E-1E09DCBBF292
Figs 1, 7–8

Diagnosis

(1) Tegmina brown, with whitish spots: large one at apex of clavus, 2 small ones along claval suture and 3–4 on disc of membrane; external margin of clavus and 3–4 patches along costal margin, pale greenish yellow (Fig. 1A); (2) hind wings without basal red patch (Fig. 1A); (3) frons regularly rounded in dorsal view, not projecting anteriorly (Fig. 1C), dark brown with ventro-lateral angles yellow-brown (Fig. 1D); (4) anterior tibiae not laminate (Fig. 1A–B); (5) abdomen red dorsally (Fig. 1A); (6) genital segments dark brown.

Etymology

The species epithet refers to the type location, Bantimurung, and is treated as a noun in apposition.

Type material

Holotype

♂: [Coll. I.R.Sc.N.B., Sulawesi, Bantimurung, 4°54'S 119°45'E, Gift J. Constant, I.G.: 32.561] (RBINS).

Paratypes

2 ♂♂, 4 ♀♀: [Bantimurung, Sulawesi, 6.02, Coll. thierry PORION] (MHNL). Note: "6.02" on the labels of the paratypes stands for June 2002.

Description

Measurements and ratios. LT: ♂ (n = 3): 24.7 mm (23.9–25.1); ♀ (n = 4): 29.0 mm (28.0–29.5); LTg/BTg = 3.24; BV/LV = 5.4; LF/BF = 0.70.

Head. Black-brown with ventro-lateral angles of frons yellow-brown (Fig. 1D, F); vertex curved and deeply grooved transversely with lowest point in middle, with all 4 margins carinate (Fig. 1C); frons slightly visible from above (Fig. 1C), convex, smooth, subquadrate with impressed, S-shaped longitudinal groove along lateral margin, starting in front of eye (Fig. 1D); clypeus narrower and shorter than frons, brown apically (Fig. 1D); labium elongate, surpassing coxae (Fig. 1B); antennae yellow-brown with scape cylindrical, elongate; pedicel bulbous (Fig. 1F).

Thorax. Black-brown with lateral pleura of pronotum yellow-brown along ventral margin; mesonotum darker than pronotum (Fig. 1C, F); pro-, meso- and metasternites brown (Fig. 1B); pronotum with transverse wrinkles, longitudinal carina and pair of impressed points on disc; mesonotum smooth with disc wrinkled; scutellum impressed on disc and with apex elevated (Fig. 1C); tegulae dark brown with posterior half yellow-brown (Fig. 1C, F).

Tegmina. (Figs 1A–C). Dark brown with whitish spots on disc: one large, subquadrate at apex of clavus, 2 small ones along claval suture and 3–4 in a longitudinal row near nodal line; external margin of clavus and 3–4 large patches along costal margin, pale greenish yellow; costal and sutural margins subparallel, apex obliquely rounded.

Hind wings. (Fig. 1A–B). Entirely dark brown, broader than tegmina.

Legs. (Fig. 1A–B). All legs dark reddish brown with tarsi I and II darker; tibiae I and II slender; tibiae III with 4–5 lateral and 7 apical spines.

ABDOMEN. (Fig. 1A–B). Bright red dorsally with 2 basal tergites black; black-brown ventrally; genital segments dark brown.

MALE GENITALIA. Pygofer, anal tube and gonostyli dark brown; pygofer higher than long, with posterior margin sinuate in lateral view and showing apically rounded conical process in middle (Fig. 8A–B); anal tube elongate, 2.36 times longer than broad, broader at ⅔ of total length, strongly curved ventrally near base (Fig. 8A–B); lateral margins sinuate (Fig. 8B) and apical margin strongly notched in dorsal view (Fig. 8C); gonostyli (Fig. 8A) elongate with posterior half much broader and strong subbasal constriction in lateral view; strong tooth directed antero-ventrally at half length; apical margin subquadrate; phallic

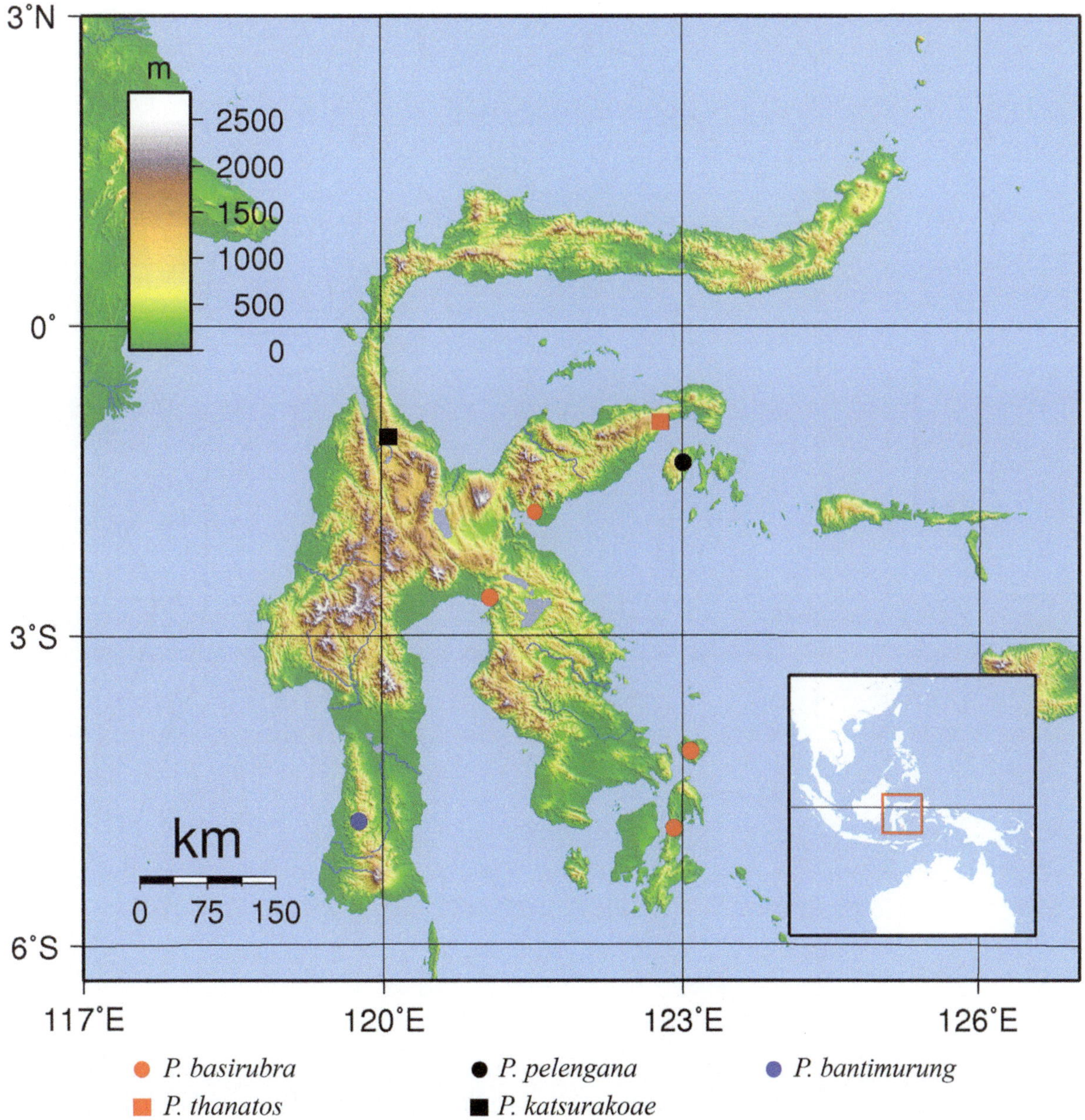

Fig. 7. Distribution map of the species of *Polydictya* Guérin-Méneville, 1844 of Sulawesi and neighbouring islands.

complex (Fig. 8D–E) with endosomal process of aedeagus laminate, slightly concave externally, showing irregular dorsal teeth along body, and apex strongly curved dorsally, hook-shaped.

Distribution

Known from South Sulawesi (Fig. 7).

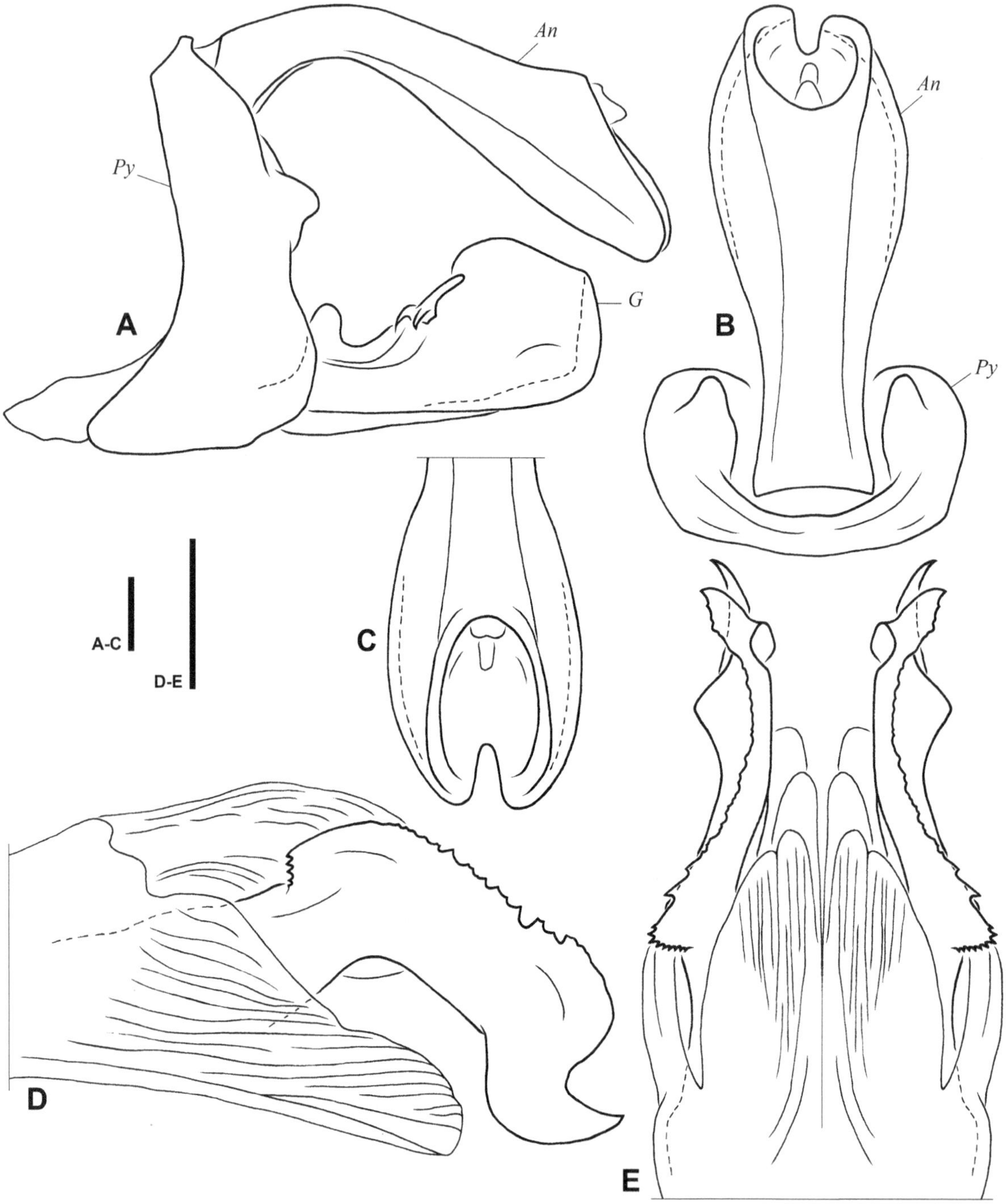

Fig. 8. *Polydictya bantimurung* sp. nov., ♂ genitalia. **A**. Pygofer, anal tube and gonostylus, left lateral view. **B**. Anal tube and pygofer, dorsal view. **C**. Apex of anal tube, postero-dorsal view. **D**. Phallic complex, lateral view. **E**. Phallic complex, dorsal view. *An* = anal tube; *Py* = pygofer; *G* = gonostylus. Scale bars = 1 mm.

Polydictya basirubra sp. nov.

urn:lsid:zoobank.org:act:10E93C3A-660E-46B5-B5C6-ECB86F8F62F8

Figs 2, 7, 9

Polydictya katsurakoae – Nagai & Porion 2002: 9 (record from Wowoni Island) [erroneous identification].

Diagnosis

(1) Tegmina brown, tinged with reddish ventrally on basal half (Fig. 2B), and with whitish markings: irregular one at apical half of clavus, 3 along claval suture, 4–5 along costal margin and 3–4 on disc of membrane (Fig. 2A); (2) hind wings with a large basal bright red patch (Fig. 2A); (3) frons regularly rounded in dorsal view, not projecting anteriorly (Fig. 2C), yellow-brown (Fig. 2D); (4) anterior tibiae not laminate (Fig. 2A–B); (5) abdomen red dorsally (Fig. 2A); (6) genital segments red.

Etymology

The species epithet is formed by the words *basis, -is* (Latin) = base, and *ruber, -bra, -brum* (adj., Latin) = red, and refers to the bright red patch at the base of the posterior wing.

Type material

Holotype

♂: [Wowoni Island, Indonesie, 09/2001, Coll. Alain Chaminade] (MHNL).

Paratypes

3 ♂♂, 2 ♀♀: 1 ♀: [Wowoni, X.02, Indonesie, Coll. Alain Chaminade] (MHNL); 1 ♂ [Wowoni Island, Indonesie, 09/01, Coll. thierry PORION] (MHNL); 1 ♂: [Malili, X.95, Sulawesi, Col. Thierry Porion] (MHNL); 1 ♀: [Buton Island, S.E. Sulawesi, Coll. thierry PORION] [Coll. T. PORION, Muséum de Lyon] (MHNL); 1 ♂: [Sulawesi Tengah: Nr. Morowali, Ranu River Area, 27.i.-20.iv.1980] [M.J.D. Brendell, B.M. 1980-280] (BMNH).

Coordinates of localities

Wowoni Island: 4°08′S 123°06′E; Buton Island: 5°03′S 122°53′E; Malili 2°38′12″S 121°5′27″E; Ranu River Area: 1°50'30"S 121°29'30"E.

Description

MEASUREMENTS AND RATIOS. LT: ♂ (n = 3): 26.7 mm (25.5–28.3); ♀ (n = 2, extrapolated): 30.0–31.1 mm; LTg/BTg = 2.61; BV/LV = 6.13; LF/BF = 0.75.

HEAD. Yellow-brown with clypeus dark brown to black-brown with paler median line (Fig. 2C–D, F); vertex curved with all margins carinate, and deeply grooved with lowest point in middle (Fig. 2C); frons slightly visible from above (Fig. 2C), convex, smooth, subquadrate, with impressed S-shaped longitudinal groove along lateral margin, starting in front of eye (Fig. 2D); clypeus narrower and shorter than frons (Fig. 2D); labium elongate, dark brown, surpassing hind coxae (Fig. 2B); antennae yellow-brown with scape cylindrical, elongate, and pedicel bulbous (Fig. 2D, F).

THORAX. (Fig. 2C, F). Dark brown with pronotum and sometimes disc of mesonotum paler yellow-brown; pronotum coriaceous with smooth median carina and 2 impressed points on anterior half of disc; mesonotum smooth with disc irregularly wrinkled and transverse wrinkles on scutellum; apex of scutellum elevated; tegulae dark brown with apex pale yellow-white.

TEGMINA. (Fig. 2A–B, E). Brown with red hue on corium, more conspicuous ventrally; 4 subquadrate to rectangular pale yellow spots along costal margin; irregular yellow-white markings on disc along claval joint and on clavus, and 2 spots on disc aligned longitudinally near nodal line; largest spot usually at apex of clavus; tegmina slightly broadening from base to apex and with apex oblique.

Hind wings. (Fig. 2A–B). Brown with antero-basal bright red patch covering about ¼ of total surface and not reaching posterior margin; broader than tegmina.

Legs. (Fig. 2A–B). All legs reddish-brown; tibiae I and II slender; tibiae III with (4–)5 lateral and 7 apical spines.

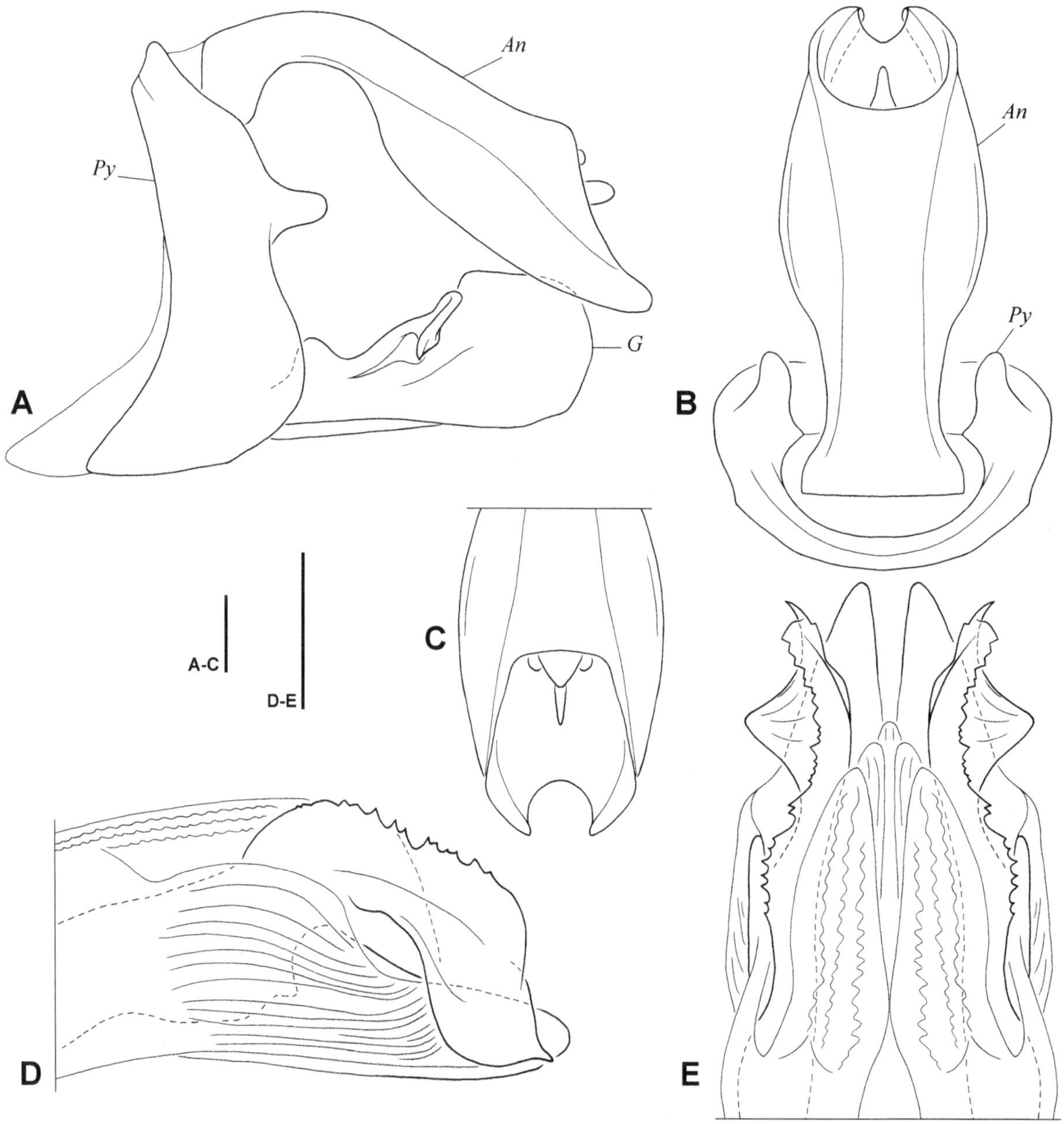

Fig. 9. *Polydictya basirubra* sp. nov., ♂ genitalia. **A**. Pygofer, anal tube and gonostylus, left lateral view. **B**. Anal tube and pygofer, dorsal view. **C**. Apex of anal tube, postero-dorsal view. **D**. Phallic complex, lateral view. **E**. Phallic complex, dorsal view. *An* = anal tube; *Py* = pygofer; *G* = gonostylus. Scale bars = 1 mm.

Abdomen. (Fig. 2A–B). Bright red with 2 basal tergites black; black ventrally; genital segments bright red.

Male genitalia. Pygofer, anal tube and gonostyli bright red; pygofer higher than long, with posterior margin sinuate in lateral view and showing digitiform process in middle (Fig. 9A–B); anal tube elongate, 2.26 times longer than broad, broader at half of total length, strongly curved ventrally near base (Fig. 9A–B); lateral margins sinuate (Fig. 9B) and apical margin strongly notched in dorsal view (Fig. 9C); gonostyli (Fig. 9A) elongate with posterior half much broader and strong subbasal constriction in lateral view; strong tooth directed antero-ventrally at half length; apical margin broadly rounded; phallic complex (Fig. 9D–E) with endosomal process of aedeagus laminate, slightly concave externally and with ventral margin laterally projecting subapically, showing irregular dorsal teeth along body and with apex pointed postero-ventrally.

Distribution

Known from Central, South and South East (Buton and Wowoni Islands) Sulawesi (Fig. 7).

Polydictya crassa Distant, 1906
Figs 3, 7

Polydictya crassa Distant, 1906b: 26 (Type in BMNH).

Polydictya crassa – Schmidt 1912: 72 (comparative note with other *Polydictya* species). — Metcalf 1947: 86 (catalogued). — Lallemand 1963: 18 (key, description). — Nagai & Porion 1996: 13; 2004: 31 (listed).

non Polydictya crassa – Nagai & Porion 2004: 7, pl. 2 fig. 13 (erroneous notes on distribution, illustration of *habitus*, misidentification of *Polydictya pelengana* sp. nov.).

Diagnosis

(1) Tegmina brown with a small, whitish spot along claval suture and 3 small, whitish spots on membrane (Fig. 3A); (2) hind wings without basal red patch (Fig. 3A); (3) frons regularly rounded in dorsal view, not projecting anteriorly (Fig. 3C), brown (Fig. 3D); (4) anterior tibiae slightly laminate (Fig. 3B); (5) abdomen not black dorsally (brown in the single specimen known, but it is possible that the original colour faded due to preservation in a fluid, e.g., ethanol) (Fig. 3A); (6) genital segments brown (same remark as under 5).

Type material examined

Holotype

♀: [S. Celebes] [crassa Dist. Type] [Distant Coll. 1911—383] [Type] [BMNH(E) #651905] (BMNH).

Measurements and ratios

LT (extrapolated): ♀ (n = 1): 30.0 mm; LTg/BTg = 2.72; BV/LV = 4.46; LF/BF = 0.79.

Distribution

The species was described from southern Sulawesi (“S Celebes”) without specific locality.

Polydictya katsurakoae Nagai & Porion, 1996
Figs 4, 7

Polydictya katsurakoae Nagai & Porion, 1996: 13 (listed), 28 (description), pl. 2, fig. 29 (*habitus* illustrated).

Polydictya katsurakoae – Chew Kea Foo *et al.* 2010: 59 (comparative note).

non Polydictya katsurakoae – Nagai & Porion 2002: 9 (erroneous record from Wowoni Island based on specimens of *P. basirubra* sp. nov.).

Diagnosis

(1) Tegmina brown, tinged with reddish ventrally, and with whitish spots: large one at apex of clavus, 3 along claval suture, 3–4 along costal margin and 3 on disc of membrane (Fig. 4A–B); (2) hind wings without basal red patch (Fig. 4A); (3) frons regularly rounded in dorsal view, not projecting anteriorly (Fig. 4C), yellow-brown (Fig. 4D); (4) anterior tibiae not laminate (Fig. 4B); (5) abdomen red dorsally (Fig. 4A); (6) genital segments red.

Type material examined

Holotype

♀: [Palolo, VI.91, C. Sulawesi, Col. Thierry Porion] [*Polydictya kasturakoae* n. sp., Holotype ♀, Nagai-Porion 1996] [EUM type No. 736] (EUM)

Paratype

1 ♀: [Palolo, VI.91, C. Sulawesi, Indonesia, Col. Thierry Porion] [*Polydictya kasturakoae* n. sp., Paratype ♀, Nagai et Porion 1994] (MHNL).

Coordinates of type locality

Palolo: 1°39'S 119°54'E.

Notes

The holotype and paratype were erroneously stated to be males in the original description (Nagai & Porion 1996).

Nagai & Porion (1996) stated that the paratype was deposited in MNHN but it is currently in MHNL as most specimens of Thierry Porion's collection.

Measurements and ratios

LT (extrapolated): ♀ (n = 2): 31.2–31.6 mm; LTg/BTg = 2.86; BV/LV = 7.56; LF/BF = 0.72.

Distribution

Known from Central Sulawesi (Fig. 7).

Polydictya pelengana sp. nov.
urn:lsid:zoobank.org:act:CD1E96F8-8487-4ED1-BC2F-820CC17B37CA
Figs 5, 7, 10

Polydictya crassa – Nagai & Porion 2004: 7 (notes), pl. 2 fig. 13 (*habitus* illustrated) [erroneous identification].

Diagnosis

(1) Tegmina dark olivaceous brown with small, whitish spots: 2–4 along claval suture, 4 along costal margin and 2 on disc of membrane (Fig. 5A); (2) hind wings without basal red patch (Fig. 5A); (3) frons slightly projecting anteriorly in dorsal view (Fig. 5C), yellow-brown (Fig. 5D); (4) anterior tibiae slightly laminate externally (Fig. 5A–B); (5) abdomen black dorsally (Fig. 5A); (6) genital segments black.

Etymology

The species epithet refers to the type locality, Peleng Island.

Type material

Holotype

♂: [Coll. I.R.Sc.N.B., Indonesia, Peleng Isl., iii.2004, exchange S. Jakl, I.G.: 31.970] (RBINS).

Paratypes

1 ♂, 5 ♀♀: 1 ♂: same data as holotype (RBINS); 1 ♀: [Coll. I.R.Sc.N.B., Indonesia, Peleng Isl., ix.2012, gift from B. Kneubühler, I.G.: 32.427] (RBINS); 1 ♂: [Peleng, Apr. 2003] (MHNL); 1 ♂: [Peleng, 3.03, Coll. Alain Chaminade] (MNHL); 1 ♂: [Peleng, III.03, Coll. thierry PORION] [*Polydictya crassa* ♂, Coll. thierry Porion] (MHNL); 1 ♂: [Peleng, III.03, Coll. thierry PORION] (MHNL); 1 ♂: [Peleng Isld, 3.03, Coll. thierry PORION] (MHNL). Note: "3.03" and "III.03" on the labels of the paratypes stand for March 2003.

Coordinates of type locality

Peleng Island: 1°24′S 123°10′E.

Description

MEASUREMENTS AND RATIOS. LT: ♂ (n = 5): 28.7 mm (27.2–30.0); ♀ (n = 1): 29.7 mm; LTg/BTg = 2.80; BV/LV = 4.57; LF/BF = 0.71.

HEAD. Pale yellow-brown (Fig. 5C–D); vertex strongly curved and deeply grooved with lowest point in middle and with all margins carinate (Fig. 5C); frons slightly visible from above (Fig. 5C), strongly convex, smooth, subquadrate with sides projecting laterally basally and with impressed C-shaped groove along lateral margin, starting in front of eye (Fig. 5D); clypeus subtriangular, narrower and shorter than frons, black-brown (Fig. 5D); labium elongate, black-brown, surpassing hind coxae (Fig. 5B); antennae pale yellow-brown with scape cylindrical, slightly elongate and pedicel strongly bulbous (Fig. 5D, F).

THORAX. (Fig. 5C, F). Dark brown to black-brown; pronotum irregularly wrinkled with smooth longitudinal carina and 2 strongly impressed points on anterior half of disc; mesonotum with disc wrinkled and sides smooth; scutellum impressed basally, elevated apically and slightly transversely winkled; tegulae brown with apex pale yellow-brown.

TEGMINA. (Fig. 5A–B, E). Dark brown to black-brown with olivaceous green hue on basal half, more conspicuous ventrally; veins and veinlets yellow-brown, maybe olivaceous green in living specimens; 3–4 small, yellow-white spots along costal margin, 2–4 along claval joint and 2 longitudinally aligned on disc near nodal line; costal and sutural margin subparallel; apex oblique with apico-sutural angle broadly rounded.

HIND WINGS. (Fig. 5A–B). Dark brown to black-brown with a basal milky-white elongate patch along costal vein; broader than tegmina.

LEGS. (Fig. 5A–B). Dark reddish brown to nearly black; anterior legs darker; tibiae I slightly foliaceous externally; tibiae II slender; tibiae III with 4–5 lateral and 7 apical spines.

ABDOMEN. (Fig. 5A–B). Tergites black with membranous joints greenish; sternites brown; genital segments black-brown.

MALE GENITALIA. Pygofer, anal tube and gonostyli black-brown; pygofer higher than long, with posterior margin sinuate in lateral view (Fig. 10A–B); anal tube elongate, 1.95 times longer than broad, broader at ⅗ of total length, strongly curved ventrally near base (Fig. 10A–B); lateral margins sinuate (Fig. 10A) and apical margin excavate in dorsal view (Fig. 10C); gonostyli (Fig. 10A) elongate with posterior half broader and strong subbasal constriction in lateral view; strong tooth directed antero-ventrally at half length; apical margin rounded; phallic complex (Fig. 10D–E) with endosomal process of aedeagus laminate, slightly concave externally and with ventral margin laterally projecting subbasally; dorsal margin rounded in lateral view and nearly smooth, showing few slightly marked teeth near base and near apex.

Distribution

Known from Peleng Island off Central Sulawesi (see map Fig. 7).

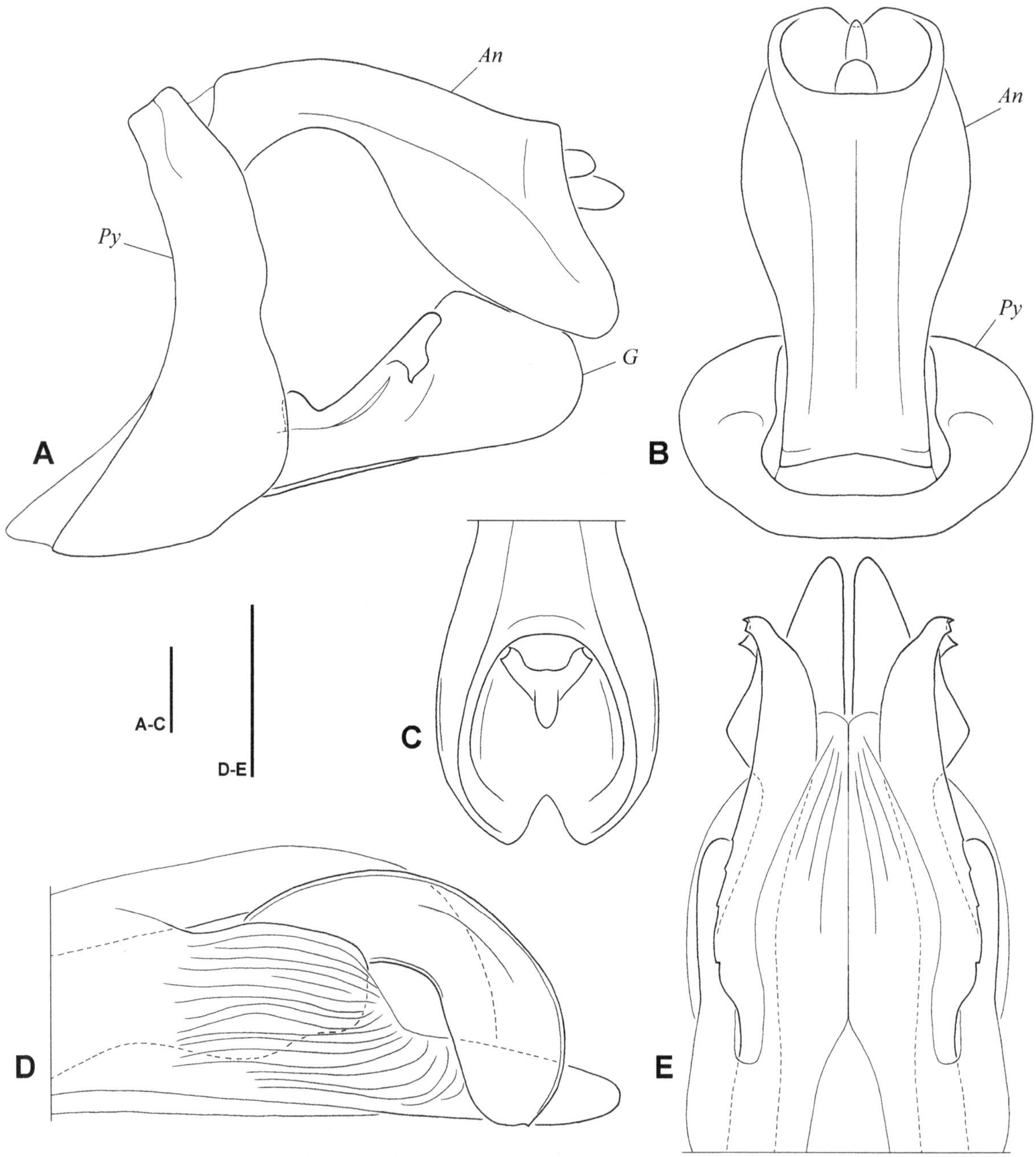

Fig. 10. *Polydictya pelengana* sp. nov., ♂ genitalia. **A**. Pygofer, anal tube and gonostylus, left lateral view. **B**. Anal tube and pygofer, dorsal view. **C**. Apex of anal tube, postero-dorsal view. **D**. Phallic complex, lateral view. **E**. Phallic complex, dorsal view. *An* = anal tube; *Py* = pygofer; *G* = gonostylus. Scale bars = 1 mm.

Polydictya thanatos Chew Kea Foo, Porion & Audibert, 2010
Figs 6, 7, 11.

Polydictya thanatos Chew Kea Foo, Porion & Audibert, 2010: 59 (described); pl. 4 figs 1-3 (*habitus* and details of head).

Diagnosis

(1) Tegmina black with 3 irregular transverse white bands before apex of clavus, one small, white line on costal cell at nodal line, and scutellar margin of clavus whitish (Fig. 6A); (2) hind wings without

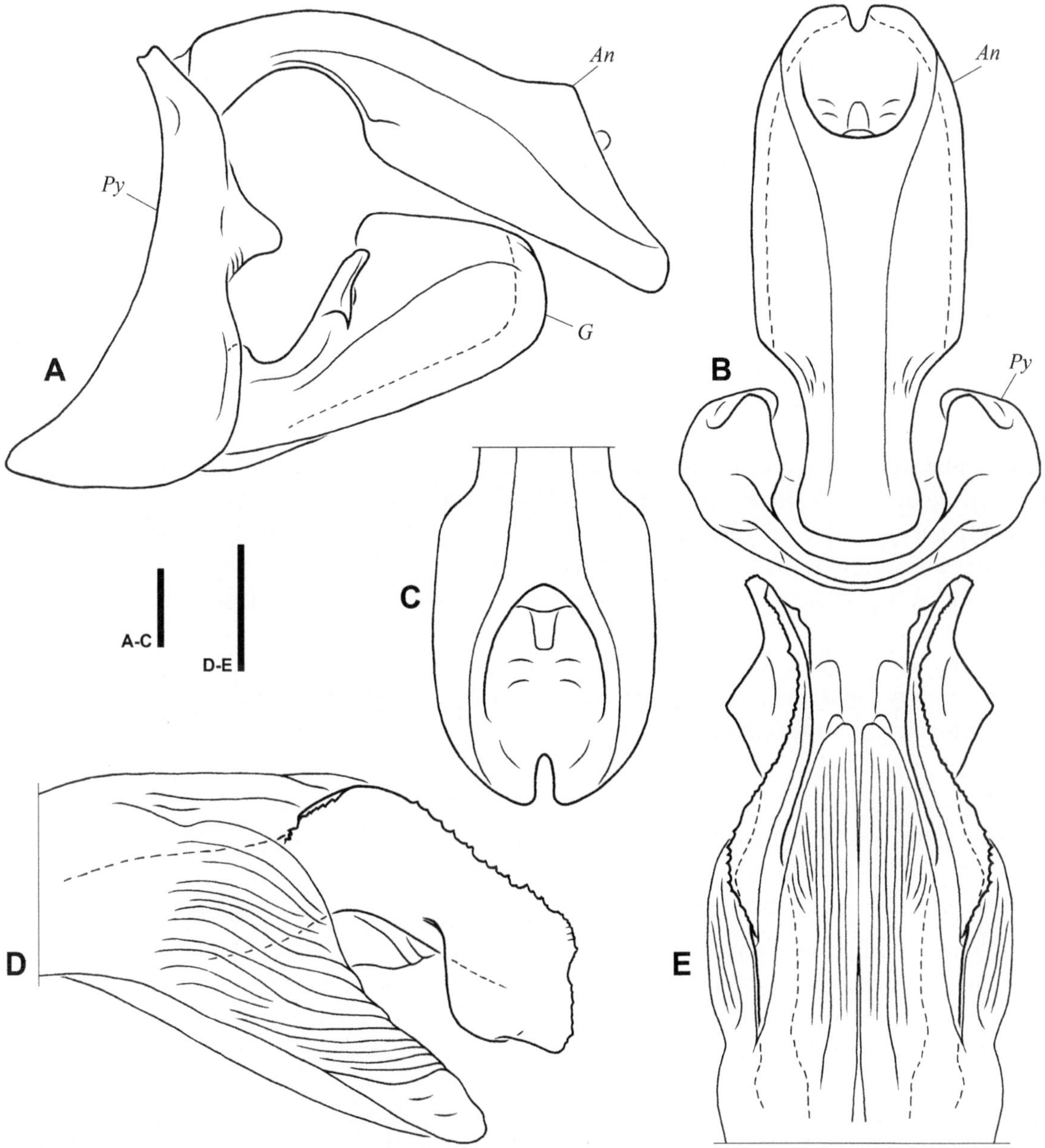

Fig. 11. *Polydictya thanatos* Chew Kea Foo , Porion & Audibert, 2010, male genitalia. **A**. Pygofer, anal tube and gonostylus, left lateral view. **B**. Anal tube and pygofer, dorsal view. **C**. Apex of anal tube, postero-dorsal view. **D**. Phallic complex, lateral view. **E**. Phallic complex, dorsal view. *An* = anal tube; *Py* = pygofer; *G* = gonostylus. Scale bars = 1 mm.

basal red patch (Fig. 6A); (3) frons regularly rounded in dorsal view, not projecting anteriorly (Fig. 6C), dark red (Fig. 6D); (4) anterior tibiae not laminate (Fig. 6A–B); (5) abdomen black dorsally (Fig. 6A); (6) genital segments black.

Type material examined

Holotype

♂: [C.E. Sulawesi, 15 Kms E. Luwuk, 300-600 m IV-2008] [*Polydictya thanatos* n. sp. S. Chew et T. Porion 2009, Holotype ♂] [Coll. Porion, 2010.54, Muséum de Lyon] (MHNL).

Paratype

1 ♂ [Indonesie, E-Sulawesi, Luwuk, 2.2009] [*Polydictya thanatos* n. sp. Paratype Mâle, S. Chew et T. Porion 2009] [Museum Paris MNHN (EH) 17722] (MNHN).

Coordinates of type locality

Luwuk: 0°55'48.32"S 122°47'45.96"E

Note

The description (Chew Kea Foo *et al.* 2010) mentions three more paratypes (two in Stanislav Jakl's collection and one in Thierry Porion's), which were not examined in the present study.

Measurements and ratios

LT: ♂ (n = 1): 27.5 mm; LTg/BTg = 2.93; BV/LV = 6.25; LF/BF = 0.79.

Male genitalia

Pygofer, anal tube and gonostyli black; pygofer higher than long, with posterior margin sinuate in lateral view and showing digitiform process in middle (Fig. 11A–B); anal tube elongate, 2.35 times longer than broad, strongly curved ventrally near base (Fig. 11A–B); sides subparallel on half of length in middle (Fig. 11B), apical margin notched in dorsal view (Fig. 11C); gonostyli (Fig. 11A) elongate with posterior half broader and strong subbasal constriction in lateral view; strong tooth directed antero-ventrally at half length; apical margin rounded; phallic complex (Fig. 11D–E) with endosomal process of aedeagus laminate, slightly concave externally and with ventral margin ventro-laterally projecting subapically; dorsal margin slightly curved laterally near base with irregular teeth, straight in lateral view and showing irregular teeth to apex; sinuate in dorsal view.

Distribution

Known from Central Sulawesi (Fig. 7).

Discussion

Considering that five of the six species of *Polydictya* from Sulawesi were described in the last twenty years, it seems likely that more species await discovery. The distribution map shows that large areas within Sulawesi and a number of neighbouring islands remain unexplored in terms of *Polydictya*. More field work should be conducted in those areas to document the insect fauna and make observations on the host-plants and biology of those insects. Although they are actively searched for by collectors, the feeding habits, larval stages and eggs remain unknown for all species treated in the present paper.

Acknowledgements

I thank all the curators listed above for the access to the material under their responsibility, Messrs Cédric Audibert, Joël Clary and Harold Labrique for their kindness and help during my visit at MHNL and Messrs Mick Webb and Max Barclay during my visit at BMNH. This study has benefitted from funds from the European Union (Synthesys Project – Grant GB-TAF-1642) for a visit to the collection of BMNH in 2011 and from MHNL for a visit to its collections. I also thank my colleagues, Dr Patrick Grootaert and Mr Pol Limbourg (RBINS), and Prof. Thierry Bourgoin (MNHN) for their permanent support, Mr Thierry Porion for interesting discussions and Mrs Megan M. Critser for reviewing the language.

References

Atkinson E.T. 1885. Notes on Indian Rhynchota. No. 4. *Journal of the Asiatic Society of Bengal* 54: 127–158.

Bourgoin T. 2014. FLOW (Fulgoromorpha Lists on The Web): A world knowledge base dedicated to Fulgoromorpha. V.8, updated [2 Sep. 2014].

Chew Kea Foo S., Porion T. & Audibert C. 2010. Cinq nouveaux Fulgoridae asiatiques (Hemiptera : Fulgoromorpha). *Les cahiers du Musée des Confluences - Etudes scientifiques* 1: 51–64.

Constant J. 2004. Révision des Eurybrachidae (I). Le genre Amychodes Karsch, 1895 (Homoptera: Fulgoromorpha: Eurybrachidae). *Bulletin de l'Institut royal des Sciences naturelles de Belgique* 74: 11–28.

Constant J. 2009. A new species of *Polydictya* from Sumatra and notes on *P. chantrainei* Nagai et Porion, 2004 (Hemiptera: Fulgoromorpha: Fulgoridae). *Annales zoologici* 59 (3): 293–296.

Constant J. 2010. A new species of *Polydictya* from Lombok (Hemiptera, Fulgoromorpha, Fulgoridae). *Nouvelle Revue d'Entomologie* 26 (2) (2009): 155–161.

Constant J. & Pham H.T. 2008. A new species of *Polydictya* from Vietnam (Hemiptera, Fulgoromorpha, Fulgoridae). *Nouvelle Revue d'Entomologie* (N.S.) 25 (1): 27–31.

Distant W.L. 1888. An enumeration of the Rhynchota received from Baron von Müller, and collected by Mr. Sayer in New Guinea during Mr. Cuthbertson's expedition. *Transactions of the Entomological Society of London* 1888: 475–489.

Distant W.L. 1906a. *The fauna of British India, including Ceylon and Burma. Rhynchota*. Vol. 3. Taylor & Francis, London.

Distant W.L. 1906b. Rhynchotal notes xxxviii. *Annals and Magazine of Natural History (ser. 7)* 18: 18–32.

Gerstaecker C.E.A. 1895. Ueber einige bemerkenswerthe Fulgorinen der Greifswalder zoologischen Sammlung. *Mittheilungen aus dem Naturwissenschaftlichen Verein für Neu-Vorpommern und Rügen* 27: 1–50.

Guérin-Méneville F.E. 1844. Insectes. *In*: Cuvier G.L.C.F.D. *Iconographie du règne animal.*

Jacobi A. 1910. 12. Hemiptera. 7. Homoptera. *In*: Sjöstedt Y. (ed.) *Wissenschaftliche ergebnisse der Schwedischen Zoologischen Expedition nach dem Kilimandjaro, dem Meru und den Umgebenden Massaisteppen Deutsch-Ostafrikas 1905–1906, unter leitung von Prof. Dr. Yngve Sjöstedt.* Vol. 2: 97–136. P. Palmquists aktiebolag, Stockholm.

Karsch F.A.F. 1890. Afrikanische Fulgoriden. *Berliner Entomologische Zeitschrift* 35 (1): 57–70.

Kirkaldy G.W. 1902. Memoirs on Oriental Rhynchota. *The Journal of the Bombay Natural History Society* 14: 46–58.

Krikaldy G.W. 1907. "Current criticism." *The Entomologist* 40: 58–60.

Lallemand V. 1959. Révision des espèces africaines de la famille Fulgoridae (Super-famille Fulgoroides – sous-ordre des Homoptères). *Publicaçoes culturais da companhia de diamantes de Angola* 41: 37–124.

Lallemand V. 1963. *Révision des Fulgoridae (Homoptera).* Deuxième partie. *Faunes asiatique et australienne.* Mémoires de l'Institut royal des Sciences naturelles de Belgique (2e série) 75, Institut royal des Sciences naturelles de Belgique, Brussels.

Melichar L. 1903. *Homopteren-Fauna von Ceylon.* F. L. Dames, Berlin.

Metcalf Z.P. 1947. *General Catalogue of the Homoptera.* Fascicle IV. *Fulgoroidea.* Part. 9. *Fulgoridae.* North Carolina State College, Raleigh (U.S.A.).

Nagai S. & Porion T. 1996. *Fulgoridae 2: Catalogue illustré des faunes asiatique et australienne.* Sciences Nat, Compiègne.

Nagai S. & Porion T. 2002. *Fulgoridae 2.* Supplement 1. *New Fulgoridae from South-East Asia.* Hillside Books, Canterbury.

Nagai S. & Porion T. 2004. *Fulgoridae 2. supplement 2. Nouveaux Fulgoridae d'Asie du Sud-Est.* Hillside Books, Canterbury.

Schmidt E. 1907. Beitrag zur Kenntnis der Fulgoriden. Die Arten des Genus Myrilla Distant. *Stettiner Entomologische Zeitung* 68: 113–116.

Schmidt E. 1912. Diagnosen neuer Fulgoriden-Gattungen und Arten nebst einigen Bemerkungen. *Stettiner Entomologische Zeitung* 73: 67–102.

Stål C. 1866. *Hemiptera Africana.* Vol. 4. *Hemiptera Homoptera Latr.* Ex officina Norstedtiana, Stockholm.

Walker F. 1851. *List of the specimens of Homopterous insects in the collection of the British Museum.* Part.2. British Museum (Natural History), London.

4

A new species of *Casmaria* H. Adams & A. Adams, 1853 (Gastropoda, Cassidae) from the Philippines identified by molecular data

Alexander FEDOSOV [1], Baldomero M. OLIVERA [2], Maren WATKINS [2] & Varvara BARKALOVA [3]

[1] A.N. Severtsov Institute of Ecology and Evolution, Russian Academy of Sciences, 119071, Leninsky prospect 33, Moscow, Russia
[2] Department of Biology, University of Utah, Salt Lake City, UT 84112, USA
[3] Moscow State University, Faculty of Biology, 1-12, Leninskie Gory, Moscow, 119991 Russia

Abstract. The genus *Casmaria* H. Adams & A. Adams, 1853 (family Cassidae) is widespread in the tropical Indo-Pacific and has been documented from some Atlantic localities as well. Two *Casmaria* species, *C. erinaceus* (Linnaeus, 1758) and *C. ponderosa* (Gmelin, 1791), are common in Indo-Pacific shallow-water sandy bottom communities and are characterized by high morphological variability; both species encompass multiple, often sympatric forms of uncertain status. In the present study we carry out a phylogenetic analysis of some Philippine *Casmaria* morphs and demonstrate that one of the distinctive morphs earlier assigned to *Casmaria ponderosa* is in fact a different species, which we describe as *Casmaria boblehmani* sp. nov. The smooth form of *Casmaria ponderosa*, *C. ponderosa ponderosa*, and the solid nodulose form, widely called "form *nodulosa*" despite being strikingly different in shell morphology, are shown to be conspecific. Studied specimens of these two morphs even from different localities share the same haplotype of the CO1 gene. In light of these new data on the morphological variability of *Casmaria* species, we discuss criteria of species delimitation in the genus *Casmaria* and possible affinities of *Casmaria boblehmani* sp. nov. within the genus.

Keywords. Polymorphic species, CO1, Cassidae, *Casmaria*, Indo-Pacific.

Introduction

The genus *Casmaria* H. Adams & A. Adams, 1853, or bonnet shells, is a distinctive group of the family Cassidae, easily recognizable by a set of shell characters, among which are a closed umbilicus and the complete lack of shell sculpture in many species. The genus is widely distributed in shallow-water marine communities of the tropical zone. Its members are abundant on sandy bottoms of the Indo-Pacific from Madagascar to Hawaii and eastern French Polynesia, and are also known from some Atlantic localities, mainly in the Caribbean province (Clench 1944; Abbott 1968).

Despite the fact that the number of species and subspecies currently attributed to the genus *Casmaria* hardly exceeds a dozen (Kreipl 1997; Buijse *et al.* 2013; WoRMS (www.marinespecies.org, accessed 10 Feb 2014)), the taxonomy of the genus remains problematic. Traditionally, diverse Indo-Pacific *Casmaria* forms were lumped into two polymorphic and extremely widespread species, *C. erinaceus* (Linnaeus, 1758) and *C. ponderosa* (Gmelin, 1791), consistent with the general wide perception of species in the group established by Abbott (1968). In these two species a number of morphologically distinct forms were named that occur in restricted localities, mostly from the periphery of Indo-Pacific; recently, most of these were given the status of full species (Buijse *et al.* 2013).

In the central Indo-Pacific both *Casmaria erinaceus* and *C. ponderosa* are represented by several sympatric forms, some of them strikingly different in appearance. Firstly, each of the two *Casmaria* species encompasses a smooth, usually thin-shelled form (*C. erinaceus vibex* and *C. ponderosa ponderosa*) and a more robust form, bearing nodules around the periphery (C. *erinaceus erinaceus* and *C. ponderosa nodulosa*) (Kreipl 2008; Beu *et al.* 2012). Moreover, specimens of both *C. erinaceus* and *C. ponderosa* demonstrate significant variation in color pattern. The status of these sympatric forms of *C. erinaceus* and *C. ponderosa* is uncertain, and the conspecificity of the smooth and nodulose forms in both species remains unproven.

In the present study we carry out a molecular-phylogenetic analysis of a set of sympatric *Casmaria* forms from the central Indo-Pacific, mainly from the Central Philippines, in order to establish species limits in Indo-Pacific *Casmaria* species.

Material and methods

Material

Material for molecular studies was collected by night diving in mid-June 2012 at Danajon Banks, off Olango Island, Cebu Province, in the Central Philippines (seven specimens) and in September 2012 at Murray Beach, off Mun Island, Nha Trang Bay, Vietnam (three specimens). Using conventional taxonomy, nine of the specimens were initially assigned to *Casmaria ponderosa* and one specimen to *Casmaria erinaceus*. The nine specimens of *Casmaria ponderosa* represented both the smooth form, known as *C. ponderosa ponderosa*, and a solid morphotype widely identified as "*C. ponderosa nodulosa*". They also demonstrated a considerable variation in color pattern.

In the field a fragment of foot tissue was clipped from each specimen and preserved in 99% ethanol for subsequent DNA extraction.

Preparation of genomic DNA and amplification of selected genetic markers

Genomic DNA was extracted from foot tissue samples using the Gentra PUREGENE DNA Isolation Kit (Gentra Systems, Minneapolis, MN) according to the manufacturer's standard protocol. The 12S, 16S and CO1 fragments were amplified using genomic DNA as a template for polymerase chain reaction (for details of the PCR reactions, see Fedosov *et al.* (2011)). The resulting PCR products were purified by gel electrophoresis and the DNA fragments recovered from agarose.

Table 1. Summary of specimens used for molecular analysis.

Species	Isolate	BOLD ID	Museum catalogue number	CO1 GenBank	16S GenBank	12S GenBank	Locality data
C. erinaceus	Casm.1			KC505344	KC505353	KC505363	Philippines, off Olango Is, Denajon Bank, 10°17.8'N 124°05.3'E, 15-25 m
C. erinaceus	IM-2007-33621	TONO119-13	MNHN IM-2007-33621	KC970029			Philippines, Panglao Is, PANGLAO 2004 Stn. R3, 9°33.0'N 123°46.5'E, 5-24 m
C. ponderosa	Casm.2			KC505343	KC505352	KC505362	Philippines, off Olango Is, Denajon Bank, 10°17.8'N 124°05.3'E, 15-25 m
C. ponderosa	Casm.3			KC505342	KC505351	KC505361	Philippines, off Olango Is, Denajon Bank, 10°17.8'N 124°05.3'E, 15-25 m
C. ponderosa	Casm.4			KC505341	KC505350	KC505360	Philippines, off Olango Is, Denajon Bank, 10°17.8'N 124°05.3'E, 15-25 m
C. ponderosa	Casm.8			KC505340	KC505349	KC505359	Philippines, off Olango Is, Denajon Bank, 10°17.8'N 124°05.3'E, 15-25 m
C. ponderosa	Casm.V1			KC505339	KC505348	KC505358	Vietnam, Nha-Trang Bay, off Mun Is, 12°10.084'N, 109°17.771'E, 6-12m
C. ponderosa	Casm.V2			KC505338	KC505347	KC505357	Vietnam, Nha-Trang Bay, off Mun Is, 12°10.084'N, 109°17.771'E, 6-12m
C. ponderosa	Casm.V3			KC505337	KC505346	KC505356	Vietnam, Nha-Trang Bay, off Mun Is, 12°10.084'N, 109°17.771'E, 6-12m
C. boblehmani sp. nov.	Casm.5		Holotype ANSP 453749	KC505336	KC505355	KC505365	Philippines, off Olango Is, Denajon Bank, 10°17.8'N 124°05.3'E, 15-25 m
C. boblehmani sp. nov.	Casm.7		Paratype 2 ANSP 453750	KC505345	KC505354	KC505364	Philippines, off Olango Is, Denajon Bank, 10°17.8'N 124°05.3'E, 15-25 m
C. boblehmani sp. nov.	IM-2007-33625	TONO121-13	MNHN IM-2007-33625	KC970027			Vanuatu, SANTO 2006 Stn. FR10, 15°36.7'S 167°05.8'E, 5–25 m
C. boblehmani sp. nov.	IM-2007-33628	TONO120-13	MNHN IM-2007-33628	KC970028			Vanuatu, SANTO 2006 Stn. FR03, 15°36.2'S 167°06.3'E, 3-32 m
OUTGROUPS							
Distorsio reticularis				HQ834113	HQ833981	HQ833859	
Linatella caudata				JF693368	HQ833982	HQ833860	
Erronea errones				HQ834112	HQ833980	HQ833858	
Semicassis pyrum				AF550508			

Characterization of clones encoding 12S, 16S and CO1 sequences

The purified DNA fragments were annealed to the pNEB206A vector using the USER Friendly Cloning kit (New England BioLabs, Inc., Beverly, MA) following the manufacturer's suggested protocol, and the resulting product transformed into competent cells. The nucleic acid sequences of the resulting 12S, 16S and CO1-encoding clones were determined according to the standard protocol for automated sequencing at the Health Sciences Center Core Sequencing Facility, University of Utah. Sequences of six individual 12S, 16S and CO1-encoding clones from each specimen were compared to ensure sequence consensus.

CO1 sequences of an additional three *Casmaria* specimens collected during recent expeditions of the Muséum National d'Histoire Naturelle, Paris (MNHN) to the Philippines (PANGLAO 2004 - one specimen) and Vanuatu (SANTO 2006 - two specimens) were contributed by Philippe Bouchet and Nicolas Puillandre (MNHN). For sequence GenBank accession numbers and detailed locality data see Table 1.

Phylogenetic analysis

CO1, 12S and 16S sequences of three caenogastropod species, *Distorsio reticularis* (Linnaeus, 1758) (Tonnoidea: Personidae), *Linatella caudata* (Gmelin, 1791) (Tonnoidea: Ranellidae) and *Erronea errones* (Linnaeus, 1758) (Cypraeidae), accessed in GenBank were used as outgroups. *Semicassis pyrum* (Lamarck, 1822) (Cassidae) was used as a within family outgroup in CO1 analysis. Sequences were aligned for each gene independently using MUSCLE (Edgar 2004). The accuracy of automatic alignments was then confirmed by eye using BioEdit version 7.0.0.0 (Hall 1999). Hypervariable regions of the 12S and 16S rRNA genes were excluded from further analyses. Parameters of the substitution model were estimated during the analysis (six substitution categories, a gamma-distributed rate variation across sites approximated in four discrete categories and a proportion of invariable sites) independently for each codon position of the CO1 gene. Bayesian Analysis was performed in two parallel runs in MrBayes (Huelsenbeck *et al.* 2001), each consisting of six Markov chains of 10,000,000 generations with a sampling frequency of one tree each 10,000 generations and with the chain temperature set to 0.02. When log-likelihood scores were found to stabilize, consensus trees were calculated after omitting the first 25% of trees as burn in.

Abbreviations used in text

ANSP = Academy of Natural Sciences, Philadelphia, USA
MNHN = Muséum National d'Histoire Naturelle, Paris, France

Results

Phylogenetic analysis

Phylogenetic trees constructed after analysis of three molecular markers (Fig. 1 A-C) are congruent and are characterized by high support of terminal groupings (PP values indicated in bold). The monophyly of *Casmaria* is supported in analyses of all three markers, although this result requires further confirmation after the analysis of a more representative dataset. In all three single-gene trees the studied specimens of the genus *Casmaria* form three distinct groups; the within-group K2P genetic distance computed for the CO1 gene averaged 0.53%, and was always lower than the minimal between-group K2P genetic distance (6.3%). This difference corresponds to the so-called "barcode gap" and suggests a specific status for each of the groups (Puillandre *et al.* 2012).

The first grouping, represented by two specimens on the CO1-based tree and a single specimen on the 12S and 16S trees, corresponds to *Casmaria erinaceus*. Specimens preliminarily identified as *Casmaria ponderosa* formed two separate, well-supported clades on the phylogenetic trees. One of these clades, including four smooth Philippine specimens and three nodulose specimens from Vietnam, corresponds to

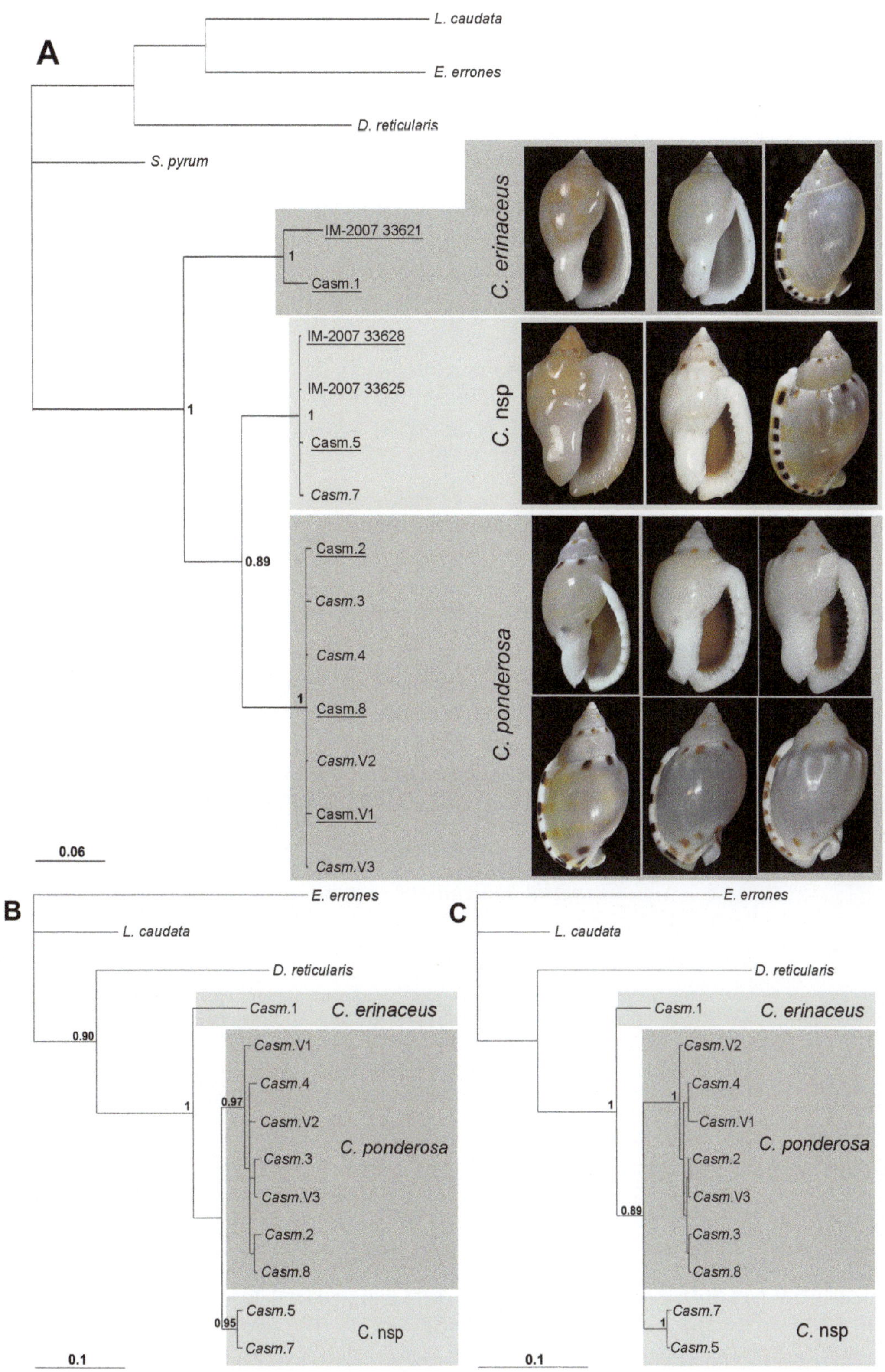

Fig. 1. Bayesian phylogenetic trees of studied members of the genus *Casmaria* obtained with standard genetic markers. Posterior probabilities when greater than 0.80 are indicated for each node. **A**. Phylogenetic tree based on the CO1 gene; for specimens underlined, vouchers are illustrated on the right. **B**. Phylogenetic tree based on the 16S rRNA gene. **C**. Phylogenetic tree based on the 12S rRNA gene.

Casmaria ponderosa. Smooth specimens represent the typical form, *Casmaria ponderosa ponderosa* and correspond to the type material of this species (Buijse *et al.* 2013). It is noteworthy that four specimens of this grouping, two smooth Philippine specimens (Casm.4, Casm.8) and two nodulose Vietnamese specimens (Casm.V2, Casm.V3), had identical CO1 sequences. The four specimens forming the second clade, two nodulose Philippine specimens and two additional specimens from Vanuatu on the CO1-based tree, represent a distinct morphotype, morphologically similar to so-called *C. ponderosa nodulosa*, but differing in details of shell structure and coloration (see below), which were previously widely attributed to *Casmaria ponderosa*. Our results recognize this morphotype as an undoubted distinct species separate from both *C. erinaceus* and *C. ponderosa*; its description is given below.

Taxonomic account

Family Cassidae Latreille, 1825
Subfamily Phaliinae Beu, 1981
Genus *Casmaria* H. Adams & A. Adams, 1853

Type species

Buccinum vibex Linnaeus, 1758 (= *Casmaria erinaceus* (Linnaeus, 1758)) (by subsequent designation in Harris 1897)

Note on the synonymy and relevant types of *Casmaria ponderosa*

Casmaria ponderosa has a long and complex synonymy, which has recently been comprehensively analyzed by Beu *et al.* (2012). All relevant type specimens were tracked down or proven to be lost. The syntype of *Cassis torquata* Reeve, 1848, illustrated earlier by Abbott (1968), was designated as the lectotype and also as the neotype of *Buccinum ponderosum* Gmelin, 1791, *B. nodulosum* Gmelin, 1791, *Phalium quadratum* Link, 1807, *Buccinum biarmatum* Dillwyn, 1817, *Buccinum pantherina* Dillwyn, 1817, and *Cassis tenuilabris* Menke, 1828 (Beu *et al.* 2012: 78). For this reason, none of these names can apply to the new species we describe.

Casmaria boblehmani sp. nov.
urn:lsid:zoobank.org:act:FA0142B4-BEF4-4F08-8BFD-C0870F39DC88
Fig. 2

Casmaria ponderosa – Robin 2008: 145, fig. 5 — Buijse *et al.* 2013: 13, fig. 25-2 [middle specimen], pl. 6, figs 6, 8, 13, pl. 7, figs 1, 5, 8-10, 12-15, 17, pl. 8, fig. 1, ?pl. 10, fig. 11.

Etymology

This species is named to honor I. Robert (Bob) Lehman, a great scientist, mentor and teacher. One of the authors (BMO) has been the grateful direct beneficiary of his remarkable generosity.

Type material

Holotype

Molecular voucher Casm.5, 36.3 mm, Danajon Banks, off Olango Island, Cebu, Philippines, 10°17.8' N, 124°05.3' E, 10-25 m, coll. local diver, Jun. 2012, ANSP-453749 (Fig. 2A-C).

Paratypes

1 (35.1 mm) and 2 (34.8 mm, molecular voucher Casm. 7), same locality as holotype, ANSP-453750 (Fig. 2D-E). The holotype and paratype 2 provided the two DNA sequences labeled Casm. 7 and Casm. 5, respectively, in Fig. 1.

Fig. 2. *Casmaria boblehmani* sp. nov. **A-C**. Holotype, 36.3 x 19.5 mm, Danajon Banks, Olango Is, Philippines. **D-E**. Paratype 2, 35.1 x 21.7 mm, same locality as holotype. **F**. 27.7 x 18.8 mm, Tip of Cape York, Queensland, Australia. **G**. 38.6 x 25.4 mm, off Woka Island, Molucca Sea, Indonesia. **H-I**. ANSP 276463, 41.5 x 24.2 mm, Akuilan Is, Fiji. **K-L**. ANSP 266122, 36.5 x 24.6 mm, Atiue, Tahiti, Society Islands.

Other material examined

PHILIPPINES: 34.6 mm, off Masbate Island, Central Philippines, coll. local diver, BO collection; 34.9 mm, off Calituban Island, Bohol, 9 m, on mud, BO collection; 44.9 mm, off Cuyo Island, Palawan, 10-25 m, BO collection; 32.2 mm, Palawan Island, unknown locality, 10-15 m, BO collection. AUSTRALIA: 27.7 x 18.8 mm, tip of Cape Yorke, Queensland, BO collection (Fig. 2F). INDONESIA: 38.6 x 25.4 mm, off Woka Island, Molucca Sea, BO collection (Fig. 2G). VANUATU: juv., Expedition "SANTO 2006", Stn. FR10, Belmoul Lagoon, 15°36.7' S, 167°05.8' E, 5-25 m, MNHN IM-2007-33625; juv., Expedition "SANTO 2006", Stn. FR03, Belmoul Lagoon entrance, 15°36.2' S, 167°06.3' E, 3-32 m, MNHN IM-2007-33628 (Fig. 1A). FIJI: 41.5 x 24.2 mm, off Akuilan Island, ANSP 276463 (Fig. 2H-I). SOCIETY ISLANDS: 36.5 x 24.6 mm, Atiue, Tahiti, ANSP 266122 (Fig. 2K-L).

Description

Adult shell small to medium-sized, solid, glossy, ovate-elongate with very prominent terminal varix, studied specimens 34.8-36.3 mm high. Proportions of shell vary considerably, ratio of shell breadth to height 0.58-0.7. Protoconch of about 4 smooth, evenly convex whorls. Teleoconch of 4-5 whorls; early spire whorls evenly convex, with no sculpture. Last adult whorl and in most specimens penultimate whorl with row of glossy nodules (6 to 12 on last whorl), weakly developed in some specimens (Fig. 2D-E), prominent in others (Fig. 2 G). Due to nodules, late whorls distinctly shouldered; shoulder situated at mid-height of penultimate whorl and adapical quarter of last adult whorl. Aperture ovate, outer lip strongly thickened, drawn upwards, with wide, shallow sinus in upper side. Interior of outer lip with distinct dentition. Apertural varix very prominent, thick, with ten brownish-black blotches on abapertural face; adapertural face with seven or eight well-developed spines directed transversally to apertural plane. Their positions correspond approximately with black blotches on abapertural face. Columellar shield solid, white, heavily callused, with numerous spiral wrinkles in some specimens. Last adult whorl of pinkish, tan, or pale brown background color, with indistinct brown spiral stripes and a row of brown to blackish-brown spots below suture. Siphonal canal with black blotch. Interior of aperture pale brown. Tip of protoconch brown.

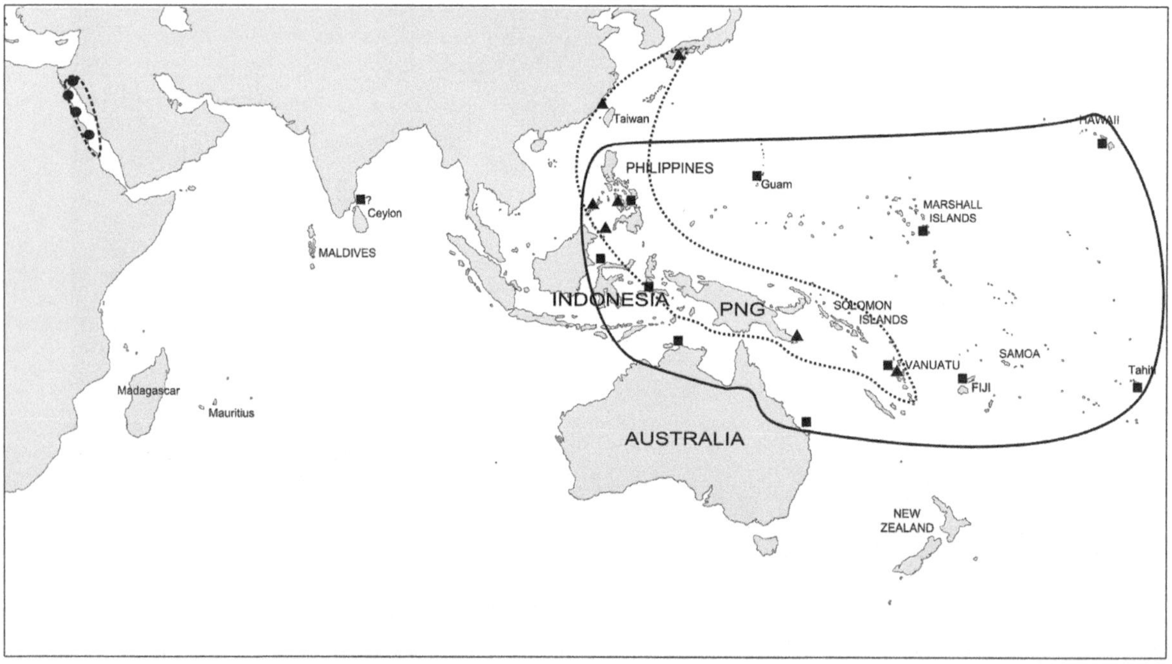

Fig. 3. Distributions of *Casmaria turgida* (short-dashed line and triangles), *C. unicolor* (dashed line and filled circles) and *C. boblehmani* sp. nov. (solid line and squares) in the Indo-Pacific.

Table 2. Summary of morphological differences among *Casmaria* species. Abbreviations: OL – outer lip; TV – terminal varix; n.f. – nodulose form; r – rarely. Number / direction of spines on OL: f – forward; tr – transverse to apertural plane. Position of spines on OL: a 1/3 – anterior third. Number of blotches on TV: N+n – N major blotches + n strokes; N se – number of subequal blotches.

Species	Smooth form	Nodulose form	Number / direction of spines on OL	Position of spines on OL	Number of blotches on TV	Dentition within	Row of blotches at suture	Row of blotches on shell base	Color pattern	Individual characters
C. erinaceus	x	x	3-5 / f	a 1/3	7-14 se	-	-	-	no, zigzags, spiral rows of dots	
C. kayae	x	x	3-5 / f	a 1/3	9-12 se	-	-	-	zigzags	
C. ponderosa	x	x	7-9 / tr	a 1/2 - 3/4	5 + 1-2	+	+	+	no	
C. turgida	x	-	8-11 / tr	a 1/2 - 3/4	5 +1-2 or 10-11 se	-	+	-	zigzags	shell length usually exceeds 60 mm
C. boblehmani	?	x	8-9 / tr	a 1/2 - 3/4	8-10 se	+	+	r	no, spiral bands	
C. unicolor	x	x	10-12 / tr	a 2/3 - 4/5	11-12 se	-	+	+ (n.f.)	no, zigzags, spiral rows of dots	inflated upper whorl profile giving the shell a wider, more stepped spire outline
C. cernica	x	-	6-7 / tr	a 1/2	5 + 1-2	-	+	+	spiral rows of blotches	narrow and narrowly varicate shell
C. atlantica	x	-	6-7 / tr	a 1/2	5 + 1-2	-	+	-	spiral bands	narrow and narrowly varicate shell
C. perryi	x	-	6-7 / tr	a 1/2	5 + 1-2	-	+	-	spiral bands	narrow and narrowly varicate shell
C. beui	x	-	>9	a 2/3	10-11 se	r	+	+	spiral rows of blotches, zigzags	shell very small, wide and thin
C. kalosmodix	x	-	?-	-	5, 10-11 se	-	+	r	axial lines + spiral bands	narrow axial lines over entire shell surface

Remarks

Specimens that are most appropriately assigned to *Casmaria boblehmani* sp. nov. are found at sites in eastern Indonesia, northern Australia (Northern Territory and Queensland), Vanuatu and the South Pacific. The largest specimen (51.8 mm) that we can confidently assign to *C. boblehmani* sp. nov. is known from the Philippines (Buijse *et al.* 2013: pl. 6, fig. 13). In the Philippines *C. boblehmani* sp. nov. occurs syntopically with *C. ponderosa*, *C. erinaceus* and *C. turgida*. All four species co-occur in sandy patches at depths of 1 to 30-40 meters.

Characters allowing the unequivocal recognition of *Casmaria boblehmani* sp. nov. and the delimitation of the *Casmaria* species described to date are summarized in Table 2.

Casmaria ponderosa and *C. boblehmani* sp. nov. can be distinguished from each other by the number and arrangement of brown blotches on the dorsal side (abapertural face) of the terminal varix: *C. ponderosa* has five major blotches and two to three intermediate strokes, while *C. boblehmani* sp. nov. has nine to ten blotches of equal or subequal size, evenly distributed along the varix of most specimens. Another character, although less stable, is a row of spots on the shell base, posterior to the siphonal canal. These are always distinct in *C. ponderosa*, but are missing in most specimens, or in rare cases present but indistinct, in *C. boblehmani* sp. nov.

Specimens of *Casmaria erinaceus* can easily be distinguished from both *C. ponderosa* and *C. boblehmani* sp. nov. by the absence of dentition on the inner surface of the outer lip and by the presence of only three or four (very rarely five) spines restricted to the anterior edge of the outer lip, directed forward. The anterior edge of the outer lip is drawn slightly forward and abaxially. Both *C. ponderosa* and *C. boblehmani* sp. nov. have distinct dentition on the inner surface of the outer lip and have six or more spines directed transversely to the apertural plane, located along the anterior and medial edges of the outer lip. Most of these characters were mentioned by Abbott (1968) and Buijse *et al.* (2013) as allowing discrimination between *C. erinaceus* and *C. ponderosa*.

Casmaria boblehmani sp. nov. can be readily distinguished from *Casmaria turgida* (Reeve, 1848). The latter species is rather conservative in morphology throughout its range (Buijse *et al.* 2013) and has a shell with evenly convex whorls, notable because of the characteristic zigzag color pattern. Besides that, *C. turgida* exceeds all co-occurring *Casmaria* forms in size. Unfortunately, *C. turgida* is missing in our molecular dataset. However, the conservative morphology of this species and its sympatry on a small scale with *C. erinaceus*, *C. ponderosa* and *C. boblehmani* sp. nov. at a number of Indo-Pacific localities prove that it is a distinct species.

Despite the fact that *C. ponderosa*, *C. boblehmani* and *C. turgida* in most cases can be distinguished from each other confidently by the combination of characters discussed above, these characters may intergrade to a certain extent, creating somewhat intermediate forms that may be difficult to interpret. Buijse *et al.* (2013: pl. 6, figs 6, 8, pl. 9, fig. 15) depicted a few specimens of intermediate appearance that we are not certain where to assign. All these specimens have smooth shells intermediate between *C. ponderosa* and *C. turgida* in size (48-51 mm).

Among the other *Casmaria* species not occurring sympatrically with *C. boblehmani* sp. nov., *C. unicolor* (Pallary, 1926) is probably the most similar. As in other *Casmaria* species, *C. unicolor* varies significantly in sculpture and encompasses both smooth and nodulose forms, and the nodulose form closely resembles *C. boblehmani* sp. nov. (see Buijse *et al.* 2013: pl. 22). These two forms seem to differ in the number of spines on the outer lip. *C. boblehmani* sp. nov. possesses seven to eight spines, while *C. unicolor* has 10-12, and they are located along the entire length of the lip in many specimens, reaching the adapical angle of the aperture. Apart from that, *C. unicolor* can be distinguished by its more inflated upper (adapical) whorl profile and more clearly distinguished, weakly flattened sutural ramp, giving it a wider, more stepped spire outline. *Casmaria unicolor* is restricted to the Red Sea (Fig. 3), and the widely distant distributions of this species and *C. boblehmani* sp. nov. suggest that these two forms are sufficiently divergent to be considered as separate species.

The three species *Casmaria cernica* (G. B. Sowerby III, 1888), *C. perryi* (Iredale, 1912) and *C. atlantica* Clench, 1944 are extremely similar to each other. They are characterized by being narrower in outline and with a narrower terminal varix than *C. boblehmani* sp. nov. Besides that, all three mentioned species share the same arrangement of dark blotches on the dorsal side of the apertural varix as in *C. ponderosa*: five major blotches with one or two intermediate strokes between each major pair.

In comparison with *Casmaria boblehmani* sp. nov., *C. kalosmodix* (Melvill, 1883) has a much larger, far more elongate and more narrowly varicate shell with narrow axial color lines that are not present in other *Casmaria* species, and with no spines on the outer lip. As in *C. boblehmani* sp. nov., the blotches on the abapertural face of the varix in *C. kalosmodix* are narrow, but in the latter species they are grouped in 3-4 clusters, separated by wide interspaces.

Another species, *Casmaria kayae* Buijse, Dekker & Verbinnen, 2013, in general resembles *C. erinaceus*. *C. boblehmani* sp. nov. can easily be distinguished from *C. kayae* by the number and position of spines on the outer lip and by the color pattern. *C. kayae* has three or four (five in a few specimens) spines on the anterior portion of the lip, and these are directed forward. Many specimens have a distinctive zigzag color pattern, quite different from the usual coloration of *C. boblehmani* sp. nov.

Finally, *C. beui* Buijse, Dekker & Verbinnen, 2013, from Western Indian Ocean, can easily be distinguished from *C. boblehmani* sp. nov. by its small size, its smooth, thin, relatively very wide shell with distinctively inflated whorls, and by its characteristic color pattern, resembling that of *C. turgida*.

Distribution

Casmaria boblehmani sp. nov. is widely distributed across the Pacific Ocean: from the Philippines, eastern Indonesia, northern Australia, to Fiji and the Society Islands, and (based on specimens illustrated by Buijse *et al.* 2013) to Hawaii (Fig. 3). Possibly reaches India (based on the specimen illustrated by Robin 2008).

Discussion

Traditionally, characters of shell shape and sculpture are given more taxonomic weight than coloration, especially in groups with highly variable shells. In particular, in the genus *Casmaria* the presence or absence of nodules on the shoulder dramatically changes the appearance of the shell; ever since the original description by Linnaeus this has confused the species-level taxonomy of the genus. Our data demonstrate that smooth (typical) and nodulose specimens of *C. ponderosa*, even from distant localities, may share the same haplotype of the CO1 gene, suggesting that (1) these forms are conspecific, and therefore (2) the presence or absence of nodules is not a character of taxonomic significance. This is the first evidence independent of traditional conchological data that confirms the conspecificity of distinct morphotypes of one cassid species. The nodulose form of *C. erinaceus* was missing from our dataset, but we expect that it will prove to be conspecific with the smooth form, as with the two forms of *C. ponderosa*. In general, our results support the hypothesis of species delimitation in Cassidae largely based on the assumption that high within-species morphological variability underlies the rise of multiple morphotypes, while these demonstrate no genetic structure.

Identification of a new species morphologically close to *Casmaria ponderosa* urged us to review the morphology of Indo-Pacific *Casmaria* species in search of morphological traits that would be consistent with our molecular data, allowing the unmistakable delimitation of Indo-Pacific *Casmaria* species or, at least, of the forms that co-occur in the Philippines. While *C. erinaceus* and *C. kayae* can be easily separated based on the number and arrangement of spines on the outer lip, delimitation of the other species is less straightforward and is largely based on shell proportions and coloration. In particular, the number and arrangement of the dark blotches on the abapertural face of the terminal varix is consistent with our hypothesis of species delimitation in *Casmaria* and seems to be conservative enough to serve as a good diagnostic character. However, although these characters work well in most cases, they vary from individual to individual and can intergrade, giving rise to rare specimens of an intermediate appearance. Interpretation of these intermediate forms requires further molecular studies.

The morphological characters shown here to be taxonomically significant were largely ignored in prior taxonomic diagnoses of the group, and other overlooked *Casmaria* species may well exist, distinguishable by more subtle details of shell coloration and morphology and detectable by molecular analysis. Further larger-scale molecular phylogenetic studies on *Casmaria* and closely related cassid genera are necessary to reveal the hidden diversity of this group.

Acknowledgements

We are grateful to Philippe Bouchet, Nicolas Puillandre and Barbara Buge (MNHN) for the access to MNHN sequences and their barcode collection, and to Gary Rosenberg and Paul Callomon for loaning specimens from the ANSP collections. We are also grateful to Andreia Salvador (NHMUK), Elaine Charwat (Linnean Society of London) and Igor Muratov (Natal Museum of South Africa) for sending photographs of types in the genus *Casmaria*. We thank Terry Merritt for assistance in preparation of the manuscript, Patrice Showers Corneli for her review of the manuscript, and Desiree Beaudry and Thong Nguyen for their help in preparation of the figures. We are grateful to Alan Beu, Philippe Bouchet and one anonymous referee for their comments that helped to amend the manuscript, and we thank Alan Beu for his generous help in editing the final text. The specimens were collected on field trips supported in part by International Cooperative Biodiversity Groups (ICBG) grant #1U01TW008163. This work was partly supported by a grant from the Russian Foundation of Basic Research, RFBR-14-04-31048 (PI A. Fedosov).

References

Abbott R.T. 1968. The helmet shells of the world (Cassidae). Part 1. *Indo-Pacific Mollusca* 2 (9): 7-198.

Beu A., Bouchet P. & Tröndlé J. 2012. Tonnoidean gastropods of French Polynesia. *Molluscan research* 32: 61-120.

Buijse J.A., Dekker H. & Verbinnen G. 2013. On the identity of *Casmaria* species (Gastropoda, Cassidae), with descriptions of two new species. *Acta Conchyliorum* 14: 3-93.

Clench W. J. 1944. The genera *Casmaria, Galeodea, Phalium* and *Cassis* in the Western Atlantic. *Johnsonia* 1 (16): 1-16.

Edgar R. C. 2004. MUSCLE: multiple sequence alignment with high accuracy and high throughput. *Nucleic Acids Research* 32: 1792-1797.

Fedosov A. E., Watkins M., Showers Corneli P., Heralde F. M. Concepcion G. P. & Olivera B. M. 2011. Phylogeny of the genus *Turris*: correlating molecular markers with radular anatomy and shell morphology. *Molecular Phylogenetics and Evolution* 59: 263-270.

Hall T.A. 1999. BioEdit: a user-friendly biological sequence alignment editor and analysis program for Windows 95/98/NT. *Nucleic Acids Symposium Series* 41: 95-98.

Harris G.F. 1897. *Catalogue of the Tertiary Mollusca in the Department of Geology, Part I. The Australasian Tertiary Mollusca.* British Museum (Natural History), London.

Huelsenbeck J.P., Ronquist F. & Hall B. 2001. MrBayes: Bayesian inference of phylogenetic trees. *Bioinformatics* 17: 754-755.

Kreipl K. 1997. *Recent Cassidae*. Verlag Christa Hemman, Wiesbaden, Germany.

Kreipl K. 2008. Family Cassidae. *In*: Poppe G.T. (ed.) *Philippine Marine Mollusks*. Conchbooks, Hackenheim.

Puillandre N., Lambert A., Brouillet S. & Achaz G. 2012. ABGD, Automatic Barcode Gap Discovery for primary species delimitation. *Molecular Ecology* 21 (8): 1864-1877.

Robin A. 2008. *Encyclopedia of Marine Gastropods. Xenophora and Conchbooks*. Hackenheim, Germany.

5

Afroneutria, a new spider genus of Afrotropical Ctenidae (Arachnida, Araneae)

Daniele POLOTOW [1] & Rudy JOCQUÉ [2]

[1] Instituto Butantan, Laboratório Especial de Coleções Zoológicas, Av. Vital Brasil, 1500, Butantã, São Paulo, CEP 05503-000, Brazil.

[2] Royal Museum for Central Africa, Tervuren, Belgium.

Abstract. The new ctenid genus *Afroneutria* is proposed to include five species: *Afroneutria velox* (Blackwall, 1865) comb. nov. (type species), *A. immortalis* (Arts, 1912) comb. nov., *A. erythrochelis* (Simon, 1876) comb. nov., *A. quadrimaculata* sp. nov. and *A. hybrida* sp. nov., all collected in central to southeast Africa. *Afroneutria* species can be distinguished from the remaining Ctenidae by the presence of a synapomorphic laminar projection on the embolus tip. The genus can also be distinguished by the presence of the following set of characters: elongated cup-shaped median apophysis and elongated and laminar embolus on the male palp; large lateral projections on the epigynum and the spermathecae divided into two parts, dorsal and ventral, in the vulva.

Keywords. Africa, *Ctenus*, systematics, taxonomy

Introduction

Ctenidae Keyserling, 1877 is composed of nocturnal and cursorial spiders, common in the lower strata of tropical forests of South and Central America, Asia and Africa. To date, the family comprises 491 species, distributed in 40 genera (World Spider Catalog 2014). Currently, most of the Ctenidae species are placed in the genus *Ctenus* Walckenaer, 1805. Based on a new diagnosis and recent cladistic analysis, *Ctenus* is considered polyphyletic as currently delimited (Silva 2003; Simó & Brescovit 2001; Brescovit & Simó 2007; Polotow & Brescovit 2009, 2014).

According to the phylogenetic hypothesis proposed by Polotow & Brescovit (2014), several *Ctenus* species need to be allocated to new genera, including a clade formed by *Ctenus velox* Blackwall, 1865, *C. immortalis* Arts, 1912 and *C. erythrochelis* Simon, 1876. They appear as a well-supported group, not related to the type species of the genus, *Ctenus dubius* Walckenaer, 1805 (Polotow & Brescovit 2014: fig. 3, clade D).

The aim of this paper is to establish the new Afrotropical ctenid genus *Afroneutria* to accommodate five species: *Afroneutria velox* (Blackwall, 1865) comb. nov. (type species), *A. immortalis* (Arts, 1912) comb. nov., *A. erythrochelis* (Simon, 1876) comb. nov., *A. quadrimaculata* sp. nov. and *A. hybrida* sp. nov., all collected in central and East Africa. We also reestablish *A. immortalis* comb. nov., previously considered a synonym of *A. velox* comb. nov.

Material and methods

Morphological observations and illustrations were made using a Wild M10 and a Wild M5 stereo microscope. The epigynum was detached from the abdomen and submerged in methyl salicylate to clear the internal structures. For SEM, specimens were cleaned ultrasonically, dried through critical point, gold coated, and then examined and photographed with a JEOL 6480 LV scanning electron microscope at the Royal Museum for Central Africa, Tervuren, Belgium (MRAC). All measurements are in millimeters.

The following abbreviations are used below:

ALE = anterior lateral eyes
AME = anterior median eyes
C = conductor
CD = copulatory ducts
Cy = cymbium
d = dorsal
E = embolus
FD = fertilization ducts
LL = locking lobes
LS = lateral sector
MA = median apophysis
MS = median sector
p = prolateral
PLE = posterior lateral eyes
PME = posterior median eyes
r = retrolateral
RTA = retrolateral tibial apophysis
S = spermathecae
Te = tegulum
Ti = tibiae
v = ventral

The material examined is deposited in the following institutions (curator in parentheses):

BMNH = The Natural History Museum, London (J. Beccaloni)
MNHN = Muséum National d'Histoire Naturelle, Paris (C. Rollard)
MRAC = Royal Museum for Central Africa, Tervuren (R. Jocqué)
OUMNH = Oxford University Museum of Natural History (Z. Simmons)
ZMB = Zoological Museum of Berlin (J. Dunlop)
ZMUC = Zoological Museum, University of Copenhagen (N. Scharff)

Results

Class Arachnida Cuvier, 1812
Order Araneae Clerck, 1757
Family Ctenidae Keyserling, 1877
Subfamily Cteninae Keyserling, 1877

Genus ***Afroneutria*** gen. nov.
urn:lsid:zoobank.org:act:59ECDA85-C86E-416E-A98D-CB1DD6AB5DA1

Type species

Ctenus velox Blackwall, 1865. Designated here.

Diagnosis

Afroneutria species can be distinguished from the remaining Ctenidae by the presence of a synapomorphic laminar projection on the embolus tip (Fig. 1B; Polotow & Brescovit 2014: character 25, fig. 3). The genus can also be distinguished from the remaining Cteninae spiders by the presence of the following set of characters: elongated cup-shaped median apophysis and elongated and laminar embolus on the male palp (Fig. 1A–B); large lateral projections on the epigynum and the spermathecae divided into two parts, dorsal and ventral, in the vulva (Fig. 1C).

Etymology

The generic name is a combination of the first part of Afrotropical and the last part of *Phoneutria* Perty, 1833, a Neotropical genus that bears superficial similarities with the genus described here. The gender is feminine.

Description

Ecribellate ctenids. Total body length (males and females) 12.40–32.00. Carapace piriform (Fig. 2A, C), brown, covered by white hairs in longitudinal median area and in lateral area; tuft of long white hairs between PME; border with white hairs; thoracic groove longitudinal, in posterior third; except *A. erythrochelis* comb. nov., which is covered by golden hairs. Chilum divided. Clypeus with long erect black bristles. Ctenid eye pattern 2-4-2, with anterior and posterior row recurved in dorsal view. Eyes round, except oval anterior lateral eyes. Chelicerae dark brown, covered by white hairs (except *A. erythrochelis* comb. nov., covered with golden hairs); promargin with three teeth, the median higher than the laterals; retromargin with four similar sized teeth; prominent basal condyle; intermarginal denticles present only in *A. immortalis* comb. nov. and *A. hybrida* sp. nov. Endites reddish brown with lighter apical area, external lateral margin excavated, distally bearing dense apical scopulae and subapical serrula. Labium dark brown, distally rounded, bearing dense apical scopula. Sternum reddish brown, oval, not extending between coxae IV. Legs reddish brown. Male legs more elongated and slender than female legs. Trochanter notched. Tarsus with claw tufts (Fig. 1D–F) composed of tenent setae, specialized in adhesion to smooth surfaces (Ramírez 2014). Abdomen oval, with tufts of white hairs and numerous black setae on anterior dorsal area (Figs 2A–D, 4A–D, 6A–D, 8A–D, 10A–B). Males and females with specific ventral coloration pattern on abdomen (Figs 2B, D; 4B, D; 6B, D; 8B, D; 10B). Male palp (Figs 1A–B, 3A–C, 5A–C, 7A–B, 9A–C, 11A–B, 13A–D, 14A–B, 15A–D): tibia with short VTA; short RTA with three projections, except *A. erythrochelis* comb. nov. with two projections and *A. hybrida* sp. nov. with truncated tip; cymbium with elongated tip; subtegulum prolateral, except *A. erythrochelis* comb. nov. with retrolateral subtegulum; tegulum oval; median apophysis cup-shaped and elongated; embolus elongated with retrolateral area hyaline and complex tip; locking lobes present; conductor short and partially sclerotized. Epigynum (Figs 3D–E, 5D–E, 7C–D, 9D–E, 12A–D): divided in median field and two lateral fields; median field with anterior area narrow and posterior area swollen and large; lateral

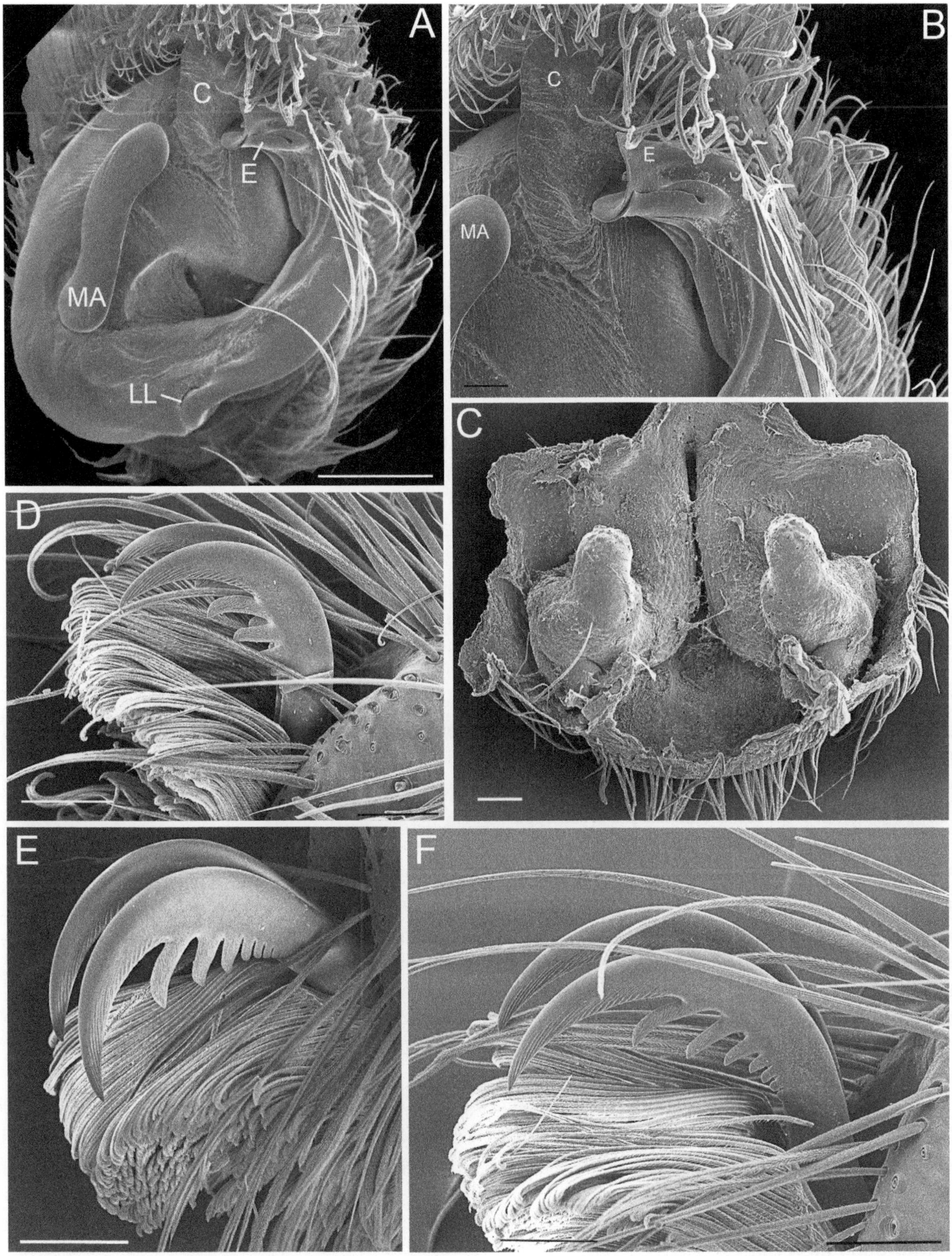

Fig. 1. *Afroneutria velox* (Blackwall, 1865) comb. nov.. — **A–B**. Male palp, ventral view. — **C–F**. ♀. **C**. Epigynum, dorsal view. **D–F**. Tarsus, lateral view. **D**. Leg III. **E**. Leg I. **F**. Leg IV. Scale bars = 200 µm.

field with lateral projections; lateral projections covering part of median field; copulatory opening with lighter area; spermathecae divided in two parts, ventral and dorsal; copulatory duct ending in ventral part; fertilization duct arising of dorsal part (Figs 1C, 3D–E, 5D–E, 7C–D, 9D–E).

Composition

Five species: *Afroneutria velox* comb. nov., *A. immortalis* comb. nov., *A. erythrochelis* comb. nov., *A. quadrimaculata* sp. nov. and *A. hybrida* sp. nov.

Distribution

Central and southeast Africa (Figs 16A–B).

Afroneutria velox (Blackwall, 1865) comb. nov.
Figs 1A–F, 2A–D, 3A–E, 12D, 13A–B, 16A

Ctenus velox Blackwall, 1865: 336.
Phoneutria melanogastra Bösenberg & Lenz, 1895: 36, pl. 1, fig. 14.
Ctenus marshalli F.O. Pickard-Cambridge, 1898: 26, pl. 4, figs 12–13.

Ctenus velox – F.O. Pickard-Cambridge 1898: 15, pl. 3, figs 1–2. — Benoit 1977: 700, fig. 2a–d.
Ctenus melanogaster – Strand 1907: 694. — Arts 1912: 196, pl. 1, fig. 14a–b. — Benoit 1977: 700, fig. 2a–d (synonymised with *C. velox*).
Ctenus melanogastra – Hyatt 1954: 884.
Ctenus marshalli – Hyatt 1954: 884 (synonymised with *C. melanogaster*).

Note

Blackwall (1865) indicated one female as the type material used in the description of *Ctenus velox*, but two adult females were found in the OUMNH collection and, according to the label, both were designated as types. Based on the original description we here designate the smaller specimen as the lectotype and the other female as the paralectotype.

Diagnosis

The males of *Afroneutria velox* comb. nov. differ from males of the other species of the genus by the laminar and truncated tip and the additional subdistal projection of the embolus (Fig. 1B) and the shape of the RTA (Fig. 3A–C). The females can be distinguished from the remaining species of the genus by the acuminated and short lateral projections of the epigynum and rounded dorsal spermathecae in the vulva (Fig. 3D–E).

Material examined

Type material

MOZAMBIQUE: *Ctenus velox*, female lectotype (designated here), 1 female paralectotype, next to confluence of Shiré River with Zambezi River (approximate coordinates: 17°40’24.0” S, 35°17’51.2” E), OUMNH 1554.

Additional material

TANZANIA: Kipatimu, 1 ♂, 1965, T.H. Hofman coll. (MRAC 137443); Kyela, 1 ♀, 9 Nov. 1991, R. Jocqué coll. (MRAC 173465); 1 ♀, 27 Nov. 1991, R. Jocqué coll. (MRAC 173191); Coast Province, Utete, Kirengoma Forest Reserve (8°20–23’ S, 38°54–58’ E), 2 ♂♂, 8 Nov.–1 Dec. 1989, J. Kingdon & Frontier Tanzania coll. (ZMUC); Coast Province, Bagamoyo, Zaraninge (Kiono) Forest Reserve, 20 km SW of Sadaani (6°04’ S, 38°35’ E), 2 ♂♂, 1 ♀, Mar. 1990, Frontier Tanzania coll. (ZMUC).

Fig. 2. *Afroneutria velox* (Blackwall, 1865) comb. nov. Habitus. **A**. ♂, dorsal view. **B**. ♂, ventral view. **C**. ♀, dorsal view. **D**. ♀, ventral view. Scale bars = 1 mm.

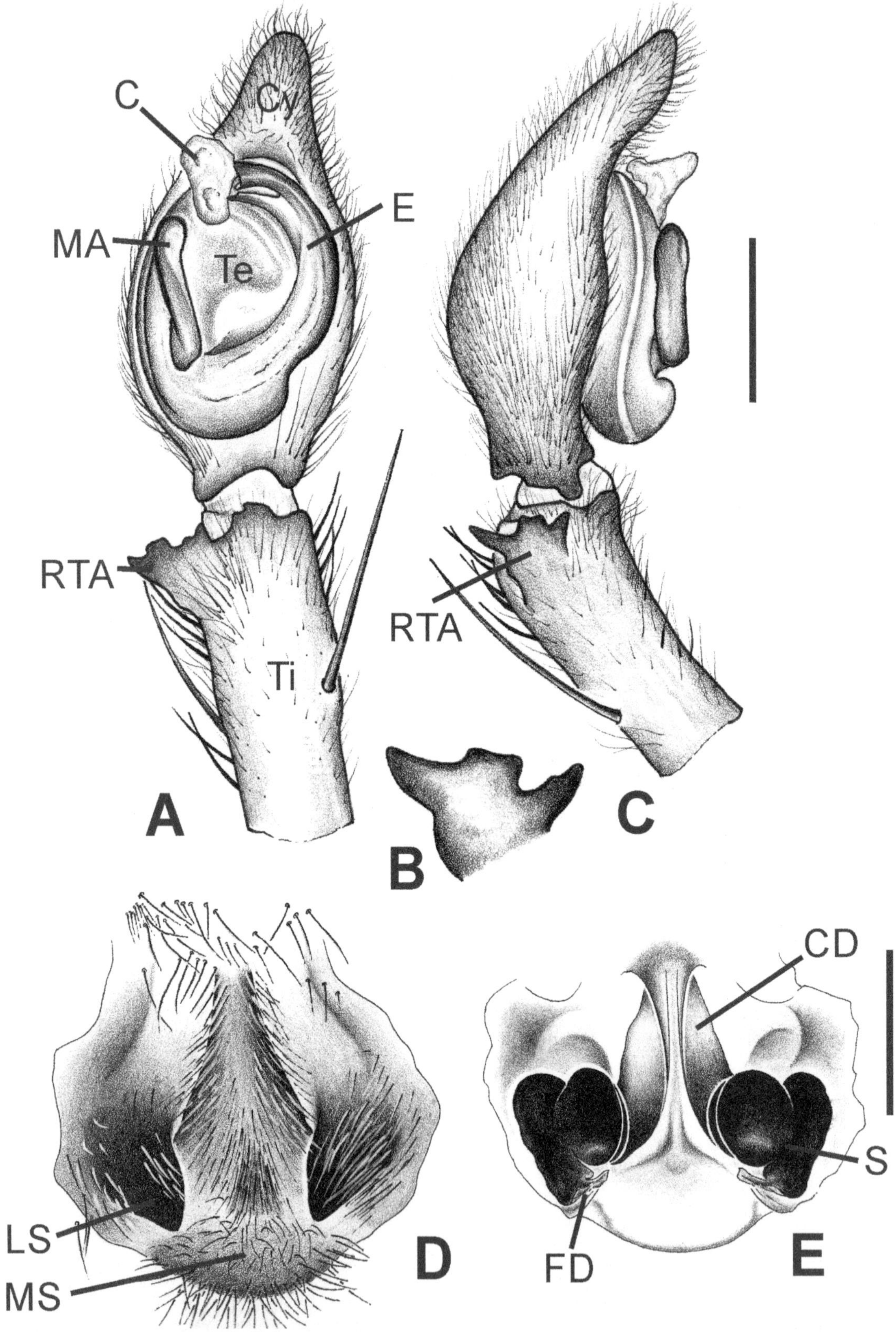

Fig. 3. *Afroneutria velox* (Blackwall, 1865) comb. nov. — **A–C**. Male palp. **A**. Ventral view. **B**. Retrolateral view. **C**. Retrolateral tibial apophysis, lateral view. — **D–E**. Female genitalia. **D**. Epigynum, ventral view. **E**. Vulva, dorsal view. Scale bars = 1 mm.

MALAWI: Chintheche, 1 ♂, 1 ♀, 1–20 Nov. 1977, R. Jocqué coll. (MRAC 155299).

DEMOCRATIC REPUBLIC OF THE CONGO: Elisabethville, Katanga, 1 ♀, 30 Jan. 1931, Bredo coll. (MRAC 15545); Buttgenbach, Lubumbashi, 1 ♀ (MRAC 25506).

Type material not examined

EAST AFRICA: *Phoneutria melanogastra*, 2 male syntypes, 1 female syntype, deposited in Zoologisches Museum Hamburg.

ZIMBABWE: *Ctenus marshalli*, holotype, ♂, Umfuli River, also known as Mupfure River (approximate coordinates: 17°55'13.3" S, 29°56'40.7" E), Mashonaland, BMNH.

Description

Male (MRAC 155299)

Total length 12.30. Carapace 6.20 long and 4.60 wide. Eye diameter: AME 0.32, ALE 0.18, PME 0.60, PLE 0.56. Leg measurements: I: femur 8.40 / patella 3.20 / tibia 8.70 / metatarsus 8.60 / tarsus 3.30 / total 32.20; II: 8.10 / 3.30 / 7.40 / 7.60 / 2.80 / 29.20; III: 6.80 / 2.60 / 5.70 / 6.40 / 2.30 / 23.80; IV: 9.30 / 2.60 / 8.90 / 11.10 / 3.30 / 35.20. Leg formula 4123. Leg spination: tibia I and II v2-2-2-2-2, r1-1, p1-1; III and IV v2-2-2, r1-1, p1-1; metatarsus I, II and III v2-2-2, p1-1-1, r1-1-1; IV v1-1-1-1-1-2, r1-1-1, p1-1-1. Coloration: venter of abdomen black, with two white spots placed transversely behind epigastric furrow (Fig. 2B). Palp: cymbium elongated, longer than tibia; conductor rectangular; median apophysis slightly curved (Figs 1A–B, 3A–C, 13A–B).

Female (MRAC 173465)

Total length 18.60. Carapace 8.10 long and 6.40 wide. Eye diameter: AME 0.36, ALE 0.24, PME 0.60, PLE 0.48. Leg measurements: I: femur 6.80 / patella 3.10 / tibia 6.30 / metatarsus 6.00 / tarsus 2.20 / total 24.40; II: 6.30 / 2.90 / 5.40 / 5.50 / 2.10 / 22.20; III: 5.20 / 2.30 / 4.50 / 5.10 / 2.10 / 19.20; IV: 7.50 / 2.60 / 6.60 / 8.40 / 2.30 / 27.40. Leg formula 4123. Leg spination: tibia I and II v2-2-2-2-2, r0, p0; III and IV v2-2-2, r1-1, p1-1; metatarsus I and II v2-2-2, r0, p0; III v2-2-2, r1-1-1, p-1-1-1; IV v1-1-1-1-1-2, r1-1-1, p1-1-1. Coloration: ventral abdomen black, with two white spots placed transversely behind epigastric furrow (Fig. 2C–D). Epigynum: triangular median field and short lateral projections of lateral fields of epigynum (Fig. 3D). Dorsal spermathecae short and round; ventral spermathecae with rebordered lateral margins (Figs 1C, 3E, 12D).

Variation

Two ♂♂: total length 10.60–12.40, carapace length 6.00–6.30, femur I length 7.10–8.70. Three ♀♀: total length 16.40–18.70, carapace length 7.00–8.10, femur I length 6.20–6.50. At least two ♂♂ present venter of abdomen black in front, red towards back, with two spots placed transversely behind epigastric furrow.

Distribution

Tanzania, Malawi, Mozambique, Democratic Republic of the Congo and Zimbabwe (Fig. 16A).

Afroneutria immortalis (Arts, 1912) revalidated, comb. nov.
Figs 4A–D, 5A–E, 12A, 14A–B, 16A

Ctenus immortalis Arts, 1912: 202, pl. 2, fig. 25.

Ctenus velox – Benoit 1977: 700, fig. 2a–d (misidentified and wrongly synonymised with *C. immortalis*).
Ctenus immortalis – Polotow & Brescovit 2014: 335, 338, 348, 355.

Fig. 4. *Afroneutria immortalis* (Arts, 1912) comb. nov. Habitus. **A**. ♂, dorsal view. **B**. ♂, ventral view. **C**. ♀, dorsal view. **D**. ♀, ventral view. Scale bars = 2 mm.

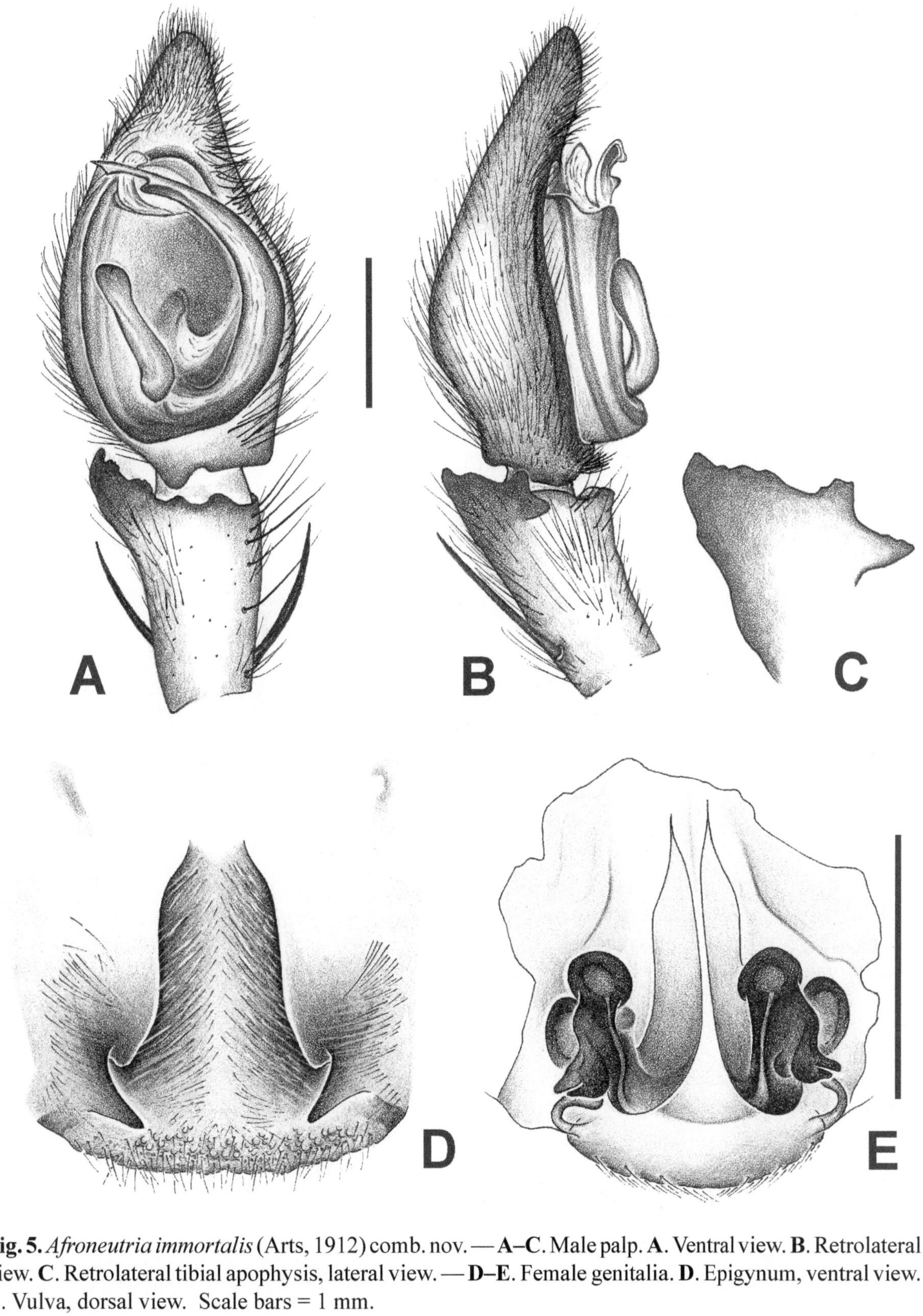

Fig. 5. *Afroneutria immortalis* (Arts, 1912) comb. nov. — **A–C**. Male palp. **A**. Ventral view. **B**. Retrolateral view. **C**. Retrolateral tibial apophysis, lateral view. — **D–E**. Female genitalia. **D**. Epigynum, ventral view. **E**. Vulva, dorsal view. Scale bars = 1 mm.

Diagnosis

The males of *Afroneutria immortalis* comb. nov. differ from those of the other species of the genus by the elongated tip and short additional laminar projection of the embolus on the male palp (Figs 5A, 14A; Polotow & Brescovit 2014: fig. 8f, indicated by an arrow) and the shape of the RTA (Fig. 5C; Polotow & Brescovit 2014: fig. 8d). The females can be distinguished by the acute lateral spurs (Figs 5D, 12A) and straight dorsal and ventral spermathecae with a swollen apical area (Fig. 5E).

Material examined

Type material

TANZANIA: *Ctenus immortalis*, 1 syntype, ♂, Langenburg (currently Tukuyu), Fülleborn coll., ZMB 9884; syntypes, ♀♀, Dar-es-Salam, Stuhlmann coll., ZMB 9883; syntypes, ♀♀, Dar-es-Salam, Regner coll., ZMB 9881.

KENYA: *Ctenus immortalis*, syntypes, 1 ♀, 2 immature, Kibwezi, Schütter coll., ZMB 9882 and 9885.

Additional material

KENYA: Coast Province, Arabuko Sokoke National Park, 1 ♀, 25 Sep. 2001, D. Van den Spiegel coll. (MRAC 211539); Tsavo, Taita West National Park, Taita Hills, Mwachora, pitfalls, 1 ♂, 20 Jun. 1999, D. Van den Spiegel coll. (MRAC 208828); Tsavo, Taita Discovery Center, 1 ♂, 27 Mar. 2000, R. Jocqué & C. Warui coll. (MRAC 209543); Watamu, 1 ♂, 1977, J.L. Cloudsley-Thompson coll. (MRAC 148774).

TANZANIA: Amani, Amani Forest Reserve, 1 ♀, Dec. 1978, K.M. Howell coll. (MRAC 159481); Mkomazi Game Reserve, Ibaya Hill, 1 ♂, 24 Nov. 1994, A. Russell-Smith coll. (MRAC 213883); Coast Province, Kisarawe, Kasimzumbwi Forest Reserve, 6°57' S, 39°03' E, 1 ♀, Jan.–Feb. 1991, Frontier Tanzania coll. (ZMUC); Coast Province, Mafia Island, 7°53' S, 39°50' E, 2 ♀♀, Oct.–Nov. 1990, Frontier Tanzania coll. (ZMUC).

Description

Male (MRAC 213883)

Total length 13.40. Carapace 7.20 long and 5.60 wide. Eye diameter: AME 0.32, ALE 0.18, PME 0.60, PLE 0.56. Leg measurements: I: femur 8.60 / patella 3.40 / tibia 8.90 / metatarsus 8.90 / tarsus 3.40 / total 33.20; II: 8.10 / 3.30 / 7.40 / 7.60 / 2.80 / 29.20; III: 6.80 / 2.60 / 5.70 / 6.40 / 2.30 / 23.80; IV: 9.50 / 2.80 / 9.10 / 11.30 / 3.50 / 36.20. Leg formula 4123. Leg spination: tibia I and II v2-2-2-2-2, r1-1, p1-1; III and IV v2-2-2, r1-1, p1-1; metatarsus I, II and III v2-2-2, p1-1-1, r1-1-1; IV v1-1-1-1-1-2, r1-1-1, p1-1-1. Coloration: venter of abdomen black, with two white spots in front of spinnerets (Fig. 4B). Palp: cymbium elongated, longer than tibia; conductor triangular; median apophysis slightly curved, with acute tip (Figs 5A–C, 14A–B).

Female (MRAC 211539)

Total length 16.40. Carapace 7.00 long and 5.40 wide. Eye diameter: AME 0.36, ALE 0.24, PME 0.60, PLE 0.48. Leg measurements: I: femur 6.80 / patella 3.00 / tibia 6.40 / metatarsus 6.00 / tarsus 2.20 / total 24.40; II: 6.30 / 2.90 / 5.40 / 5.50 / 2.10 / 22.20; III: 5.20 / 2.40 / 4.40 / 5.10 / 2.00 / 19.10; IV: 7.50 / 2.60 / 6.60 / 8.40 / 2.30 / 27.40. Leg formula 4123. Leg spination: tibia I and II v2-2-2-2-2, r0, p0; III and IV v2-2-2, r1-1, p1-1; metatarsus I and II v2-2-2, r0, p0; III v2-2-2, r1-1-1, p-1-1-1; IV v1-1-1-1-1-2, r1-1-1, p1-1-1. Coloration: venter of abdomen black, with two white spots in front of spinnerets (Fig. 4D). Epigynum: triangular median field and elongated lateral projections of lateral field (Figs 5D, 12A). Dorsal and ventral spermathecae elongated (Fig. 5E).

Variation

Six ♂♂: total length 10.40–13.60, carapace length 5.30–7.20, femur I length 4.70–8.10. Two ♀♀: total length 15.70–16.40, carapace length 7.00–7.60, femur I length 6.80–7.90. Several ♀♀ with four white spots on venter of abdomen, two behind epigastric furrow, two in front of spinnerets.

Distribution

Kenya and Tanzania (Fig. 16A).

Afroneutria erythrochelis (Simon, 1876) comb. nov.
Figs 6A–D, 7A–D, 15C–D, 16B

Phoneutria erythrochelis Simon, 1876: 222.
Phoneutria auricularis Karsch, 1879: 347.
Ctenus johnstoni F.O. Pickard Cambridge, 1898: 21, pl. 3, fig. 7.
Ctenus carsoni F.O. Pickard Cambridge, 1898: 24, pl. 3, figs 4–5.
Ctenus aureopubescens Strand, 1906: 82.

Ctenus erythrochelis – Simon 1897: 107, figs 98, 101. — Lessert 1915: 44, pl. 2, fig. 33.
Phoneutria auricularis – Arts 1912: 188, pl. 1, fig. 1.
Ctenus johnstoni – Arts 1912: 188, pl. 1, fig. 1.
Ctenus carsoni – Arts 1912: 188, pl. 1, fig. 1.
Ctenus aureopubescens – Benoit 1977: 698, figs 1a–d.

Diagnosis

The males of *Afroneutria erythrochelis* comb. nov. differ from those of the other species of the genus by the swollen retrolateral area of the cymbium, the curved median apophysis, the retrolateral base of the embolus and by the laminar projection of the embolus extending beyond its tip (Figs 7A–B, 15C–D). The females can be distinguished by the short lateral projections of the epigynum and the dorsal spermathecae curved ventrally, forming a spiral in the vulva (Figs 7C–D, 12C).

Material examined

Type material

ANGOLA: *Phoneutria erythrochelis*, holotype, ♂, Landana, MNHN.

ANGOLA: *Phoneutria auricularis*, syntypes, 2 ♂♂, 2 ♀♀, 2 immature, Benguela, Chinchoxo, Falkenstein coll., ZMB 2563.

Additional material

ANGOLA: Distrito do Lunda, Cuango, 9°10' S, 17°58' E, 1 ♂, Sep. 1972, C. Cabral coll. (MRAC 167914).

BURUNDI: Bujumbura, 3°23' S, 29°21' E, 1 ♀, 1959, G. Marlier coll. (MRAC 114120); Kanyania, 3°51' S, 29°44' E, 1 ♂, Jul. 1947, Dames de Marie coll. (MRAC 58236); 35 km SE of Kanyania, 3°51' S, 29°44' E, 1 ♂, 1948, Dames de Marie coll. (MRAC 60880).

DEMOCRATIC REPUBLIC OF THE CONGO: Congo da Lemba, 5°12' S, 13°42' E, 1 ♂, 1 ♀, 1912, R. Mayné coll. (MRAC 16303); Ekoli, terr. Opola, 0°22' S, 24°16' E, 2 ♀♀, 1952, Meews coll. (MRAC 73889, 73890); Funda Biabo, 9°50' S, 25°33' E, 1 ♂, 2 Apr. 1914, L. Charliers coll. (MRAC 25490); Katako-Kombe, 3°22' S, 24°24' E, 1 ♀, 10 Apr. 1952, Dr. Fontaine coll. (MRAC 72460); Chutes Guillaume, Kwango, 7°42' S, 17°18' E, 1 ♂, 1936, G. Schmitz coll. (MRAC 21546); Lomami, 9°24' S, 24°46' E, 1 ♀, Dr. Rodhain coll. (MRAC 25496); Lukula, 6°09' S, 26°39' E, 1 ♂, Aug. 1937, E. Dartevelle

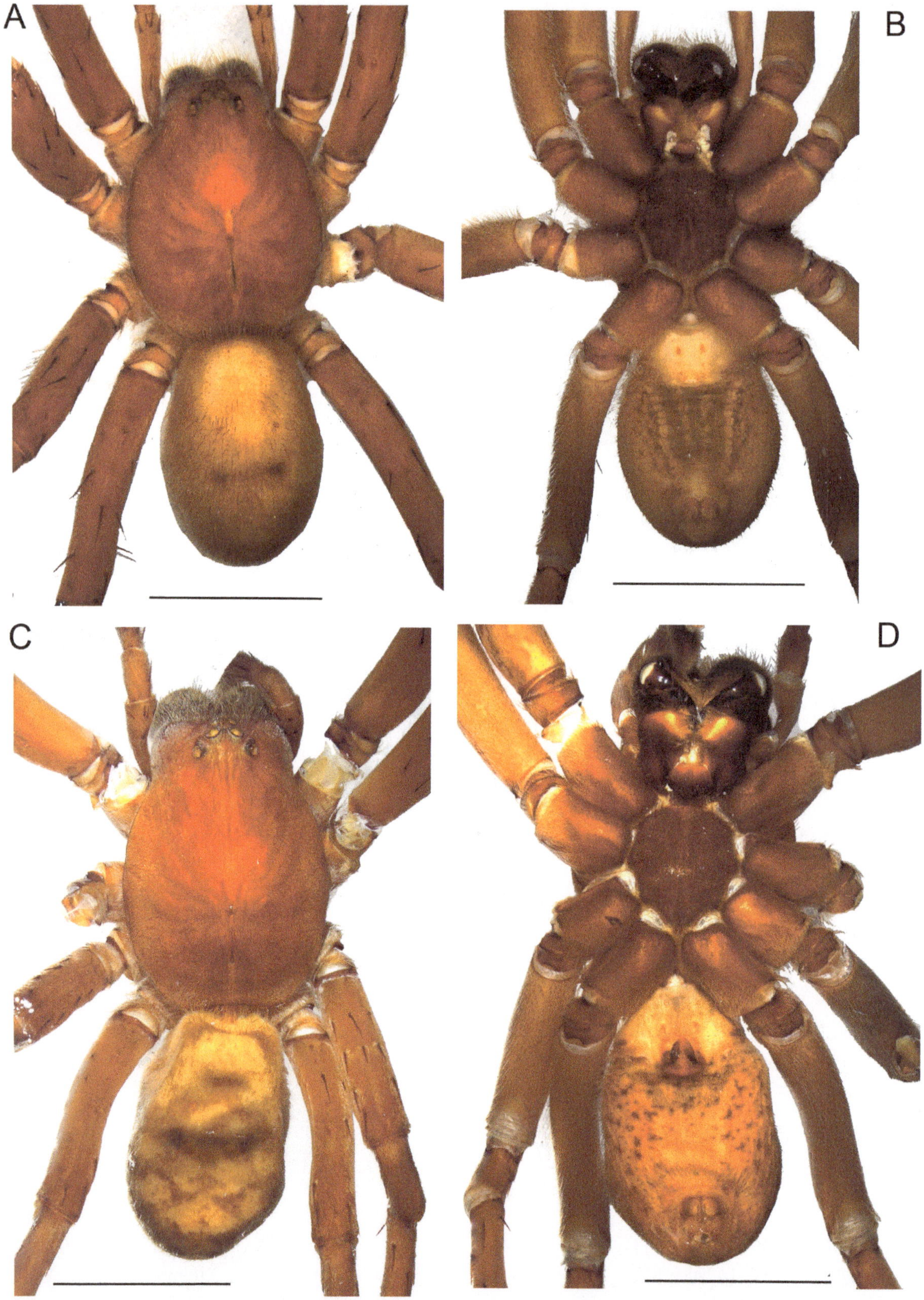

Fig. 6. *Afroneutria erythrochelis* (Simon, 1876) comb. nov. Habitus. **A.** ♂, dorsal view. **B.** ♂, ventral view. **C.** ♀, dorsal view. **D.** ♀, ventral view. Scale bars = 10 mm.

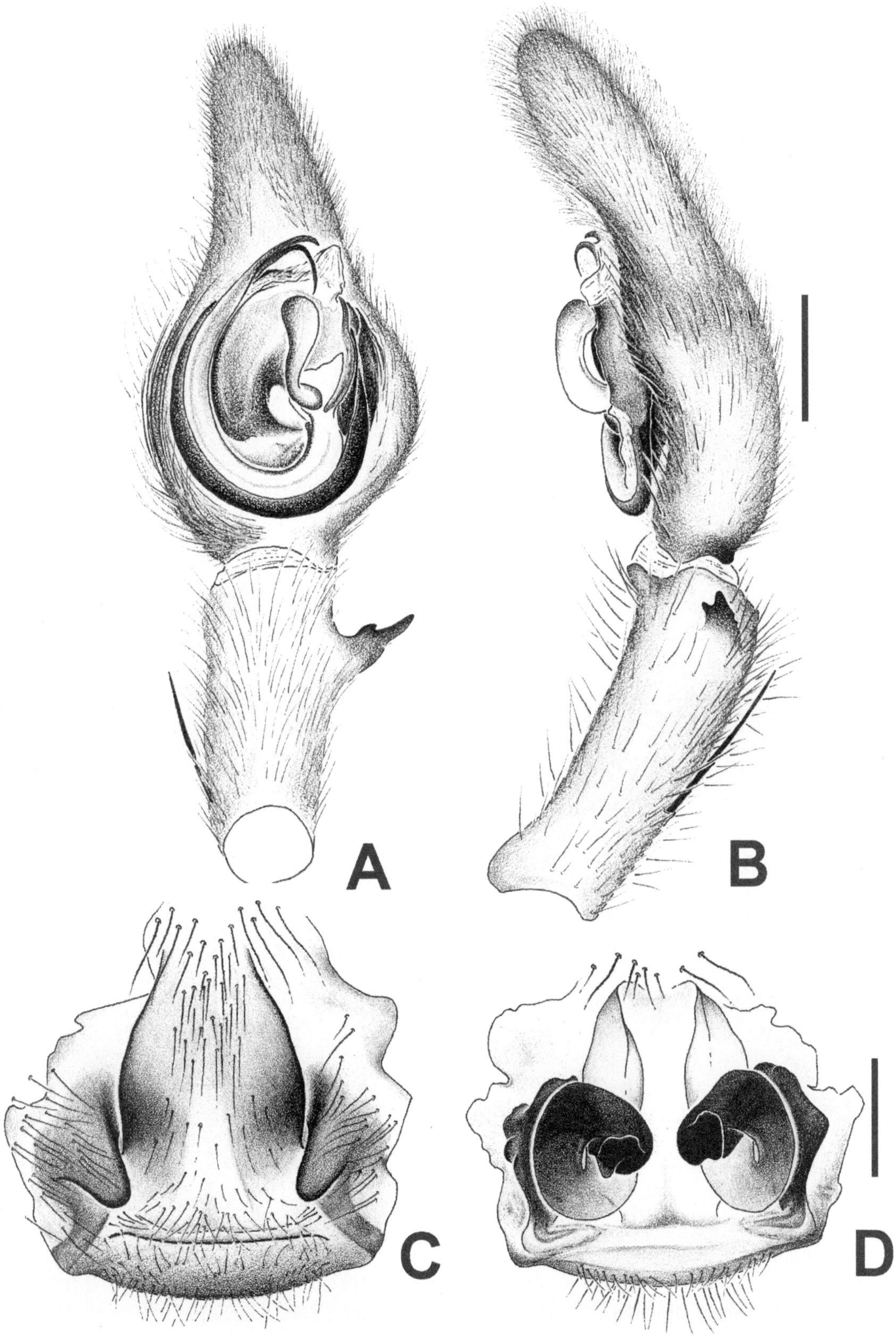

Fig. 7. *Afroneutria erythrochelis* (Simon, 1876) comb. nov. — **A–B**. Male palp. **A**. Ventral view. **B**. Retrolateral view. — **C–D**. Female genitalia. **C**. Epigynum, ventral view. **D**. Vulva, dorsal view. Scale bars = 1 mm.

coll. (MRAC 21487); Camp de Lukula, 6°09' S, 26°39' E, 2 ♂♂, 1911, Dr. Daniel coll. (MRAC 21539); Lukula, 6°09' S, 26°39' E, 1 ♀, 1953, Dr. Wauthier coll. (MRAC 74606); Kivu, Rugari, 2°02' S, 28°56' E, 1 ♂, 1937, R.P. Feyss coll. (MRAC 21518); Kisantu, 5°08' S, 15°06' E, 1 ♀, R.F. Gillet coll. (MRAC 25488); Kisantu, 5°08' S, 15°06' E, 1 ♀, R.P. Vanderyst coll. (MRAC 25500); Tumba, 5°27' S, 14°38' E, 1 ♂, 22 Jul. 1951, M. Bequaert coll. (MRAC 69745); Boma, 5°51' S, 13°03' E, 1 ♂, 1926, H. Schouteden coll. (MRAC 25498); Boma, 5°51' S, 13°03' E, 1 ♀, Nov. 1945, J. Vrydagh coll. (MRAC 25503); Bas-Congo, Noki, 5°00' S, 14°49' E, 1 ♂, A. Cabra coll. (MRAC 21538); Leopoldville, 4°19' S, 15°19' E, 1 ♂, Feb. 1945, R. Fiasse coll. (MRAC 58214); Leopoldville, 4°19' S, 15°19' E, 1 ♀, Jul. 1954, Dr. Fontaine coll. (MRAC 78890); Matadi, 5°49' S, 13°28' E, 1 ♂, 1959, C. Crespin coll. (MRAC 114754); Popokabaka, 5°42' S, 16°34' E, 1 ♀, 1925, R.P. Vanderyst coll. (MRAC 25499); Thysville, 5°15' S, 14°52' E, 1 ♀, 1930, Dr. Vanderhaegen coll. (MRAC 25501); Kamina, 8°44' S, 25°00' E, 2 ♂♂, 1950, Buls coll. (MRAC 67365); Kapolowe, Jadotville, 10°58' S, 26°56' E, 1 ♂, 1959, J.P. Herremans coll. (MRAC 140851); Katanga, Lubumbashi, 11°40' S, 27°28' E, 2 ♀♀, Apr.–May 1966, J. Godeaux coll. (MRAC 131502); Elisabethville, 11°40' S, 27°28' E, 1 ♀, 1936, Richard coll. (MRAC 21503); Elisabethville, 11°40' S, 27°28' E, 1 ♂, Sep. 1932, Ch. Seydel coll. (MRAC 21504); Elisabethville, 11°40' S, 27°28' E, 1 ♀, 1913 (MRAC 25492); Elisabethville, 11°40' S, 27°28' E, 2 ♂♂, 25 Sep. 1959, Ch. Seydel coll. (MRAC 140491); Albertville, 5°57' S, 29°12' E, 1 ♂, 1 ♀, 1959, J. Verhoustraete coll. (MRAC 115056, 115059); Ile de M'Boko, Tanganika, Fizi, 4°18' S, 28°56' E, 1 ♂, 3 Sep. 1958, N. Leleup coll. (MRAC 113231); Mulongo, 7°50' S, 26°59' E, 1 ♂, P. Gerard coll. (MRAC 12428); Kivu, Uvira, 3°24' S, 29°08' E, 2 ♀♀, Aug. 1961, R. Kiss coll. (MRAC 120068); Kivu, Uvira, Lake Tanganika, 3°24' S, 29°08' E, 1 ♀, Aug. 1961, R. Kiss coll. (MRAC 120081); Leverville, 4°50' S, 18°44' E, 1 ♂, 1928, Mme. Tinant coll. (MRAC 25493); Dilolo, 10°41' S, 22°21' E, 1 ♂, G.F. De Witte coll. (MRAC 16308); Katanga, Kisenge, 10°42' S, 23°10' E, 2 ♀♀, Dec. 1964, A. Regnard coll. (MRAC 127561); Kwamouth, 3°11' S, 16°12' E, 1 ♀, Jun. 1913, J. Maes coll. (MRAC 25494); Libenge, 3°39' N, 18°38' E, 1 ♀, C. Leontovitch coll. (MRAC 30274); Ndua, close to Bolobo, 2°10' S, 16°14' E, 1 ♀, Dec. 1951, N'gue coll. (MRAC 72204); Kalima, 2°34' S, 26°37' E, 1 ♂, R.S. Dierexens coll. (MRAC 61614); Kasongo-Lunda, 6°28' S, 16°49' E, 1 ♀, 1939, Mme. Bequaert coll. (MRAC 2178); Maniema, Kasongo, Kahuta, 4°27' S, 26°40' E, 1 ♀, 1959, P.L.G. Benoit & Segers coll. (MRAC 140375); Bambesa, 3°26' N, 25°41' E, 1 ♀, Aug. 1938, J. Vrydagh coll. (MRAC 2790).

REPUBLIC OF CONGO: Pointe Noire, 4°47' S, 11°52' E, 1 ♀, 1938, E. Dartevelle coll. (MRAC 21528).

GABON: Tchibanga, 2°55' S, 10°59' E, 1 ♀, 20 Sep. 1984, A. Souche, Pauly coll. (MRAC 168750).

MALAWI: Chisasira Forest, 25 km S of Chintheche, 11°50' S, 33°53' E, 1 ♂, 27 Sep.–14 Oct. 1977, R. Jocqué coll. (MRAC 155300); Chintheche, 11°50' S, 34°10' E, 1 ♂, 1–20 Nov. 1977, R. Jocqué coll. (MRAC 155304).

NAMIBIA: Andara-Kavango, Okavango River, 18°04' S, 21°29' E, 1 ♀, Sep. 1979, M.E. Baddeley coll. (MRAC 152797).

RWANDA: Kigali, 01°57' S, 30°04' E, 1 ♀, Sep. 1961, A. Lestrade coll. (MRAC 123159).

TANZANIA: Pugu Forest, Kisarawe, near Minala, 6°52' S, 39°05' E, 1 ♂, 1969, K.M. Howell coll. (MRAC 159272); Dar-es-Salaam, UDSM Campus, 6°48' S, 39°17' E, 1 ♂, 1970–1971, K.M. Howell coll. (MRAC 159472); Dar-es-Salaam, UDSM Campus, 6°48' S, 39°17' E, 1 ♀, Sep. 1980, K.M. Howell coll. (MRAC 169162).

ZIMBABWE: Bulawayo, 20°09' S, 28°36' E, 1 ♀, Nov. 1964–Jan. 1965, S. Bucklin coll. (MRAC 128094).

Type material not examined

MALAWI: *Ctenus johnstoni*, holotype, ♀, Zomba, Lake Nyassa, BMNH.

KENYA and DEMOCRATIC REPUBLIC OF THE CONGO: *Ctenus carsoni*, syntypes, 3 ♂♂, Mombasa and Nyanza Province (Kenya) and Kavala Island, Lake Tanganyika (Democratic Republic of the Congo), Carson coll., Pasha coll., BMNH.

GHANA: *Ctenus aureopubescens*, holotype, ♂, probably ZMB, not found.

Description

Male (MRAC 60880)

Total length 29.10. Carapace 15.10 long and 11.45 wide. Eye diameter: AME 0.55, ALE 0.25, PME 0.72, PLE 0.66. Leg measurements: I: femur 16.30 / patella 6.80 / tibia 16.10 / metatarsus 15.90 / tarsus 4.40 / total 59.50; II: 14.30 / 6.30 / 12.40 / 14.10 / 4.20 / 51.30; III: 11.50 / 5.20 / 8.80 / 10.70 / 3.40 / 39.60; IV: 15.80 / 5.70 / 13.70 / 17.45 / 4.20 / 56.85. Leg formula 1423. Leg spination: tibia I and II v2-2-2-2-2, r1-1, p1-1; III and IV v2-2-2, r1-1, p1-1; metatarsus I, II and III v2-2-2, p1-1-1, r1-1-1; IV v1-1-1-1-1-2, r1-1-1, p1-1-1. Legs covered by long, golden hairs (Fig. 6A–B). Palp: cymbium elongated, almost as long as tibia; conductor rectangular, curved; median apophysis drop-shaped, curved at base (Figs 7A–B, 15C–D).

Female (MRAC 2178)

Total length 32.00. Carapace 16.80 long and 12.96 wide. Eye diameter: AME 0.55, ALE 0.25, PME 0.70, PLE 0.65. Leg measurements: I: femur 15.35 / patella 7.50 / tibia 13.90 / metatarsus 13.75 / tarsus 4.40 / total 54.90; II: 14.25 / 6.60 / 11.60 / 12.20 / 3.80 / 48.45; III: 11.20 / 5.50 / 8.30 / 10.20 / 3.50 / 38.70; IV: 15.00 / 6.10 / 12.20 / 15.65 / 3.90 / 52.85. Leg formula 1423. Leg spination: tibia I and II v2-2-2-2-2, r0, p0; III and IV v2-2-2, r1-1, p1-1; metatarsus I and II v2-2-2, r0, p0; III v2-2-2, r1-1-1, p-1-1-1; IV v1-1-1-1-1-2, r1-1-1, p1-1-1. Legs covered by long, golden hairs (Fig. 6C–D). Epigynum: triangular median field and short lateral projections of lateral fields (Fig. 7C, 12C). Dorsal spermathecae elongated, curved, ventral spermathecae short (Fig. 7D).

Variation

Ten ♂♂: total length 19.60–29.10, carapace length 9.50–15.10, femur I length 10.50–16.30. Ten ♀♀: total length 31.50 –33.45, carapace length 14.50–18.25, femur I length 12.60–16.80.

Distribution

Angola, Burundi, Democratic Republic of the Congo, Malawi, Namibia, Republic of Congo, Rwanda, Tanzania and Zimbabwe (Fig. 16B).

Afroneutria quadrimaculata sp. nov.
urn:lsid:zoobank.org:act:0FF06727-624D-465E-ACA1-8A9CEFB12A50
Figs 8A–D, 9A–E, 12B, 15A–B, 16A

Ctenus velox – Benoit 1977: 700, fig. 2a–d (misidentification).

Diagnosis

The males of *Afroneutria quadrimaculata* sp. nov. differ from those of the other species of the genus by the laminar embolus with enlarged tip and the curved, subdistal, additional projection of the embolus (Figs 9A, 15A). The females can be distinguished by the large and elongated lateral projections of the epigynum and the curved dorsal spermathecae in the vulva (Figs 9D–E, 12B).

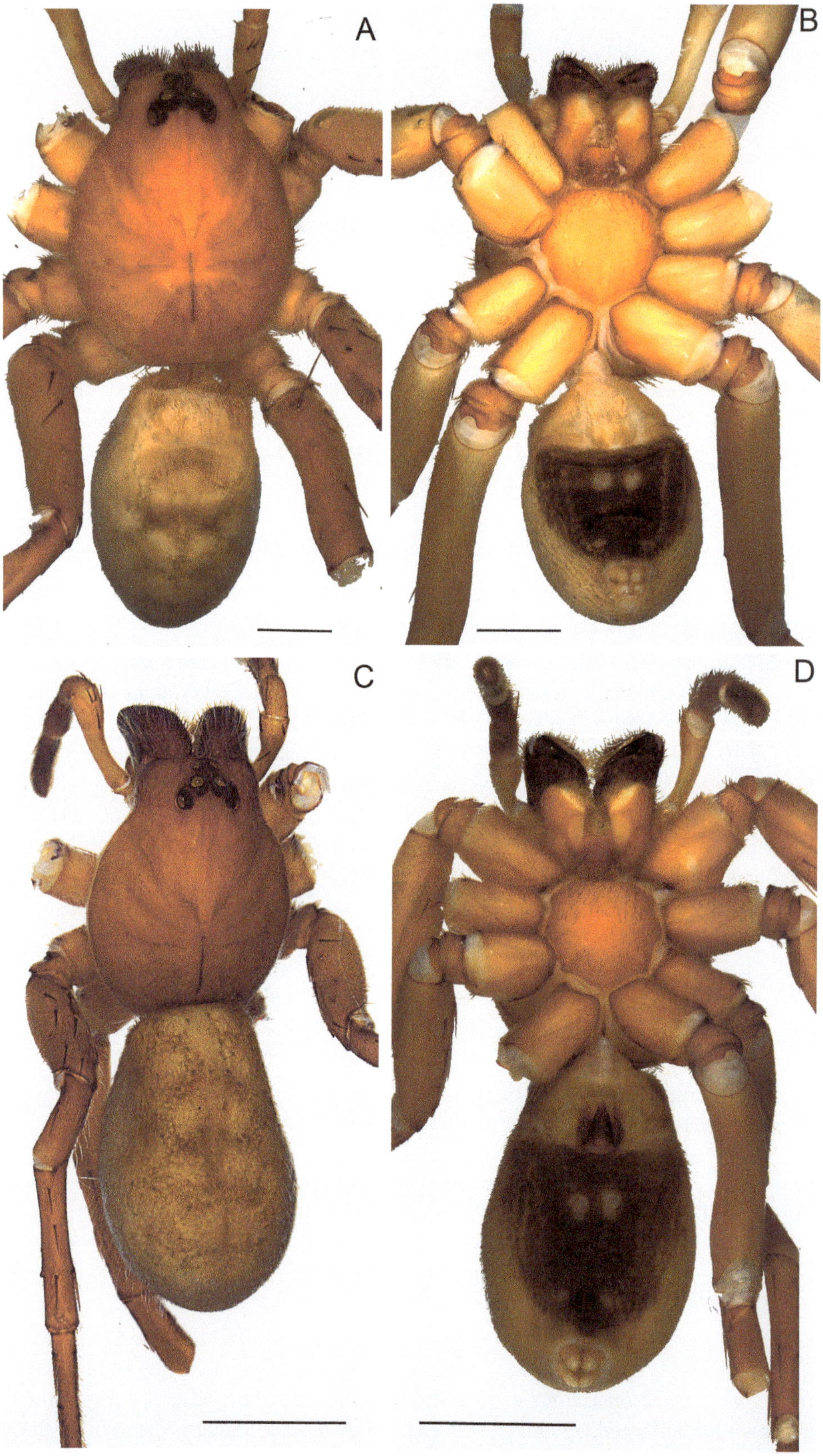

Fig. 8. *Afroneutria quadrimaculata* sp. nov. Habitus. **A**. ♂, dorsal view. **B**. ♂, ventral view. **C**. ♀, dorsal view. **D**. ♀, ventral view. Scale bars = 2 mm.

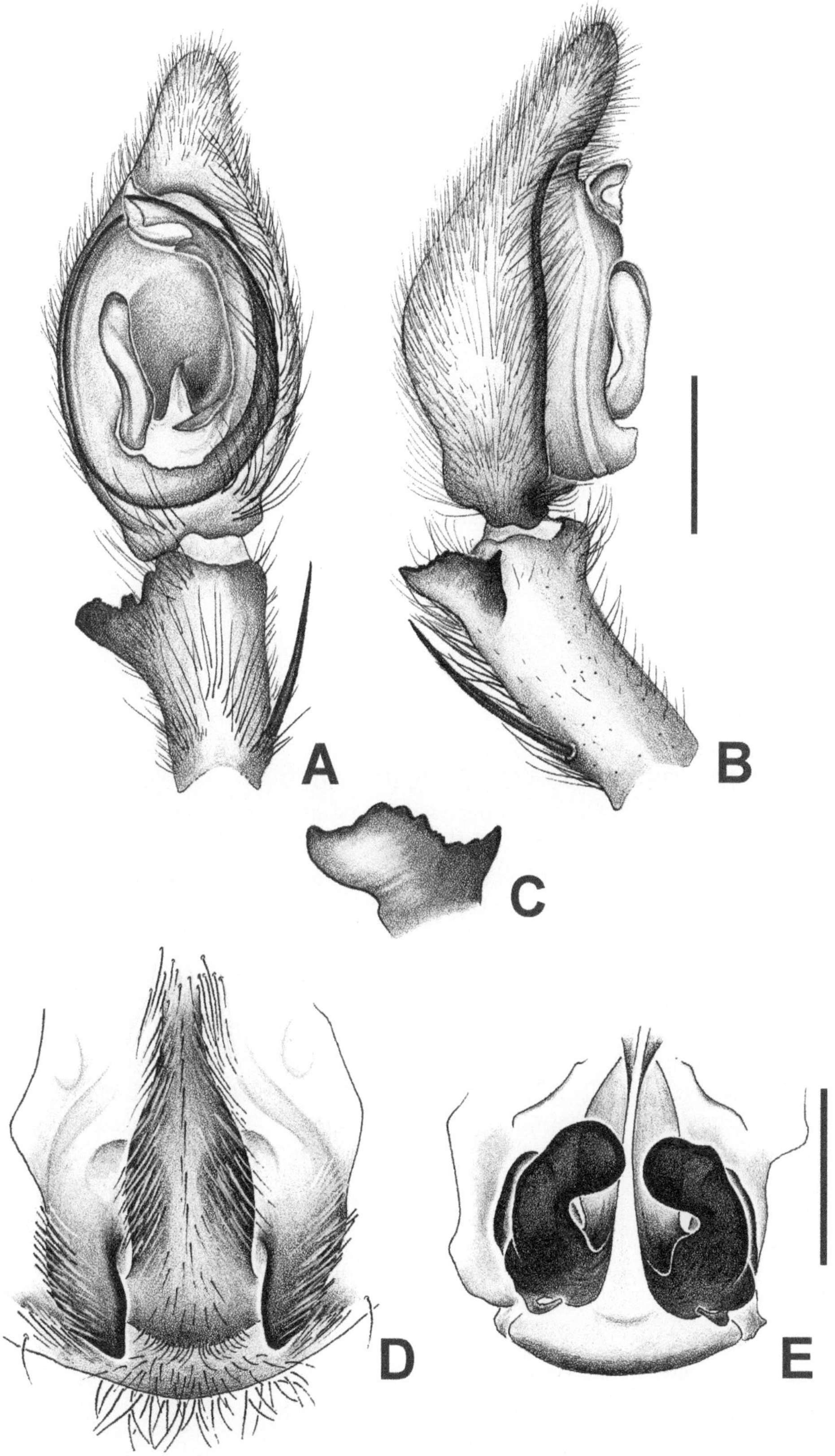

Fig. 9. *Afroneutria quadrimaculata* sp. nov. — **A–C**. Male palp. **A**. Ventral view. **B**. Retrolateral view. **C**. Retrolateral tibial apophysis, lateral view. — **D–E**. Female genitalia. **D**. Epigynum, ventral view. **E**. Vulva, dorsal view. Scale bars = 2 mm.

Etymology

The specific name is derived from Latin, *quadri-* (four) and *macula-* (spot), meaning to have four spots and making reference to the four white spots on the venter of the abdomen.

Material examined

Holotype

DEMOCRATIC REPUBLIC OF THE CONGO: ♂, Jadotville, Kasompi, Katanga, Oct. 1956, Z. Bacq coll. (MRAC 90554).

Paratype

DEMOCRATIC REPUBLIC OF THE CONGO: ♀, Kisenge, Katanga, Dec. 1964, A. Regnard coll. (MRAC 127584).

Additional material

DEMOCRATIC REPUBLIC OF THE CONGO: Kasai, Bibanga, 1 ♀, 1940, M. Bequaert coll. (MRAC 21542); Katanga, Kisenge, 1 ♂, 1965, Q. Regnard coll. (MRAC 128007); Katanga, Jadotville, Kasompi, 1 ♀, Oct. 1956 (MRAC 90555); 1 ♂, 6 Oct. 1956 (MRAC 90596); Katanga, Luiswishi Forest, 1 ♀, 2 Mar. 1974, F. Malaise coll. (MRAC 149210); Katanga, L'Upemba National Park, River Lufira (680 m), 2 ♂♂, 5 ♀♀, 15 Sep. 1947, G.F. de Witte coll. (MRAC 152022).

Description

Male (MRAC 128007)

Total length 14.90. Carapace 7.50 long and 5.90 wide. Eye diameter: AME 0.36, ALE 0.24, PME 0.48, PLE 0.44. Leg measurements: I: femur 9.00 / patella 3.40 / tibia 9.10 / metatarsus 8.80 / tarsus 3.30 / total 33.60; II: 8.10 / 3.30 / 7.30 / 7.80 / 3.00 / 29.50; III: 7.00 / 2.70 / 5.50 / 6.70 / 2.50 / 24.40; IV: 9.20 / 3.10 / 8.40 / 10.90 / 3.30 / 34.90. Leg formula 4123. Leg spination: tibia I and II v2-2-2-2-2, r1-1, p1-1; III and IV v2-2-2, r1-1, p1-1; metatarsus I, II and III v2-2-2, p1-1-1, r1-1-1; IV v1-1-1-1-1-2, r1-1-1, p1-1-1. Coloration: venter of abdomen black with four white spots, two behind epigastric furrow, two in front of spinnerets (Fig. 8B). Palp: cymbium slightly longer than tibia; conductor oval; median apophysis curved (Figs 9A–C, 15A–B).

Female (MRAC 90555)

Total length 20.10. Carapace 9.50 long and 7.40 wide. Eye diameter: AME 0.44, ALE 0.28, PME 0.56, PLE 0.52. Leg measurements: I: femur 8.70 / patella 4.40 / tibia 8.20 / metatarsus 7.60 / tarsus 2.70 / total 31.60; II: 7.80 / 4.10 / 6.80 / 6.70 / 2.70 / 28.10; III: 6.80 / 3.20 / 5.50 / 6.30 / 2.40 / 24.20; IV: 9.00 / 3.70 / 8.00 / 10.20 / 3.10 / 34.00. Leg formula 4123. Leg spination: tibia I and II v2-2-2-2-2, r0, p0; III and IV v2-2-2, r1-1, p1-1; metatarsus I and II v2-2-2, r0, p0; III v2-2-2, r1-1-1, p-1-1-1; IV v1-1-1-1-1-2, r1-1-1, p1-1-1. Coloration: as in male (Fig. 8C–D). Epigynum: with triangular median field and elongated lateral projections of lateral fields (Figs 9D, 12B). Dorsal spermathecae long, curved, ventral spermathecae short (Fig. 9E).

Variation

Six ♂♂: total length 15.00–14.10, carapace length 7.50–7.40, femur I length 9.00–7.70. Ten ♀♀: total length 24.20–16.50, carapace length 10.00–8.10, femur I length 8.90–7.50.

Distribution

Democratic Republic of the Congo (Fig. 16A).

Afroneutria hybrida sp. nov.

urn:lsid:zoobank.org:act:425F304C-8F67-4DFA-A475-A3E3115CFEB8

Figs 10A–B, 11A–C, 13C–D, 16B

Diagnosis

The males of *Afroneutria hybrida* sp. nov. differ from those of the other species of the genus by the laminar embolus with acute tip, the short additional subdistal projection of the embolus and the truncated RTA (Figs 11A–B, 13C–D).

Etymology

The specific name is derived from Latin *ibrida*, meaning a mongrel, specifically the offspring of a tame sow and a wild boar, and making reference to the similarity of the morphology to that of *Afroneutria velox* comb. nov. and *A. quadrimaculata* sp. nov.

Material examined

Holotype

TANZANIA: ♂, Kambai Forest Reserve, Muheza District, Tanga Region, 4°59' S, 38°41' E, ZMUC.

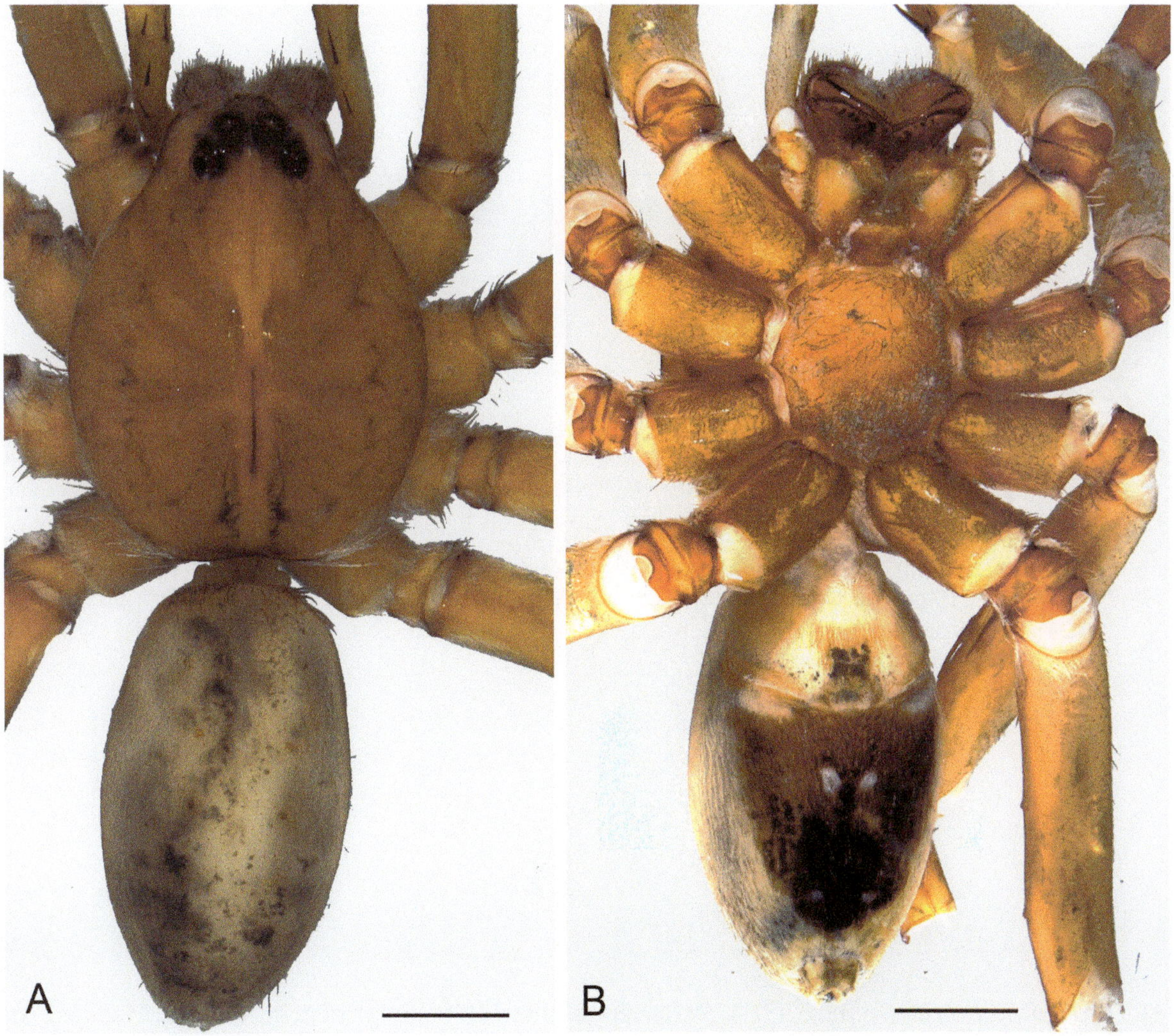

Fig. 10. *Afroneutria hybrida* sp. nov. Habitus, ♂. **A**. Dorsal view. **B**. Ventral view. Scale bars = 2 mm.

Paratype

TANZANIA: 1 ♂, same collection data as holotype, ZMUC.

Additional material

None.

Description (holotype)

Total length 12.40. Carapace 6.50 long and 5.50 wide. Eye diameter: AME 0.32, ALE 0.19, PME 0.60, PLE 0.56. Leg measurements: I: femur 9.70 / patella 3.20 / tibia 10.60 / metatarsus 11.40 / tarsus 4.70 / total 39.60; II: 8.70 / 3.10 / 8.90 / 9.50 / 3.50 / 33.70; III: 7.50 / 2.40 / 6.70 / 7.70 / 3.10 / 27.40; IV: 10.20 / 2.60 / 10.00 / 13.20 / 4.30 / 40.30. Leg formula 4123. Leg spination: tibia I and II v2-2-2-2-2, r1-1, p1-1; III and IV v2-2-2, r1-1, p1-1; metatarsus I, II and III v2-2-2, p1-1-1, r1-1-1; IV v1-1-1-1-1-2, r1-1-1, p1-1-1. Coloration: venter of abdomen black, with two white spots placed transversely behind epigastric

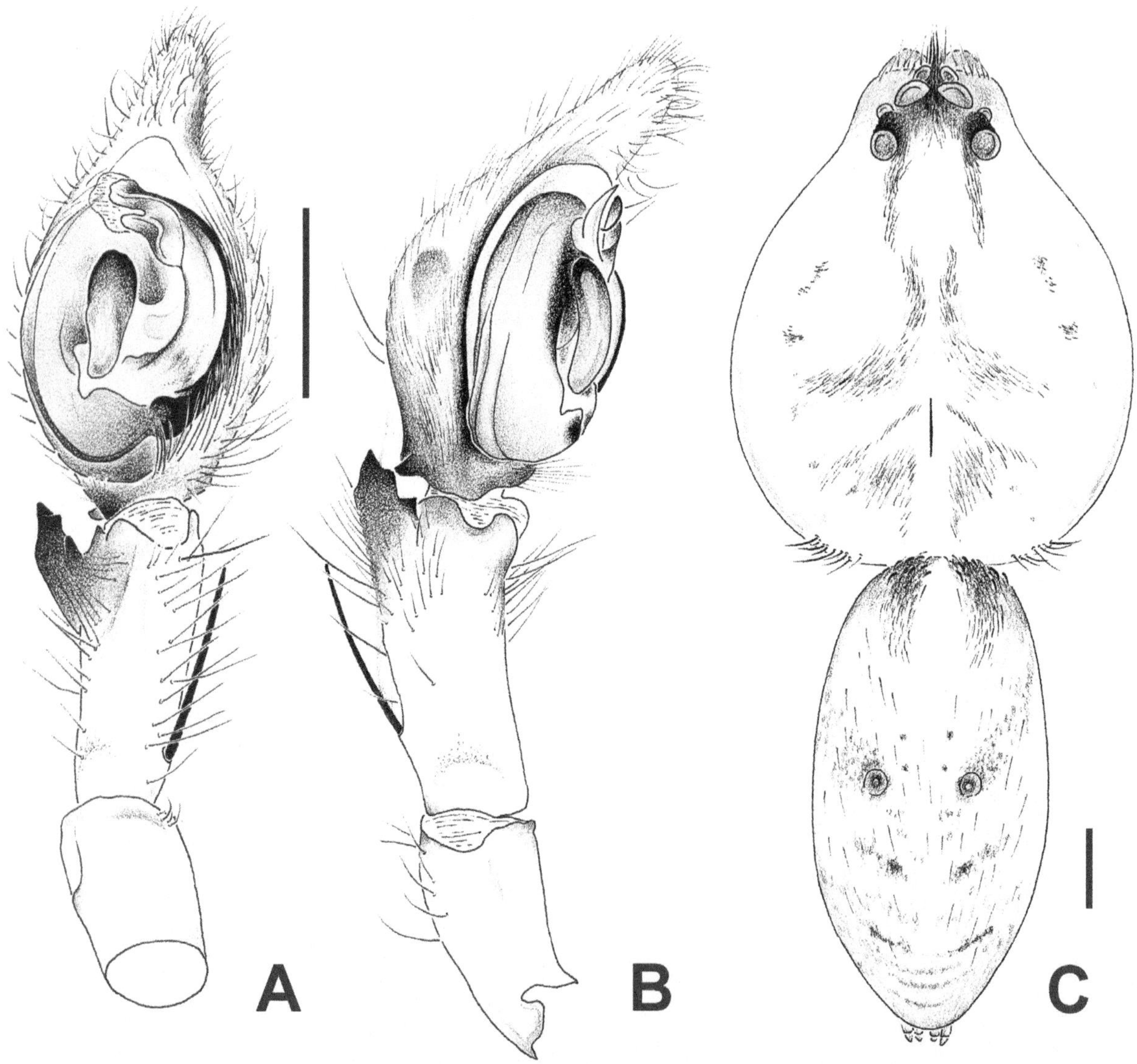

Fig. 11. *Afroneutria hybrida* sp. nov., ♂. **A**. Palp, ventral view. **B**. Palp, retrolateral view. **C**. Habitus, dorsal view. Scale bars = 1 mm.

furrow (Fig. 10B). Palp: cymbium elongated, longer than tibia; conductor oval; median apophysis straight (Figs 11A–B, 13C–D).

Distribution

Tanzania (Fig. 16B).

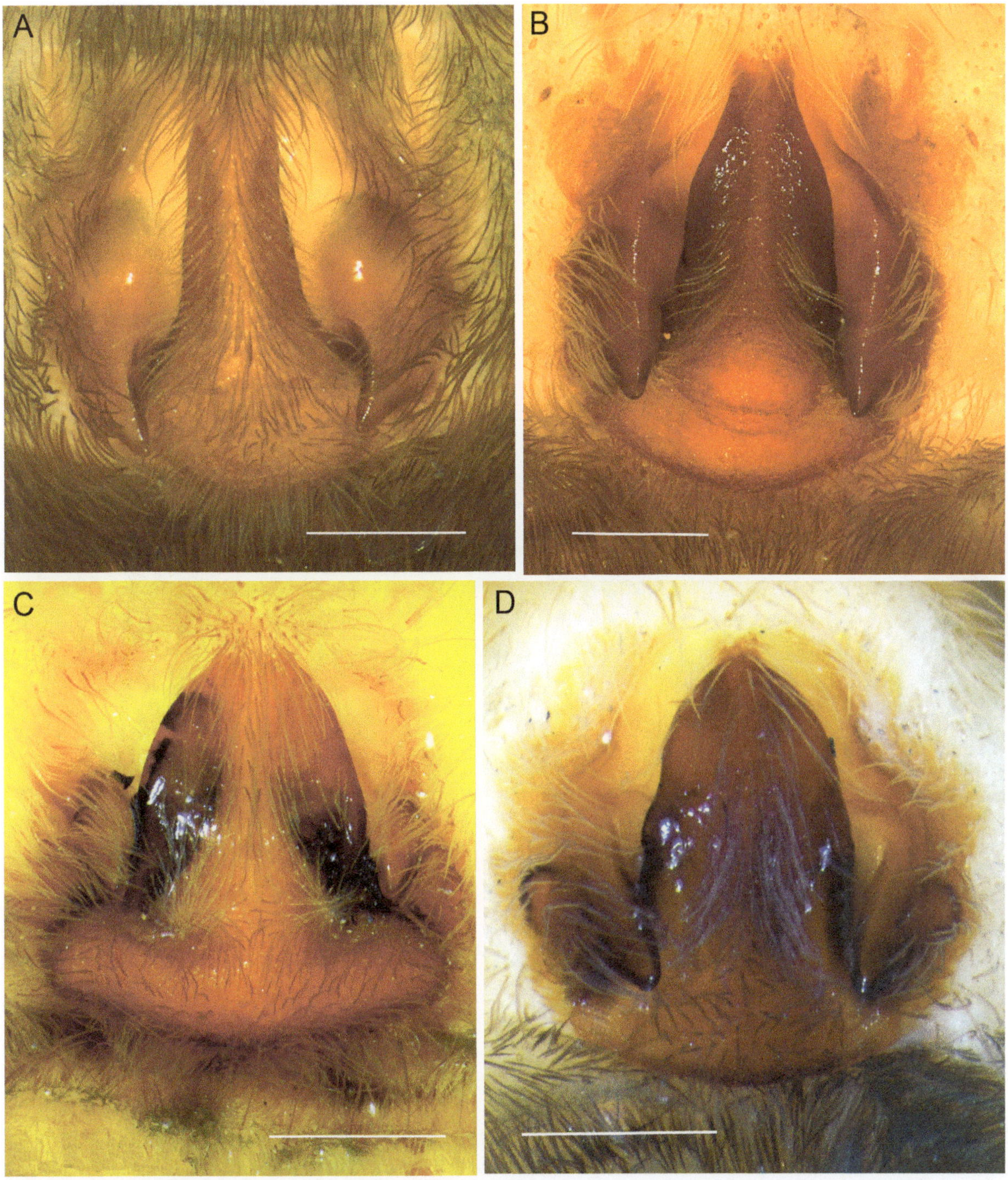

Fig. 12. Epigynum, ventral view. **A**. *Afroneutria immortalis* (Arts, 1912) comb. nov. **B**. *Afroneutria quadrimaculata* sp. nov. **C**. *Afroneutria erythrochelis* (Simon, 1876) comb. nov. **D**. *Afroneutria velox* (Blackwall, 1865) comb. nov. Scale bars = 1 mm.

Discussion

Afroneutria gen. nov. is proposed based on the results of the cladistic analysis presented in Polotow & Brescovit (2014). In that analysis, three of the species treated here were used: *Afroneutria erythrochelis* comb. nov., *A. immortalis* comb. nov. and *A. velox* comb. nov. The clade appeared as sister-group of a clade formed by *Ctenus sinuatipes* F.O. Pickard-Cambridge, 1897 and *C. curvipes* (Keyserling, 1881), both from Central America (Polotow & Brescovit 2014: fig. 3). The *Afroneutria* clade is supported by a unique elongated projection in the apical area of the embolus (Polotow & Brescovit 2014: character 25), also present in the species described here, *A. quadrimaculata* sp. nov. and *A. hybrida* sp. nov.

The Afrotropical region houses 123 species of Ctenidae, distributed in eight genera: *Africactenus* Hyatt, 1954, *Anahita* Karsch, 1879, *Apolania* Simon, 1898, *Caloctenus* Keyserling, 1877, *Ctenus* Walckenaer, 1805, *Petaloctenus* Jocqué & Steyn, 1997, *Thoriosa* Simon, 1910 and *Trogloctenus* Lessert, 1935 (Platnick 2015). *Viridasius* Simon, 1889 and *Vulsor* Simon, 1889 are excluded from the count, because they belong to another family according to Polotow *et al.* (in press). *Ctenus* contains the largest number of species (more than 70). The recent redescription of the Neotropical type species, *Ctenus dubius*, by Brescovit & Simó (2007) and cladistic analyses (Silva 2003; Polotow & Brescovit 2009) indicated that the genus is polyphyletic as currently delimited. Thus, the description of *Afroneutria* gen. nov. is helping to redefine Ctenidae and the polyphyletic *Ctenus* in the Afrotropical region.

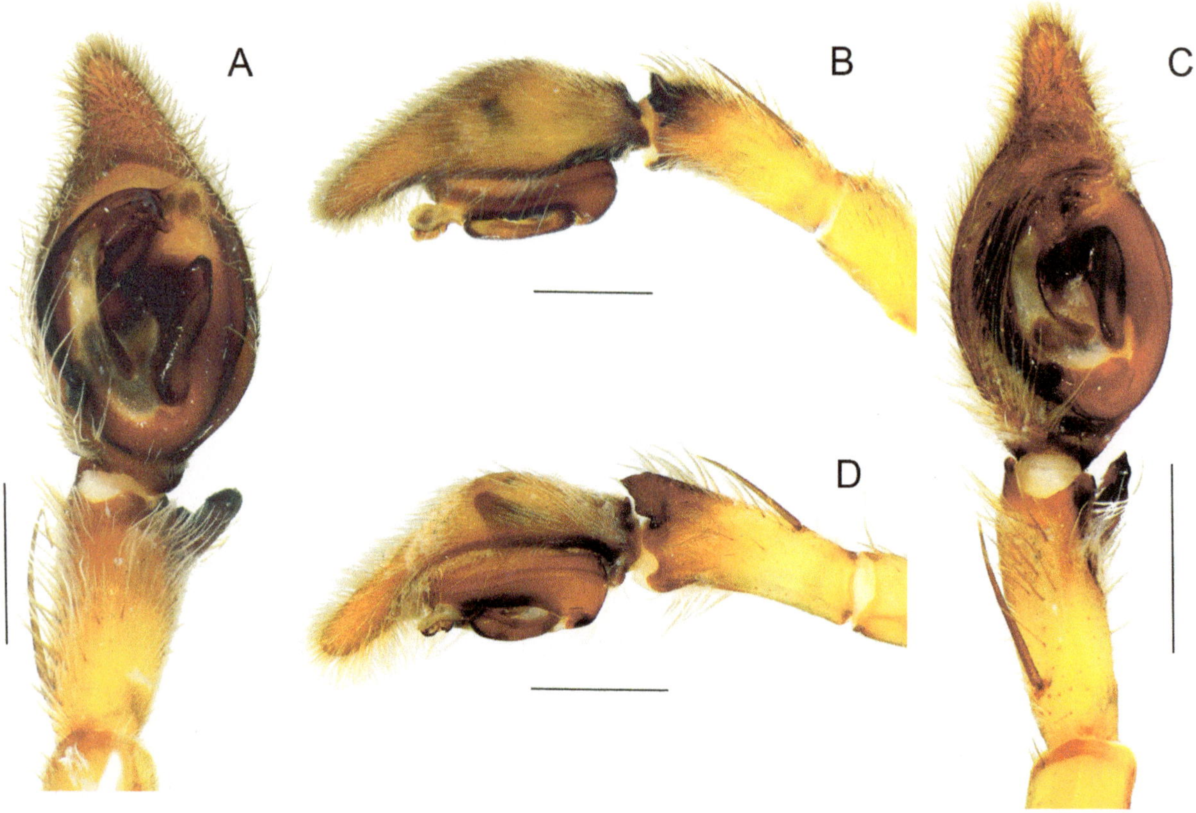

Fig. 13. Male palp. — **A–B**. *Afroneutria velox* (Blackwall, 1865) comb. nov. **A**. Ventral view. **B**. Retrolateral view. — **C–D**. *Afroneutria hybrida* sp. nov. **C**. Ventral view. **D**. Retrolateral view. Scale bars = 1 mm.

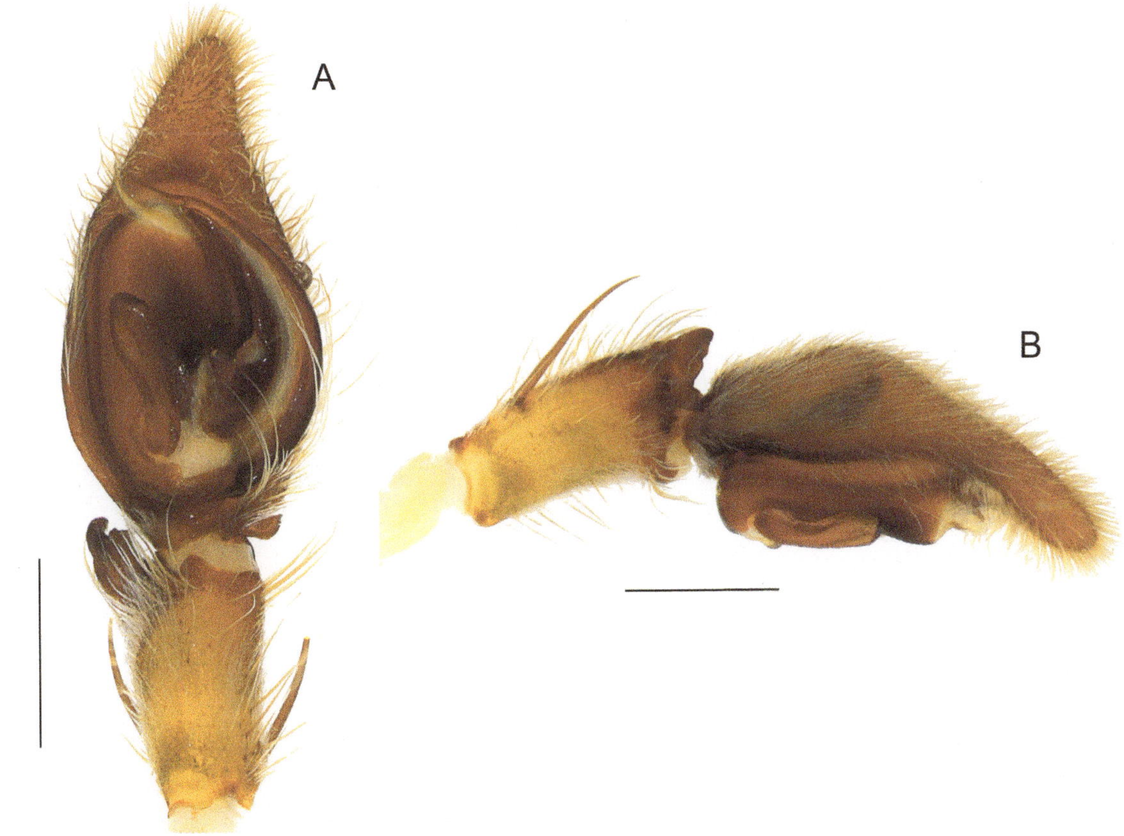

Fig. 14. *Afroneutria immortalis* (Arts, 1912) comb. nov. Male palp. **A**. Ventral view. **B**. Retrolateral view. Scale bars = 1 mm.

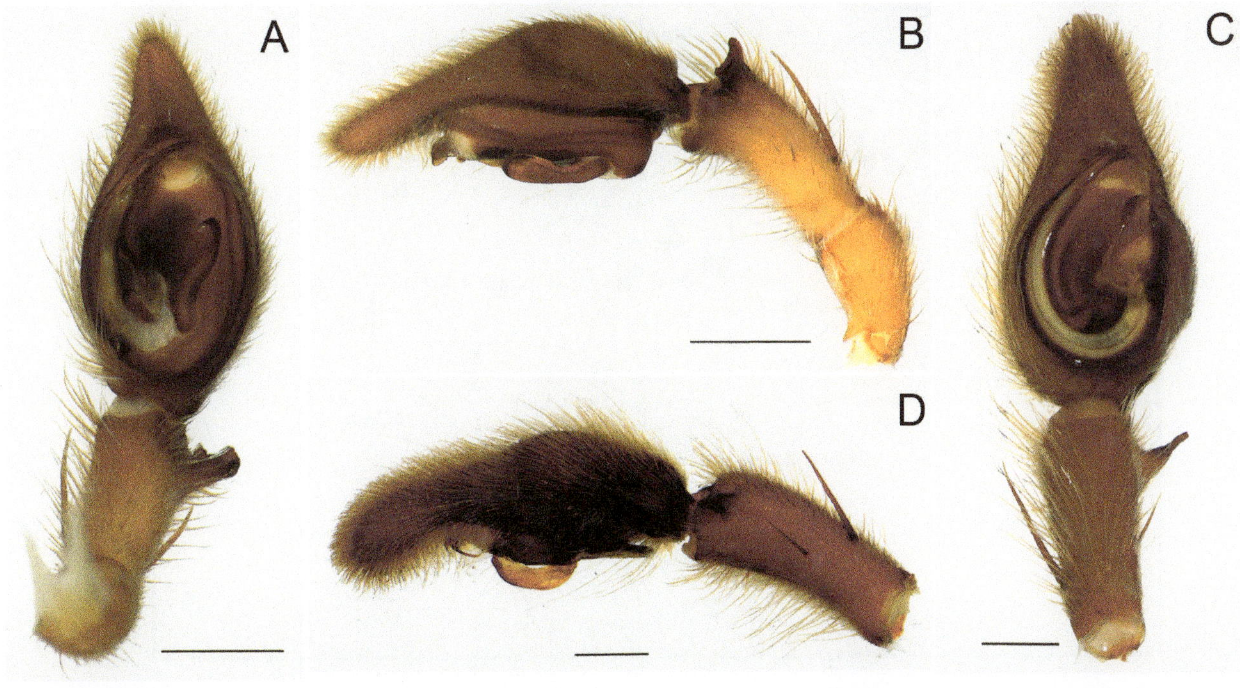

Fig. 15. Male palp. — **A–B**. *Afroneutria quadrimaculata* sp. nov. **A**. Ventral view. **B**. Retrolateral view. — **C–D**. *Afroneutria erythrochelis* (Simon, 1876) comb. nov. **C**. Ventral view. **D**. Retrolateral view. Scale bars = 1 mm.

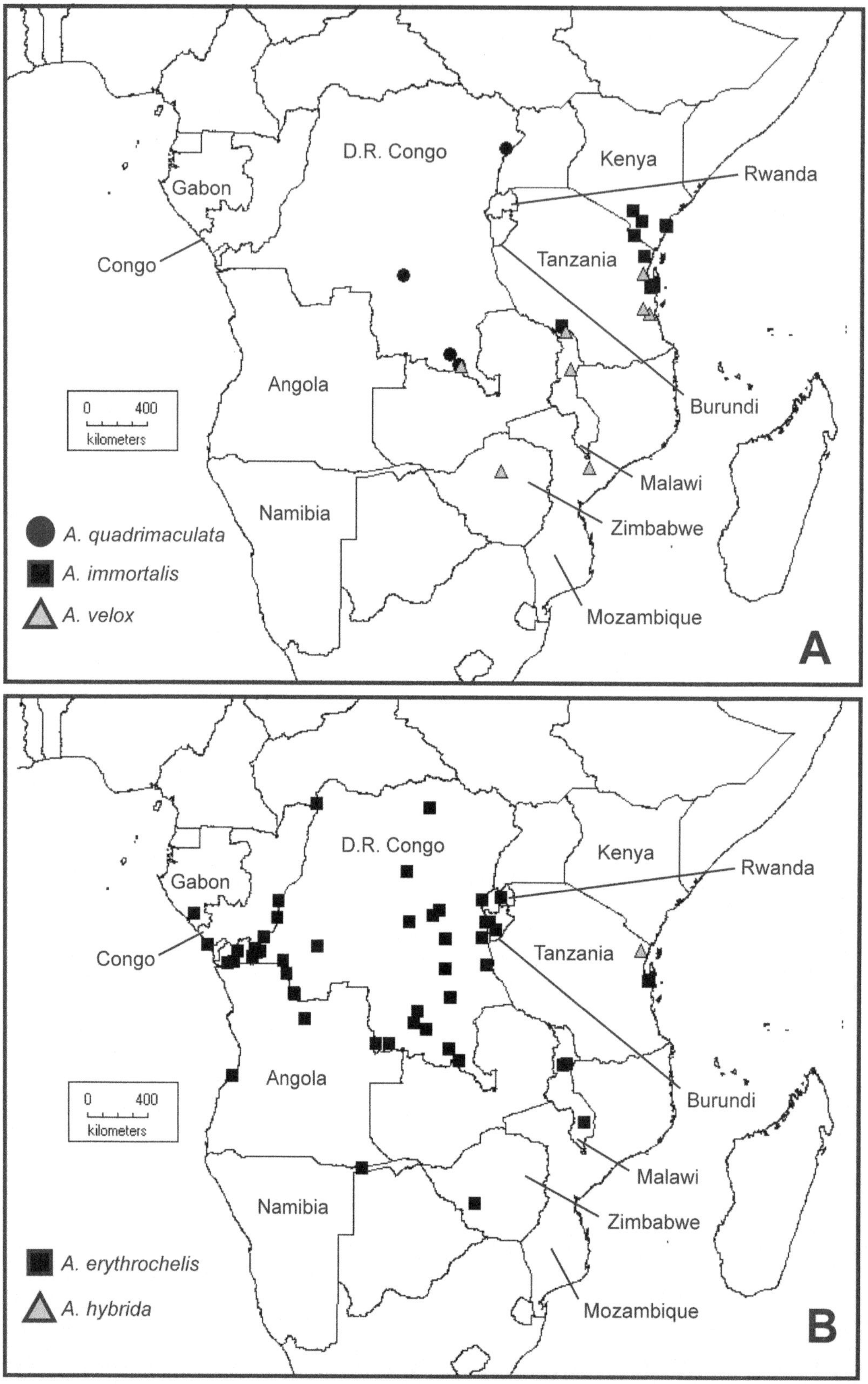

Fig. 16. Distribution records of species of *Afroneutria* gen. nov. on the African continent. **A.** Distribution of *A. velox* (Blackwall, 1865) comb. nov., *A. immortalis* (Arts, 1912) comb. nov. and *A. quadrimaculata* sp. nov. **B.** Distribution of *A. erythrochelis* (Simon, 1876) comb. nov. and *A. hybrida* sp. nov.

Acknowledgments

Financial support for this study was provided by the Fundação de Amparo a Pesquisa do Estado de São Paulo - FAPESP (06/55230-7, doctorate fellowship) and the Belgian National Focal Point to the Global Taxonomy Initiative at the Royal Museum for Central Africa, Tervuren. We would like to thank all curators who kindly lent critical specimens for this research.

References

Arts L. des. 1912. Zusammenstellung der afrikanischen Arten der Gattung *Ctenus*. *Jahrbuch der Hamburgischen Wissenschaftlichen Anstalten* 29 (2): 183–218.

Benoit P.L.G. 1977. Etudes sur les Ctenidae africains (Araneae) V. Gen. *Ctenus* Walck. – groupe *erythrochelis*. *Revue zoologique africaine* 91: 698–703.

Blackwall J. 1865. Descriptions of recently discovered species and characters of a new genus of Araneidea from the East of Central Africa. *The Annals and Magazine of Natural History*, 3rd Series, 16: 336–352.

Bösenberg W. & Lenz H. 1895. Ostafrikanishe Spinnen gesammelt von Herrn Dr. F. Stuhlmann in den Jahren 1888 und 1889. *Jahrbuch der Hamburgischen Wissenschaftlichen Anstalten* 12 (2): 27–51.

Brescovit A.D. & Simó M. 2007. On the Brazilian Atlantic Forest species of the spider genus *Ctenus* Walckenaer, with the description of a neotype for *C. dubius* Walckenaer (Araneae, Ctenidae, Cteninae). *Bulletin of the British Arachnological Society* 14: 1–17.

Hyatt K.H. 1954. The African spiders of the family Ctenidae in the collections of the British Museum (Natural History). *The Annals and Magazine of Natural History*, 12th Series, 7: 877–894.

Karsch F. 1879. West-afrikanische Arachniden, gesammelt von Herrn Stabsarzt Dr. Falkenstein. *Zeitschrift für die gesammten Naturwissenschaften* 52: 329–373.

Lessert R. de. 1915. Arachnides de l'Ouganda et de l'Afrique orientale allemande (voyage du Dr J. Carl dans la region des lacs de l'Afrique centrale). *Revue suisse de zoologie* 23: 1–80.

Pickard-Cambridge F.O. 1898. On the cteniform spiders of Africa, Arabia and Syria. *Proceedings of the Zoological Society of London* 1898: 13–32.

Polotow D. & Brescovit A.D. 2009. Revision and cladistic analysis of *Isoctenus* and description of a new neotropical genus (Araneae, Ctenidae, Cteninae). *Zoological Journal of the Linnean Society* 155: 583–614.

Polotow D. & Brescovit A.D. 2014. Phylogenetic analysis of the tropical wolf spider subfamily Cteninae (Arachnida, Araneae, Ctenidae). *Zoological Journal of the Linnean Society* 170: 333–361.

Polotow D., Carmichael A. & Griswold C.E. In press. Total evidence analysis of the phylogenetic relationship of Lycosoidea spiders (Araneae, Entelegynae). *Invertebrate Systematics*.

Ramírez M.J. 2014. The morphology and phylogeny of dionychan spiders (Araneae, Araneomorphae). *Bulletin of the American Museum of Natural History* 390: 1–374.

Silva D. 2003. Higher-level relationships of the spider family Ctenidae (Araneae: Ctenoidea). *Bulletin of the American Museum of Natural History* 274: 1–86.

Simó M. & Brescovit A.D. 2001. Revision and cladistic analysis of the Neotropical spider genus *Phoneutria* Perty, 1833 (Araneae, Ctenidae), with notes on related Cteninae. *Bulletin of the British Arachnological Society* 12 (2): 67–82.

Simon E. 1876. Étude sur le arachnides du Congo. *Bulletin de la Société zoologique de France* 1: 215–224.

Simon E. 1897. *Histoire Naturelle des Araignées.* Librairie encyclopédique de Roret, Paris.

Strand E. 1906. Tropisch-afrikanische Spinnen des Kgl. Naturalien-kabinetts in Stuttgart. *Jahreshefte des Vereins für Vaterländische Naturkunde in Württemberg* 62: 13–103.

Strand E. 1907. Afrikanische Spinnen (exkl. Aviculariiden), hauptsächlich aus dem Kapland. *Zoologische Jahrbücher, Abteilung für Systematik* 25: 557–731.

World Spider Catalog. 2014. World Spider Catalog, version 15.5. Natural History Museum Bern [online].

6

Description of a new species of *Paracrobeles* Heyns, 1968 (Nematoda, Rhabditida, Cephalobidae) from Kelso Dunes, Mojave National Preserve, California, USA

Sven BOSTRÖM [1] & Oleksandr HOLOVACHOV [2]

[1, 2] Department of Zoology, Swedish Museum of Natural History, Box 50007, SE-104 05 Stockholm, Sweden.

Abstract. A new species of *Paracrobeles*, *P. kelsodunensis* sp. nov. is described from the Kelso Dunes area, Mojave National Preserve, southern California. *Paracrobeles kelsodunensis* sp. nov. is particularly characterised by a body length of 469–626 µm in females and 463–569 µm in males; lateral field with four incisures, extending almost to tail terminus; three pairs of asymmetrical lips, separated by U-shaped primary axils with two long guarding processes, each lip usually with four tines along its margin; three long labial probolae, deeply bifurcated, with slender prongs without tines; metastegostom with a strong anteriorly directed dorsal tooth; pharyngeal corpus anteriorly spindle-shaped, posteriorly elongate bulbous with dilated lumen; spermatheca 24–87 µm long; postvulval uterine sac 60–133 µm long; vulva in a sunken area; spicules 33–38 µm long; and male tail with a 5–8 µm long mucro. The generic diagnosis is emended on the basis of recently described species and a key to the species of *Paracrobeles* is provided.

Keywords. Morphology, new species, *Paracrobeles*, SEM, taxonomy.

Introduction

The genus *Paracrobeles* was erected by Heyns (1968), with *Paracrobeles laterellus* Heyns, 1968 as type species, for a cephalobid from South Africa having: pharyngeal metacorpus developed into an elongate median bulb with a large chamber; labial probolae long, bifurcate, sharply pointed, prongs without fringe; and cephalic probolae prominent, sharply pointed without fringe. *P. laterellus* has also been reported from Namibia by Rashid *et al.* (1990). Since then three new species have been described, viz. *P. psammophilus* Navarro & Lluch, 1999 from Spain, *P. mojavicus* Taylor, Baldwin & Mundo-Ocampo, 2004 from California, USA and *P. deserticola* Abolafia, Divsalar, Panahi & Shokoohi, 2014 from Iran. *P. psammophilus* was also recorded from Italy by Orselli & Vinciguerra (2002), who presented SEM pictures of the species. A population of *Paracrobeles* from Kelso Dunes, Mojave National Preserve, southern California, is described here from studies by light and scanning electron microscope. It is close to *P. mojavicus*, but differs in some characters and is thus considered to represent a new species. The generic diagnosis is emended on the basis of recently described species and a key to the species of *Paracrobeles* is provided.

Materials and methods

During a tour in the Kelso Dunes area, Mojave National Preserve, southern California, the junior author collected several samples of sand in the rhizosphere of different species of desert plants along the southern slope of the dunes. Nematodes were recovered from sand samples using a modified Baermann funnel extraction technique. For LM, specimens were relaxed by gentle heat, fixed in cold 4% formaldehyde solution, transferred to pure glycerine by a slow evaporation method and mounted on permanent slides in glycerine with paraffin wax as support for the coverslip. For SEM, specimens were post-fixed in 1% osmium tetroxide (OsO_4) and transferred to pure acetone through an acetone/distilled water series. Specimens were critical point dried in liquid CO_2, mounted on stubs, gold-plated under vacuum to a thickness of 200 Å in an Agar High Resolution Sputter Coater Model 20, and examined in a Hitachi S-4300 SEM at an accelerating voltage of 5 kV.

Morphometric characters applied herein and their abbreviations are as defined for Cephalobidae in De Ley *et al.* (1999); terminology of the labial and cephalic region and stoma follows that of Holovachov *et al.* (2009). Type specimens are deposited in the invertebrate collections of the Department of Zoology, Swedish Museum of Natural History, Stockholm, Sweden (SMNH).

Results

Phylum Nematoda Diesing, 1861
Class Chromadorea Inglis, 1983
Order Rhabditida Chitwood, 1933
Family Cephalobidae Filipjev, 1934

Genus ***Paracrobeles*** Heyns, 1968

Type species

Paracrobeles laterellus Heyns, 1968

Diagnosis (emended after Holovachov *et al.* 2009)

Cuticle annulated, without distinctly annulated internal layer; annuli with longitudinal striation (tessellated). Lateral field with two wings (three or four incisures), ending near tail terminus in females and in males. Lip region weakly offset, consisting of six lips arranged in three pairs: one dorsal and two subventral. Pairs of lips separated by primary axils with one or two acute triangular guarding processes; secondary axils shallow. Cephalic probolae with three or four long and slender tines. Labial probolae deeply bifurcated without tines along the slender prongs. Six outer labial and four cephalic papilliform sensilla arranged in a cephaloboid manner. Amphidial aperture rounded, located on lateral lips. Stoma divided into cheilo-, gymno- and stegostom: cheilostom barrel-shaped, with strongly sclerotized bacilliform cheilorhabdia; gymnostom narrow tubular, as wide as stegostom, with weakly sclerotized plate-like gymnorhabdia; stegostom consists of a funnel-shaped prostegostom and variably shaped mesostego-, metastego- and telostegostom parts. Metastegostom tooth absent or present. Pharynx cephaloboid: pharyngeal procorpus cylindrical; metacorpus elongate bulbous; lumen of metacorpus often expanded to a large triradiate chamber with seemingly sclerotized lining; isthmus narrower than metacorpus; basal pharyngeal bulb oval, with strongly developed valves. Nerve ring encircling metacorpus, metacorpus-isthmus junction or anterior part of isthmus. Excretory pore opens at level of nerve ring. Deirids present. Female reproductive system cephaloboid; posterior part of ovary straight and relatively short; spermatheca present; postvulval uterine sac present; vulva flat with contour of body or in a depression; vagina straight or directed anteriad. Male reproductive system cephaloboid; spicules cephaloboid, with corpus and manubrium of approximately equal width; gubernaculum plate-like; cornua crurum absent. Male genital papillae: two ventrosublateral pairs located anterior to cloaca; one ventrosublateral pair located just posterior to cloacal opening; two pairs located at middle of tail length; and three pairs (lateral, subventral and subdorsal) near tail terminus; there is a midventral papilla on anterior cloacal lip. Rectum short (about as long as anal body diameter). Phasmid openings located at about one-third to half of tail length in both sexes. Female tail conoid, straight or slightly arcuate ventrad, tail terminus pointed or finely rounded; male tail conoid, slightly arcuate ventrad, tail terminus finely rounded.

Valid species

P. deserticola Abolafia, Divsalar, Panahi & Shokoohi, 2014
P. laterellus Heyns, 1968
P. mojavicus Taylor, Baldwin & Mundo-Ocampo, 2004
P. psammophilus Navarro & Lluch, 1999

Paracrobeles kelsodunensis sp. nov.

urn:lsid:zoobank.org:act:3D3142CA-44FA-422F-9B6A-8A988B3B75EC
Figs 1–2, 3A–C, I; Table 1

Diagnosis

Paracrobeles kelsodunensis sp. nov. is characterised by a body length of 469–626 µm in females and 463–569 µm in males; cuticle coarsely annulated, annuli 3.0–4.8 µm wide at midbody; lateral field with two wings, each separated by a broad groove, extending almost to tail terminus; three pairs of asymmetrical lips, pairs of lips separated by U-shaped primary axils, with two long guarding processes emerging from the first annule, each lip with four (usually) or three (rarely) tines along its margin: two long acute, with or without one shorter in the middle, and one long acute extending along the secondary axil; three long labial probolae, deeply bifurcated, with slender prongs without tines; metastegostom with a strong, anteriorly directed dorsal tooth; pharynx corpus anteriorly spindle-shaped, posteriorly elongate bulbous with dilated lumen; isthmus narrow, demarcated by a break in muscular tissue: anteriorly with heavy musculature, posteriorly further narrowing with reduced musculature; pharyngeal corpus 2.5–3.3 times isthmus length; nerve ring and excretory pore at level of metacorpus to metacorpus-isthmus junction; spermatheca 24–87 µm long; postvulval uterine sac 60–133 µm long; vulva in a depression; spicules 33–38 µm long; and male tail with a 5–8 µm long mucro.

Etymology

The new species name refers to the place where it was found.

Material examined

USA: holotype ♀, paratypes 15 ♀♀ and 9 ♂♂, SMNH Type-8715–Type 8720, 28 Mar. 2010, California, Mojave National Preserve, Kelso Dunes, soil around roots of desert plants (34°53.698' N, 115°42.155' W; 34°53.754' N, 115°42.248' W and 34°54.226' N, 115°42.200' W), *legit* O. Holovachov & P. De Ley.

Description

Adult

Body variably arcuate when killed by heat. Cuticle coarsely annulated, annuli 3.0–4.8 µm wide at midbody. Irregular longitudinal striae give the cuticle a tiled appearance. Lateral field with two wings, areolated, each separated by a broad groove, appearing as four incisures under LM, occupying about 20% of body diameter, extending almost to tail terminus in both sexes. Lip region weakly offset, carrying 6 + 4 papillae and two rounded amphid apertures. Three pairs of asymmetrical lips, one dorsal and two ventrolateral. Pairs of lips separated by U-shaped primary axils, with two long guarding processes emerging from the first annule. Each lip with four (commonly) or three (rarely – seen in one specimen so far) tines along its margin: two long acute, with or without one shorter in the middle, and one long acute extending along the secondary axil. Three labial probolae, 13.0–15.5 µm long, deeply bifurcated, with slender prongs without tines. Stoma about one lip region diameter long. Stomatal parts not clearly discernible. Cheilorhabdia oval in latero-median view; metastegostom with a strong, anteriorly directed dorsal tooth. Pharyngeal corpus anteriorly spindle-shaped, posteriorly elongate bulbous with dilated lumen; isthmus narrow, demarcated by a break in muscular tissue: anteriorly with heavy musculature, posteriorly further narrowing with reduced musculature; bulb oval, with valves. Nerve ring and excretory pore vary in position, from the level of metacorpus to metacorpus-isthmus junction, at 60–68% of neck length and at 57–68% of neck length, respectively. Deirids at level of isthmus, at 69–82% of neck length. Excretory canal cuticularised distally.

Female

Reproductive system monodelphic, prodelphic, in dextral position in relation to intestine. Ovary reflexed posteriorly at oviduct, ovary straight posterior to vulva. Spermatheca well developed. Postvulval uterine sac large, 1.4–3.1 times vulval body diameter (VBD) long. Vagina straight and perpendicular to body axis, about one-third to two-fifths of VBD. Vulva in a sunken area. Intra-uterine eggs 43–66 x 30–36 µm. Tail conoid, generally slightly curved ventrad, curved dorsad in some specimens, with 16–21 ventral annuli, non-annulated in terminal 6–12 µm, terminus minutely rounded. Rectum sigmoid, about half of ABD long. Phasmid openings located at about one-third to two-fifths of tail length.

Male

Similar to female in most respects, except for the sexual characters. Reproductive system monorchic, dextral in position; testis reflexed ventrad anteriorly. Spicules slender, paired and symmetrical, strongly curved ventrad; with oval manubrium and subcylindrical, gradually narrowing shaft. Gubernaculum plate-like. Genital papillae distributed as follows: two pairs ventrosublateral precloacal (at 5–11 µm and at 45–57 µm anterior to cloaca), one pair ventrosublateral adcloacal, a single midventral on anterior cloacal lip; two pairs (one ventrosublateral and one lateral) at midtail; three pairs (one lateral, one subventral and one dorsosublateral) near tail terminus. Phasmid openings located at about two-fifths of tail length, one to two annuli posterior to the lateral midtail papillae. Tail slightly curved ventrad, conoid, with a 5–8 µm long mucro and minutely rounded terminus.

Remarks

The population of *Paracrobeles kelsodunensis* sp. nov. from Kelso Dunes described here agrees in many respects with the description of *P. mojavicus* collected from sandy soil in a lava field, Mojave Desert, California. The new species differs from *P. mojavicus* by having a prominent, anteriorly directed, dorsal metastegostom tooth (vs no metastegostom tooth); a somewhat more anterior position of the excretory pore (at level of metacorpus or metacorpus-isthmus junction vs at level of isthmus); male tail with a 5–8 µm long mucro (*vs* male tail without mucro). Type specimens of *P. mojavicus* were examined (see Fig. 3D–I) and among them, one male (Fig. 3I) was found to have a prominent metastegostom tooth. The presence of a metastegostom tooth has not been described for any other species than *P. kelsodunensis* sp. nov., which might indicate that the population described by Taylor *et al.* (2004) is a mixture of species. One possible explanation could be a polymorphism based on food sources, and a comparison between molecular characters would probably be needed to resolve the status of the *Paracrobeles* populations from the Mojave Desert.

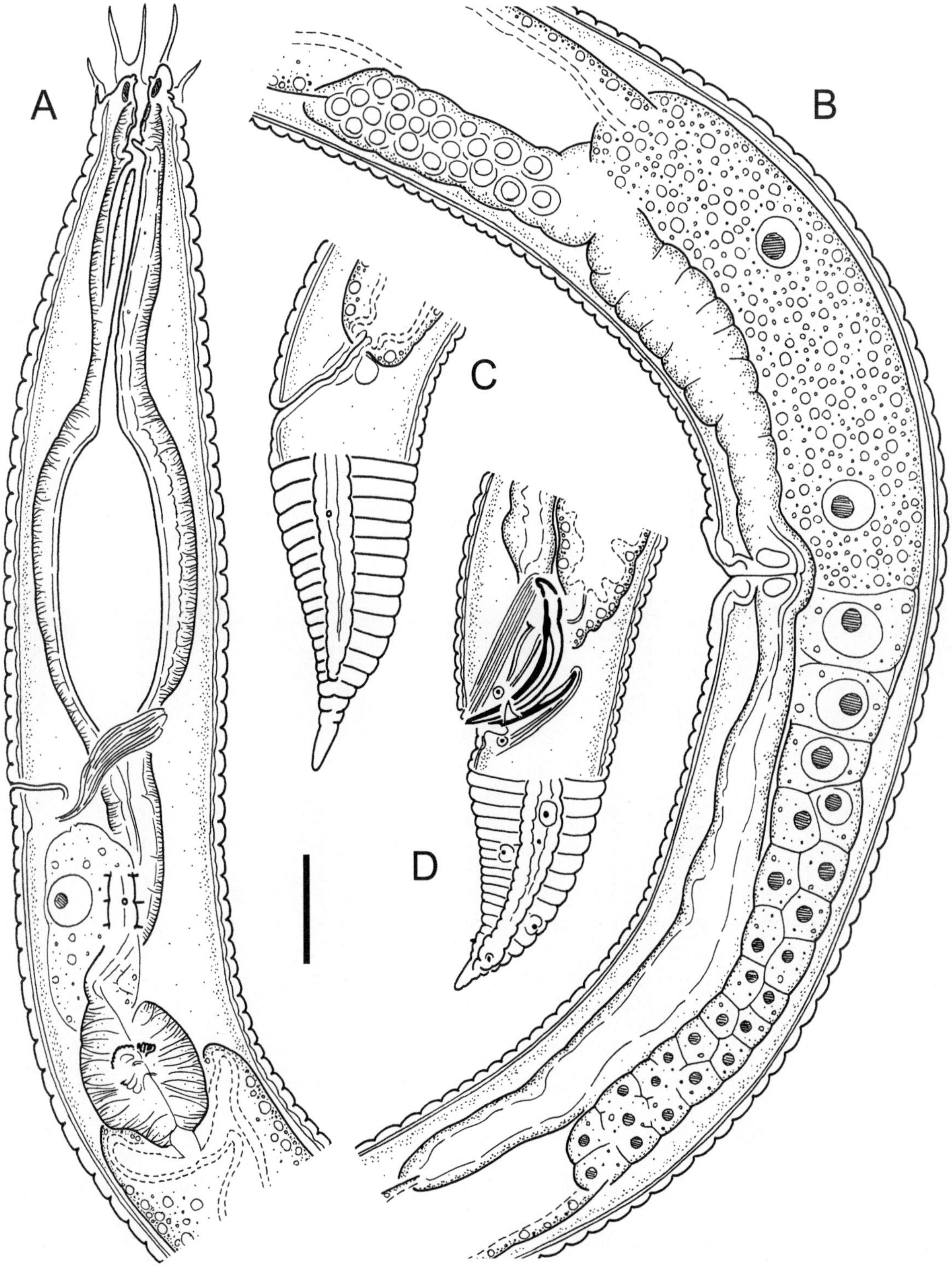

Fig. 1. *Paracrobeles kelsodunensis* sp. nov. **A**. Pharyngeal region. **B**. Female gonad. **C**. Female tail. **D**. Male tail. Scale bar = 20 µm.

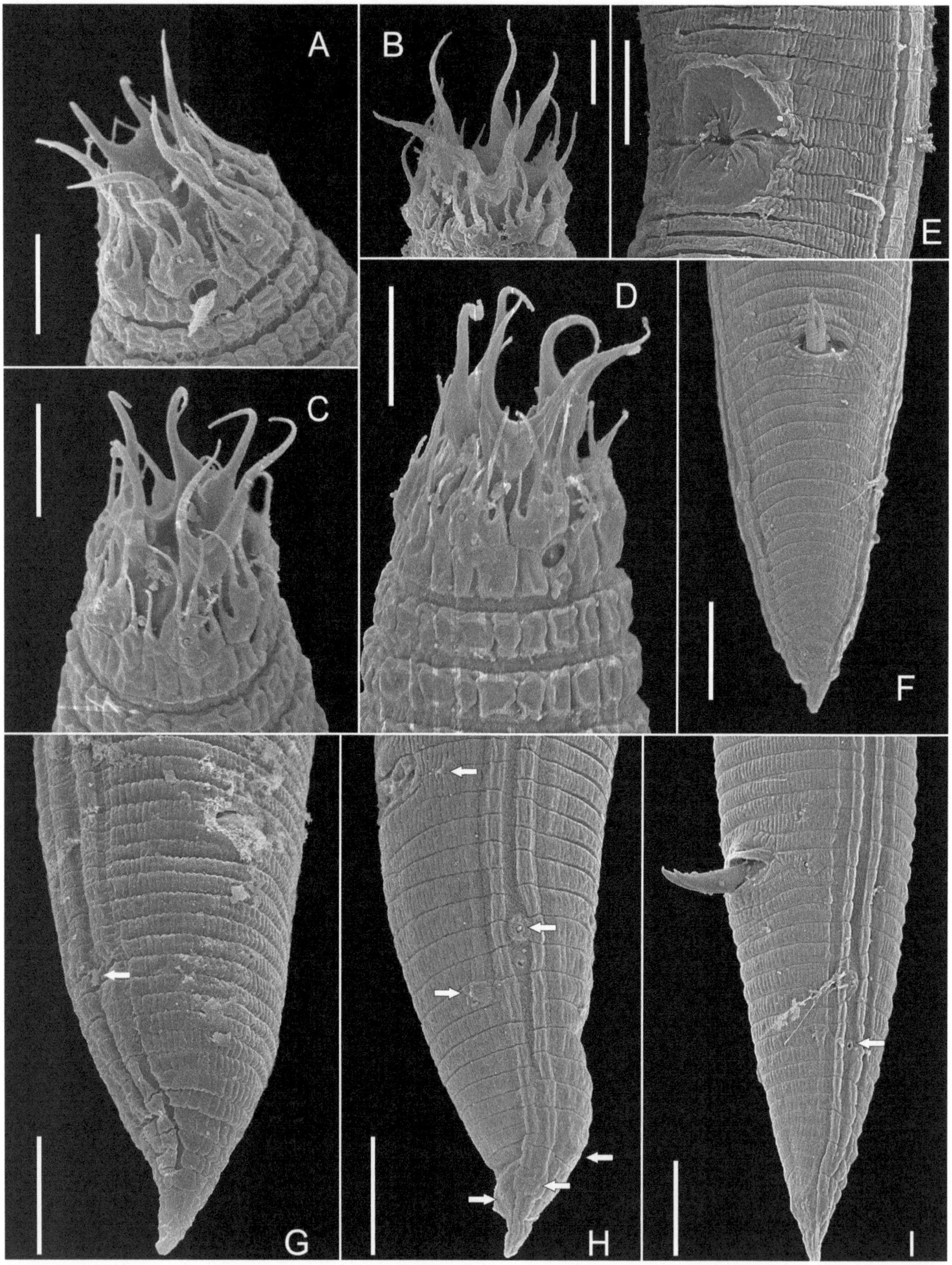

Fig. 2. *Paracrobeles kelsodunensis* sp. nov. SEM micrographs. **A**. Anterior end, left lateral view. **B**. Anterior end, semi-en face view. **C–D**. Anterior end, subventral view. **E**. Vulval region. **F**. Male tail, ventral view. **G**. Female tail, right sublateral view (arrow points at phasmid). **H**. Male tail, left lateral view (arrows point at papillae). **I**. Male tail, left lateral view (arrow points at phasmid). Scale bars: A–D = 5 μm, E–I = 10 μm.

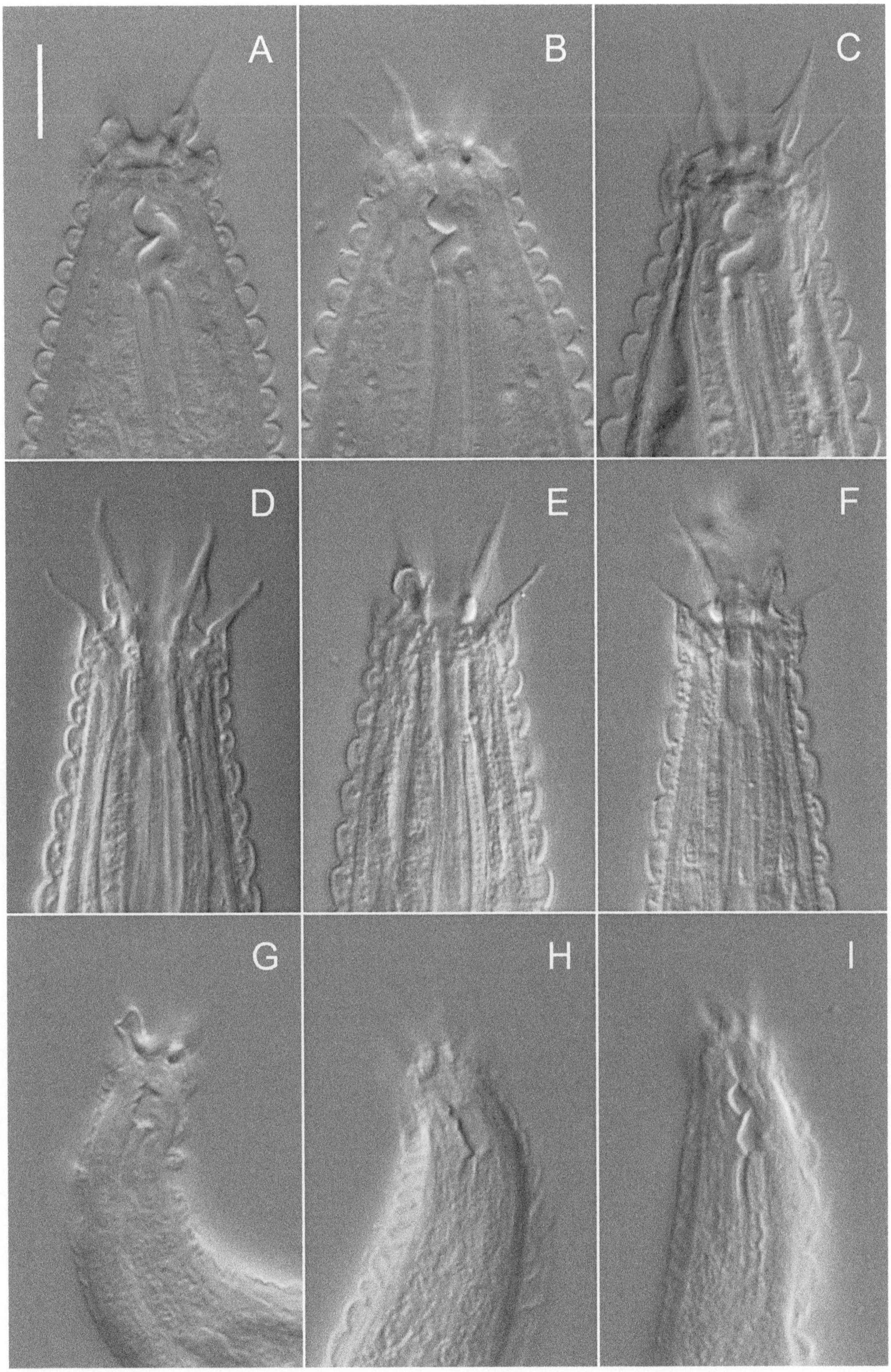

Fig. 3. A–C. *Paracrobeles kelsodunensis* sp. nov. LM micrographs. **A**. Male anterior end, ventral side to the right. **B–C**. Female anterior end, ventral side to the right. **D–H**. *Paracrobeles mojavicus* Taylor, Baldwin & Mundo-Ocampo, 2004. **D–F**. Female anterior end, ventral side to the right. **G–H**. Male anterior end, ventral side to the right. **I**. *Paracrobeles* cf. *kelsodunensis* sp. nov., male anterior end, ventral side to the right. Scale bar: A–I = 10 µm.

Table 1. Measurements (in μm) of *Paracrobeles kelsodunensis* sp. nov. from Kelso Dunes, Mojave Desert, California (presented as mean±s.d. and (range) or only range). * Number of annuli from anterior end to nerve ring, excretory pore and deirid, respectively; – indicates that data is not applicable.

	Holotype ♀	16 ♀♀ (incl. holotype)	9 ♂♂
Body length	626	538±51 (469–626)	514±33 (463–569)
Body diameter (BD)	42	43.0±3.2 (37–49)	37.5±2.3 (34–41)
Pharynx length	185	169±11.8 (150–188)	161.5±9.5 (147–173)
Tail length	66	54.5±6.4 (42–66)	54.0±3.9 (47–58)
Anal or cloacal diameter (ABD)	29	27.6±1.7 (25–31)	29.9±0.9 (28–31)
Vulva or Testis	391	334±29.1 (297–391)	265±31 (223–315)
V-A/T	2.8	2.8±0.3 (2.5–3.4)	–
a	14.9	12.5±1.0 (10.9–14.9)	13.7±1.2 (12.1–16.7)
b	3.4	3.2±0.2 (2.9–3.7)	3.2±0.2 (2.9–3.6)
c	9.5	10.0±0.6 (9.1–11.5)	9.6±1.0 (8.0–11.8)
c’	2.3	2.0±0.2 (1.6–2.3)	1.8±0.2 (1.6–2.1)
V or T (%)	62.5	62.0±1.2 (60–65)	51.1±5.1 (44–59)
Lip region diameter	18	17.7±0.4 (17–18)	17.1±0.6 (16–18)
Labial probolae	15.5	14.7±0.7 (13.0–15.5)	14.1±0.6 (13.0–14.5)
Stoma length	14.5	14.9±0.5 (14.5–15.5)	15.1±0.5 (14.5–15.5)
Corpus length	114	106.1±5.3 (100–119)	99.7±3.9 (94–105)
Isthmus length	39	35.9±3.5 (30–42)	33.4±2.3 (29–37)
Bulb length	30	29.6±1.2 (28–33)	29.9±1.0 (28–31)
Bulb diameter	22	22.3±1.0 (20–23)	21.0±0.9 (20–22)
Corpus/isthmus ratio	2.9	3.0±0.3 (2.5–3.3)	3.0±0.2 (2.7–3.3)
Nerve ring from anterior end	133	123.9±7.7 (112–142)	119.0±8.0 (109–137)
Excretory pore from anterior end	141	119.4±10.4 (102–142)	114.8±10.8 (101–133)
Deirid from anterior end	161	152.8±11.1 (135–163)	134.0±13.4 (110–155)
R_{NR}*	24	26.1±1.4 (23–29)	26.4±1.2 (24–28)
R_{EP}*	26	24.9±1.3 (22–27)	25.3±1.2 (24–27)
R_{DEI}*	31	31.8±2.2 (29–36)	30.8±2.2 (27–34)
Annuli width at midbody	4.0–4.8	3.0–4.8	3.4–4.0
Annuli width anteriorly	4.0–4.8	3.4–4.8	3.0–4.0
Vagina or Testis flexure length	14.5	13.6±0.8 (13.0–15.5)	47.7±13.0 (31–72)
Spermatheca or Spicule length	60	49.7±16.8 (24–87)	35.1±1.5 (33–38)
PUS or Gubernaculum length	129	102.1±20.4 (60–133)	19.9±1.3 (18–22)
PUS/VBD	3.1	2.6±0.5 (1.4–3.1)	–
Rectum	18	17.9±1.3 (14.5–20.5)	–
Rectum/ABD	0.6	0.6±0.1 (0.5–0.7)	–
Phasmid	20.5	19.1±1.9 (16–29)	21.5±2.5 (18–26)
Phasmid (% of tail)	31	35.5±3.4 (31–40)	40.1±2.8 (37–45)

Key to species of *Paracrobeles* (emended after Abolafia *et al.* 2014; when using the key, please also consult Table 2 for additional diagnostic information)

1. Spicules less than 40 μm long 2
– Spicules more than 40 μm long, primary axils with single guarding process 4

2. Vulva at 58–60% of body length .... ***P. laterellus*** Heyns, 1968
– Vulva at 60–71% of body length, primary axils with two guarding processes 3

3. Metastegostom tooth absent .... ***P. mojavicus*** Taylor, Baldwin & Mundo-Ocampo, 2004
– Metastegostom tooth present, anteriorly directed .... ***P. kelsodunensis*** sp. nov.

Table 2. Morphometrics of *Paracrobeles* species. Measurements in µm (except L, in mm) (emended from Abolafia *et al.*, 2014). * Measured from the illustrations or calculated from other measurements; ** times the corresponding body diameter; – indicates that data are not applicable; [1]Abolafia *et al.* (2014); [2]Heyns (1968); [3]Rashid *et al.* (1990); [4]Taylor *et al.* (2004); [5]Navarro & Lluch (1999); [6]Orselli & Vinciguerra (2002); [7]Present paper.

Species	n	L	a	b	c	c'	V (%)	Stoma length	Labial probolae	Excretory pore –ant. end.	Tail	Spermatheca or Spicules	Postvulval sac or Gubernaculum	Country
desertícola[1]	8 ♀♀	0.48–0.60	14–18	3.0–4.0	9–10	2.0–3.0	50–63	11–14	14–16	96–111	53–67	33–50	60–96 2.9–4.5**	Iran
	8 ♂♂	0.46–0.65	13–19	3.0–4.0	9–13	1.8–2	–	12–14	15–16	101–120	48–54	46–56	16–24	
laterellus[2]	4 ♀♀	0.53–0.74	17–20	3.3–4.1	11–15	2.3*	58–60	9*	10*	106*	50–60	–	1.0–1.5**	Namibia
	6 ♂♂	0.57–0.66	17–25	3.5–3.9	11–12	2.0*	–	–	–	–	50–55	32–35/43–46	17–22	
laterellus[3]	1 ♀	0.57	18	3.7	11	2.2	59	12	9*	107	50	26	46	Namibia
mojavicus[4]	17 ♀♀	0.47–0.71	13–18	2.6–3.5	9–13	1.8–2.5	62–71	13–18	11–17	115–163	41–66	38–77	54–110	California, USA
	19 ♂♂	0.45–0.75	14–19	2.7–3.7	10–12	1.7–2.3	–	13–18	11–17	103–160	43–67	28–37	17–22	
psammophilus[5]	10 ♀♀	0.36–0.55	10–17	2.9–3.7	7–11	1.9–2.7	59–63	10–14	10–12	69–117	46–53	48*	57–101	Spain
	10 ♂♂	0.42–0.53	10–18	3.1–3.6	9–12	1.6–2.2	–	10–14	–	75–107	42–53	42–57	20–39	
psammophilus[6]	34 ♀♀	0.49–0.58	13–16	2.3–2.9	8–10	2.2–2.8	59–62	10–18	11–18	114–139	58–68	30–37	65–106	Italy
	42 ♂♂	0.52–0.61	12–20	2.4–3.2	8–9	1.7–2.8	–	12–19	13–17	–	56–70	68–81	28–31	
kelsodunensis[7] sp. nov.	16 ♀♀	0.47–0.63	11–15	2.9–3.7	9–12	1.6–2.3	60–65	14–16	13–16	102–142	42–66	24–87	60–133 1.4–3.1**	California, USA
	9 ♂♂	0.46–0.57	12–17	2.9–3.6	8–12	1.6–2.1	–	14–16	13–15	101–133	47–58	33–38	18–22	

4. Lips with three tines (including the ones in secondary axils); postvulval sac not swollen .... ***P. deserticola*** Abolafia, Divsalar, Panahi & Shokoohi, 2014
– Lips with four tines (including the ones in secondary axils); postvulval sac swollen .... ***P. psammophilus*** Navarro & Lluch, 1999

Discussion

The Mojave Desert and especially the Kelso Dunes seem to be an area with a high diversity of species of the family Cephalobidae Filipjev, 1934. Representatives of this family are mostly terrestrial and bacteria-consuming nematodes with a worldwide distribution. They seem to be especially diverse and abundant in deserts, and sand dunes appear to be a suitable habitat for cephalobids. About ten new species have been described from the Mojave Desert, which might indicate a hot spot for cephalobid diversity (De Ley 2014). Hitherto one new genus and six new species, including the one described here, and one already known species have been described from the Kelso Dunes (Boström & Holovachov 2012, 2013a, 2013b, 2014). Some of the new species described from the Mojave Desert are closely related, which might further indicate that speciation processes are going on in this area. Species of the genus *Paracrobeles* are rare inhabitants of terrestrial ecosystems and have a rather restricted distribution. So far they have been found in warm, dry sandy soils in southern Africa (South Africa and Namibia), the Mediterranean (Spain and Italy), California and Iran (Abolafia *et al.* 2014; Heyns 1968; Navarro & Lluch 1999; Orselli & Vinciguerra 2002; Rashid *et al.* 1990; Taylor *et al.* 2004). The description of the new species adds morphological data important for species identification and broadens the diagnosis of *Paracrobeles*.

Acknowledgements

The second author was supported in part by an award from the NSF Partnerships for Enhancing Expertise in Taxonomy (PEET) program grant "PEET: Training the Next Generation of Nematode Taxonomists: Applying the Tools of Modern Monography Across Free-living and Parasitic Tylenchina" (DEB-0731516). Sampling was performed by O. Holovachov and P. De Ley under permit # MOJA-2010-SCI-0003 from the National Park Service, United States Department of Interior.

References

Abolafia J., Divsalar N., Panahi H. & Shokoohi E. 2014. Description of *Paracrobeles deserticola* sp. n. and *Nothacrobeles hebetocaudatus* sp. n. (Nematoda: Rhabditida: Cephalobidae) from Iran and the phylogenetic relationships of these two species. *Zootaxa* 3827 (1): 1–19.

Boström S. & Holovachov O. 2012. Description of *Chilodellus eremus* gen. n., sp. n. and *Stegelleta arenaria* sp. n. (Rhabditida: Cephalobidae) from Kelso Dunes, Mojave National Preserve, California, USA. *Journal of Nematode Morphology and Systematics* 15 (1): 21–31.

Boström S. & Holovachov O. 2013a. Description of two new species of *Nothacrobeles* Allen & Noffsinger, 1971 (Rhabditida: Cephalobidae) from Kelso Dunes, Mojave National Preserve, California, USA. *Journal of Nematode Morphology and Systematics* 16 (1): 25–34.

Boström S. & Holovachov O. 2013b. Description of one new species of *Heterocephalobellus* Rashid, Geraert & Sharma, 1985 (Rhabditida, Cephalobidae) from Kelso Dunes, Mojave National Preserve, California, USA and Monte desert, Usno, Argentina. *Journal of Nematode Morphology and Systematics* 16 (2): 161–166.

Boström S. & Holovachov O. 2014. Descriptions of species of *Stegelleta* Thorne, 1938 (Nematoda, Rhabditida, Cephalobidae) from California, New Zealand and Senegal, and a revision of the genus. *European Journal of Taxonomy* 87: 1–19.

De Ley I.T., De Ley P., Baldwin J.G., Mundo-Ocampo M. & Nadler S.A. 1999. Three new species of *Nothacrobeles* (Nemata: Cephalobidae) from the Mojave Desert, California. *Journal of Nematology* 31 (4): 482–497.

De Ley P. 2014. Past and present highlights of nematode research in the Mojave Desert. *Mojave National Preserve Science Newsletter* May 2014: 5–10.

Heyns J. 1968. *Paracrobeles laterellus* n. gen., n. sp. from South Africa (Nematoda: Cephalobidae). *Nematologica* 14: 511–514.

Holovachov O., De Ley I.T., Mundo-Ocampo M. & De Ley P. 2009. *Identification of Cephaloboidea (Nematoda)*. EUMAINE, Gent and Nematology, UC Riverside.

Navarro P. & Lluch J. 1999. *Paracrobeles psammophilus* sp. nov. (Nematoda: Cephalobidae) from El Saler, Valencia (Spain). *Journal of Zoology* 249 (4): 481–485.

Orselli L. & Vinciguerra M.T. 2002. Nematodes from Italian sand dunes. 6. Two new and three rare species of Cephalobidae (Nematoda). *Nematologia mediterranea* 30: 211–220.

Rashid F., Heyns J. & Coomans A.1990. *Paracrobeles* and *Acrobeles* species from South West Africa/Namibia with description of a new *Acrobeles* species (Nematoda: Cephalobidae). *Phytophylactica* 22 (1): 41–49.

Taylor T.M., Baldwin J.G. & Mundo-Ocampo M. 2004. *Paracrobeles mojavicus* sp. n. (Nematoda: Cephalobidae) from the Mojave Desert, California. *Journal of Nematode Morphology and Systematics* 6 (2): 151–160.

7

The new genus of stick insect *Lobofemora* from Vietnam, with the description of three new species (Phasmida: Phasmatidae: Clitumnini)

Joachim BRESSEEL[1] & Jérôme CONSTANT[2]

[1, 2] Royal Belgian Institute of Natural Sciences, O.D. Phylogeny and Taxonomy, Entomology, Vautier street 29, B-1000 Brussels, Belgium.

Abstract. The new genus *Lobofemora* gen. nov. is described from Vietnam to accommodate three new species: *L. bachmaensis* sp. nov. (Bach Ma National Park, central Vietnam), *L. bidoupensis* sp. nov. (Bidoup-Nui Ba National Park, central Vietnam) and *L. scheirei* sp. nov., the type species (Cat Tien National Park and Dong Nai Biosphere Reserve, south Vietnam). It is provisionally placed in the tribe Clitumnini Brunner von Wattenwyl, 1893. The genus is the only known Clitumnini which shows tegmina and sometimes alae in the males. Females show conspicuous foliaceous lobes on the median femora. Males and females of all species and the eggs of *L. scheirei* sp. nov. and *L. bidoupensis* sp. nov. are described and illustrated. An identification key to the species and a distribution map are provided. The definition of the tribe Clitumnini *sensu* Hennemann & Conle (2008) is slightly adapted to include the new genus and the tribal placement is discussed. The male of *L. scheirei* sp. nov. is able to stridulate by rubbing the outer margins of the tegmina against the subcostal and radial veins of the alae.

Keywords. Phasmatodea, stridulation, Global Taxonomic Initiative, South-East Asia, walking stick.

Introduction

The tribe Clitumnini Brunner von Wattenwyl, 1893 was recently re-diagnosed and revised on the generic level by Hennemann & Conle (2008). The tribe comprises fourteen genera, some of which are very species rich, e.g., *Ramulus* Saussure, 1862 (171 species, see Brock 2015) and *Entoria* Stål, 1875 (31 species, see Brock 2015). Recent expeditions to Vietnam in the frame of a Global Taxonomic Initiative project by the authors and Vietnamese colleagues have revealed three congeneric new species that could not be attributed to any known genus. Following the key of Hennemann & Conle (2008), the new genus belongs in the tribe Clitumnini showing the following typical features: small to medium sized species with body lengths ranging from 71.2 to 95.5 mm in females and 52.5 to 74.0 mm in males; elongate legs; sexual dimorphism distinct; antennae shortened with the scapus enlarged, laterally dilated and consisting of less than 28 segments; both sexes with short, roughly cylindrical cerci; males with the anal segment consisting of two movable semitergites which are minutely dentated interiorly and lack a developed sclerotised external vomer. Furthermore, the very elongate eggs and hairy appendages on the outer border of the capitulum is a character shared with three other genera of Clitumnini.

The new genus, however, violates several characters of Clitumnini; hence, it is remarkable and interesting in several aspects. The mesothorax is less than 2.5 x the combined length of head and prothorax in females. The median segment is about as long as the metanotum. Tegmina and often alae present in males, males have teeth on the mesofemora and have a granulose thorax. Females with back of the head raised and/or conical.

The new genus is described below as *Lobofemora* gen. nov. and is currently known from different localities in Central and South Vietnam. *Lobofemora scheirei* sp. nov. was found in the lowlands of Dong Nai Biosphere Reserve and Cat Tien National Park, both located in Southern Vietnam. *Lobofemora bidoupensis* sp. nov. is described from the highlands of Bidoup-Nui Ba National Park and *Lobofemora bachmaensis* sp. nov. is recorded from Bach Ma National Park in the Annamite Mountains, Central Vietnam.

Material and methods

Due to their nocturnal behaviour, most Phasmida including the genus *Lobofemora* gen. nov. were collected at night. A light-weight and water-proof head torch, Petzl MYO RXP, was used during searching the vegetation. The females were kept alive in a plastic box for producing eggs. Hatchlings of the different species were reared to adulthood by the first author and Dr Bruno Kneubühler (Switzerland). The wild caught specimens were euthanized with ethyl acetate fumes. The specimens were then stored in airtight plastic "zip"-bags in wood chips (used in rodent cages) and sprinkled with ethyl acetate (EtOAc) to prevent rotting, mould and keeping the specimens flexible. The bags are then frozen on arrival and the specimens are mounted later on.

The egg morphology follows the terminology of Clark Sellick (1997, 1998). The description of the colouration is based on live specimens.

Acronyms used for the collections

BMNH	=	Natural History Museum, London, United Kingdom
FHC	=	Frank Hennemann Collection, Hohenecken, Germany
MNHU	=	Museum für Naturkunde der Humboldt-Universität, Berlin, Germany
RBINS	=	Royal Belgian Institute of Natural Sciences, Brussels, Belgium
RMNH	=	Nationaal Natuurhistorisch Museum (Naturalis), Leiden, The Netherlands
VNMN	=	Vietnam National Museum of Nature, Vietnamese Academy of Sciences, Hanoi, Vietnam
ZMUH	=	Zoologisches Museum, Universität Hamburg, Germany
ZSMC	=	Zoologische Staatssammlung München, Germany

Results

Family Phasmatidae Gray, 1835 *sensu lato*
Subfamily Clitumninae Brunner von Wattenwyl, 1893

Tribe **Clitumnini** Brunner von Wattenwyl, 1893

We follow the definition of the tribe proposed by Hennemann & Conle (2008: 67), except for the following characters:

1. The length of the median segment: Clitumnini have the median segment strongly shortened and at most ¼ the length of the metanotum; *Lobofemora* gen. nov. has the median segment about as long as the metanotum.
2. Relative length of mesothorax: Clitumnini have the mesothorax at least 2.5 x the combined length of head and prothorax; *Lobofemora* gen. nov. has the mesothorax shorter than twice the combined length of head and prothorax.

Genus ***Lobofemora*** gen. nov.
urn:lsid:zoobank.org:act:5EBCBEF1-D0EA-4433-9B85-287709869730

Type species

Lobofemora scheirei sp. nov. by present designation.

Etymology

The name of the new genus is formed by the juxtaposition of *lobus* (Latin), meaning lobe and "femora". The name refers to the foliaceous lobes on the median legs of the females.

Diagnosis

Differing from all other know genera of Clitumnini by the following combination of characters:
1. Females with back of head more or less strongly raised, granulose and conical.
2. Males with at least some scale-like tegmina present, sometimes with developed alae.
3. Median segment about as long as or longer than metanotum.
4. Females with conspicuous enlarged lobes on mesofemora and males occasionally with dorsal teeth on the mesofemora.
5. Mesothorax less than 2.5 x the length of head and pronotum combined.
6. Semi-tergites of males very short and broad and thorax densely granulose.

Description

Male

HEAD. Longer than wide, slightly tapering posteriorly and granulose. Vertex elongated. Dorsal portion of head flattened, with two definite spines positioned at posterior margin of eyes and projecting dorsally. Eyes circular and strongly projecting hemispherically. Scapus flattened dorso-ventrally and rounded laterally. Pedicellus cylindrical.

THORAX. Pronotum shorter than head and slightly widening posteriorly. Anterior margin incurved and raised, followed by faint, median longitudinal impression, not reaching posterior margin. Central transverse impression not reaching lateral edges of pronotum. Mesonotum granulose, widening posteriorly. Fine median longitudinal line. Tegmina completely developed or present as small scales. Alae absent, micropterous or brachypterous. Metanotum shorter than median segment. Meso- and metasternum with posteriorly, between legs, Y-shaped impression, antero-laterally resulting in two minute pseudo-foramina.

ABDOMEN. Abdominal segments slightly granulose and with fine median, longitudinal line. Tergum X with median, longitudinal carina, apically split into two semi-tergites. Inner apical rim of semi-tergites armed with several spines. Apex of semi-tergites obliquely pointing downwards. Posterolateral angles rounded. Poculum rounded, slightly projecting over apex of abdominal tergum IX. Apex of poculum narrower and rounded. No sclerotised vomer present. Cerci short, not reaching apex of tergum X, incurving, circular in cross-section, apices rounded.

LEGS. Profemora slightly longer than head and thorax combined; compressed and curved basally. Anterodorsal carina with few, small serrations.

Female

HEAD. Longer than wide, slightly tapering posteriorly and strongly granulose. Vertex elongated and with a shallow longitudinal impression. Dorsal portion of head flattened, with transverse, laterally acute ridge at posterior margin of eyes. Eyes circular and projecting hemispherically. Antennae short, with scapus strongly dorso-ventrally flattened and with lateral margins rounded. Apical half with very short segments.

THORAX. Pronotum shorter than head and granulose. Anterior margin incurved and raised, followed by median longitudinal impression, not reaching posterior margin. Transverse impression in middle, not reaching lateral edges of pronotum. Mesonotum granulose, with fine median longitudinal line. Metanotum with anterior margin weakly incurved and raised, longer than wide and strongly granulose. Meso- and metasternum as in male.

ABDOMEN. Terga granulose, with fine median, longitudinal line. Tergum X with median, longitudinal carina; notched posteriorly; posterolateral angles rounded. Epiproct elongated. Subgenital plate slightly elongated, just projecting over apex of tergum X and with median longitudinal carina; apex acute. Cerci short, not reaching apex of tergum X. Gonapophyses VIII elongated, reaching apex of epiproct.

LEGS. Profemora longer than head, pronotum and mesonotum combined; compressed and curved basally. Anterodorsal carina with small serrations. Mesofemora with posterodorsal carina, with two to three lobes, anterior one being most prominent. Protibiae longer than head and thorax combined. Mesotibiae slightly longer than mesofemora. Metatibiae shorter than profemora.

Distribution

Currently known from different localities in Central and South Vietnam (see map Fig. 1).

Species included and distribution

Lobofemora bachmaensis sp. nov. [Bach Ma National Park]
Lobofemora bidoupensis sp. nov. [Bidoup Nui Ba National Park]
Lobofemora scheirei sp. nov. [Cat Tien National Park and Dong Nai Biosphere Reserve]

Keys to the species of *Lobofemora* gen. nov.

Males

1. Alae developed, projecting over posterior margin of median segment (Fig. 6D, H) .... *L. scheirei* sp. nov.
– Alae scale-like or absent (Figs 2D, H; 4D, H) 2

2. Tegmina scale-like, flattened, just projecting over anterior margin of metanotum; alae absent (Fig. 2D, H); posterodorsal carina of mesotibiae with conspicuous triangular tooth (Fig. 2B) .... *L. bachmaensis* sp. nov.
– Tegmina reaching halfway along metanotum, convex; alae scale-like (Fig. 4D, H); posterodorsal carina of mesotibiae unarmed (Fig. 4B) .... *L. bidoupensis* sp. nov.

Females

1. Mesofemora with two lobes on posterodorsal carina: more basal one prominent; more apical one present as subapical triangular spine (Fig. 3D); vertex projecting posteriorly, almost reaching halfway along pronotum length (Fig. 3E, I) .... *L. bachmaensis* sp. nov.
– Mesofemora with three lobes on posterodorsal carina: most basal one most prominent and gradually becoming smaller; most apical one present as subapical triangular spine (Figs 5D, 7D); vertex rounded or slightly elongated posteriorly, not surpassing ⅓ of pronotum length (Figs 5E, I; 7E, I) 2

2. Posterodorsal carina of profemora armed with two enlarged triangular bispinose teeth; medioventral carina of metatibiae armed with two small triangular lobes anteriorly (Fig. 5D); vertex rounded posteriorly (Fig. 5E, I) .... *L. bidoupensis* sp. nov.
– Posterodorsal carina of profemora unarmed; medioventral carina of metatibiae without triangular lobes anteriorly (Fig. 7D); vertex slightly elongated posteriorly (Fig. 7E, I)........ *L. scheirei* sp. nov.

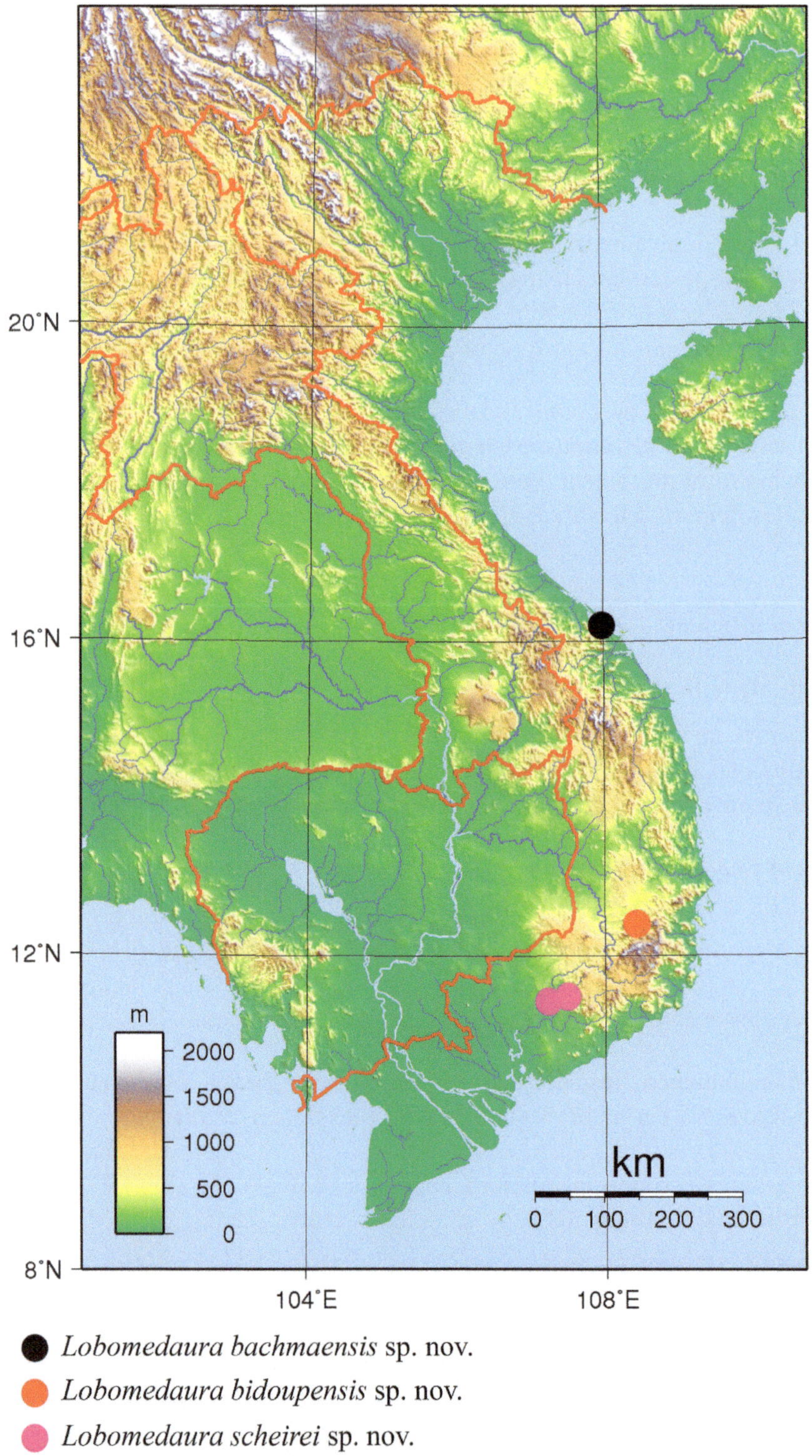

Fig. 1. Distribution map of the species of *Lobofemora* gen. nov.

Fig. 2. *Lobofemora bachmaensis* sp. nov. ♂, holotype. **A**. Habitus, dorsal view. **B**. Habitus, lateral view. **C**. Habitus, ventral view. **D**. Anterior part of body, dorsal view. **E**. Apex of abdomen, dorsal view. **F**. Apex of abdomen, lateral view. **G**. Apex of abdomen, ventral view. **H**. Anterior part of body, lateral view. D–H = not to scale.

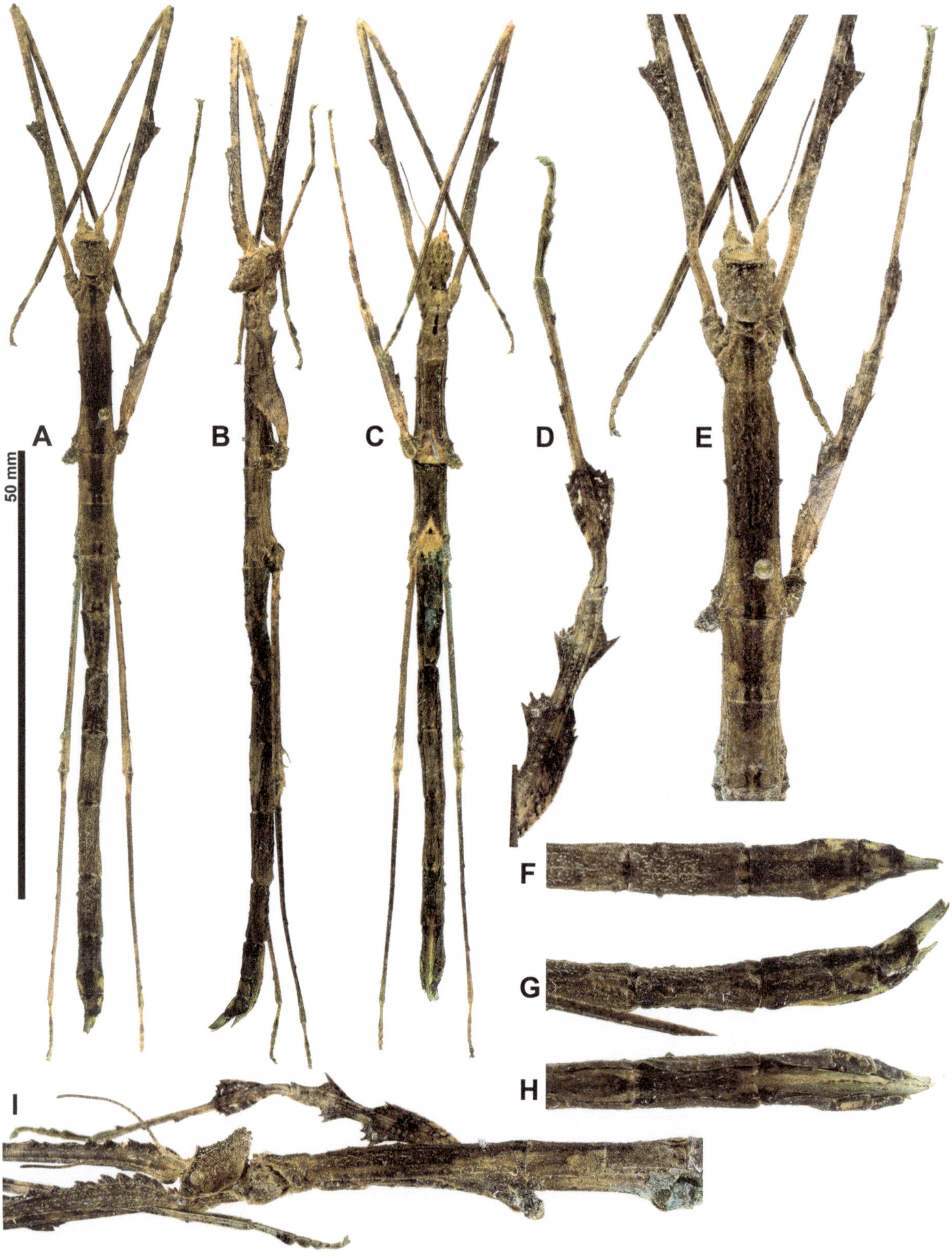

Fig. 3. *Lobofemora bachmaensis* sp. nov. ♀, paratype. **A**. Habitus, dorsal view. **B**. Habitus, lateral view. **C**. Habitus, ventral view. **D**. Right median leg, detail. **E**. Anterior part of body, dorsal view. **F**. Apex of abdomen, dorsal view. **G**. Apex of abdomen, lateral view. **H**. Apex of abdomen, ventral view. **I**. Anterior part of body, lateral view. D–I = not to scale.

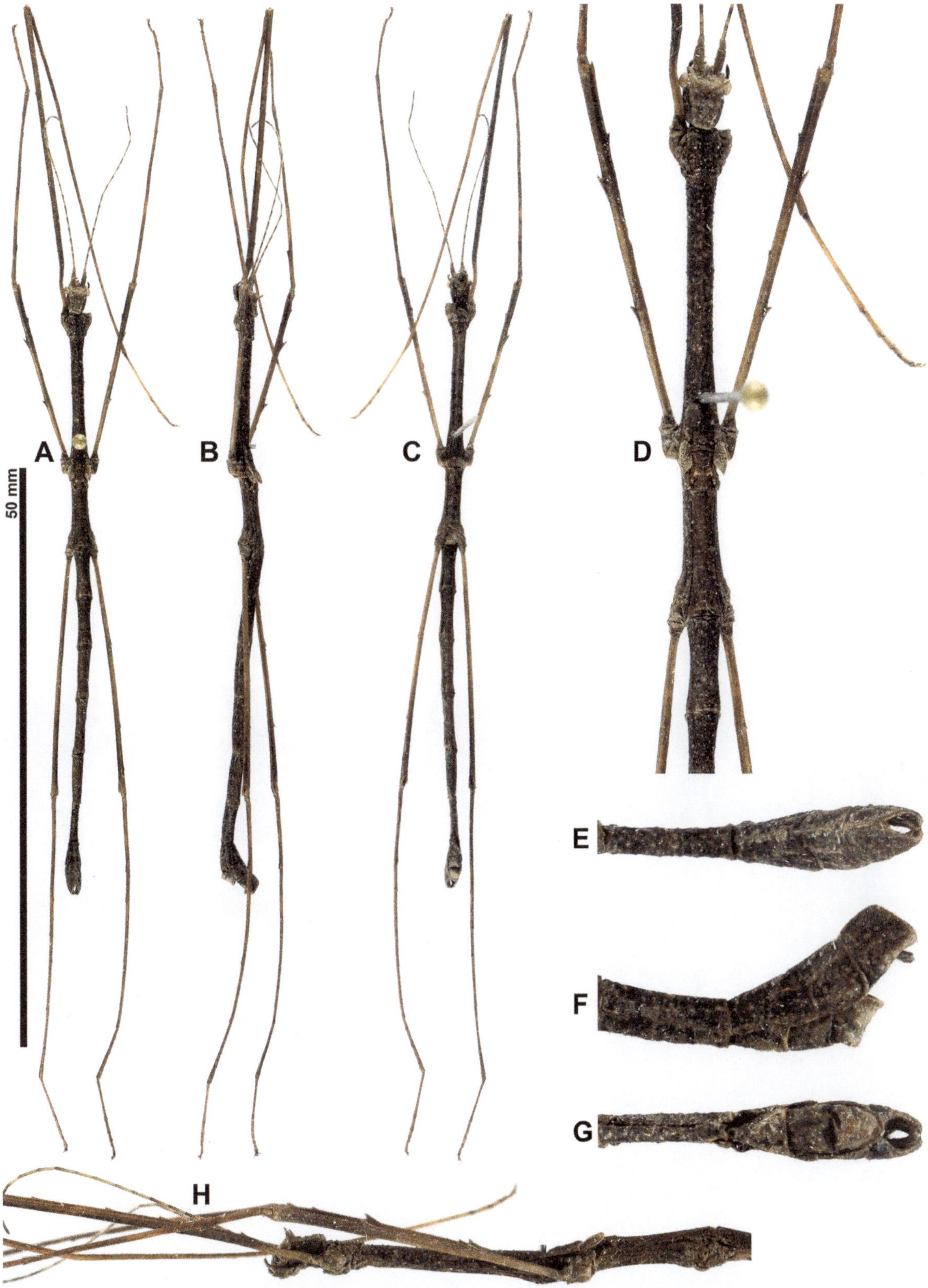

Fig. 4. *Lobofemora bidoupensis* sp. nov. ♂, holotype. **A**. Habitus, dorsal view. **B**. Habitus, lateral view. **C**. Habitus, ventral view. **D**. Anterior part of body, dorsal view. **E**. Apex of abdomen, dorsal view. **F**. Apex of abdomen, lateral view. **G**. Apex of abdomen, ventral view. **H**. Anterior part of body, lateral view. D–H = not to scale.

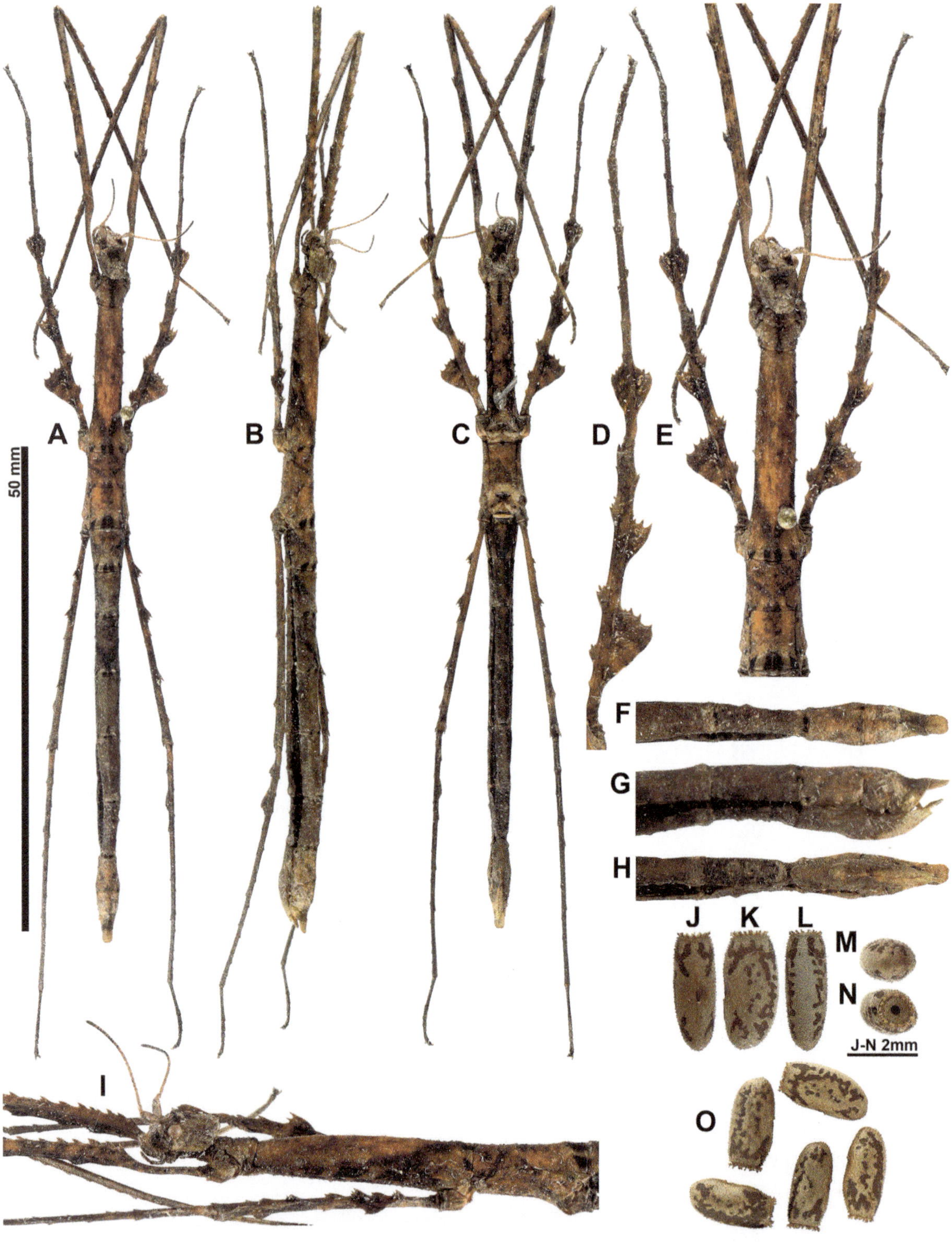

Fig. 5. *Lobofemora bidoupensis* sp. nov. **A–I**. ♀, paratype. **A**. Habitus, dorsal view. **B**. Habitus, lateral view. **C**. Habitus, ventral view. **D**. Right median leg, detail. **E**. Anterior part of body, dorsal view. **F**. Apex of abdomen, dorsal view. **G**. Apex of abdomen, lateral view. **H**. Apex of abdomen, ventral view. **I**. Anterior part of body, lateral view. **J–N**. Egg. **J**. Detail of micropylar plate. **K**. Lateral view. **L**. Ventral view. **M**. Detail of polar area. **N**. Detail of operculum. **O**. Eggs. D–I, O = not to scale.

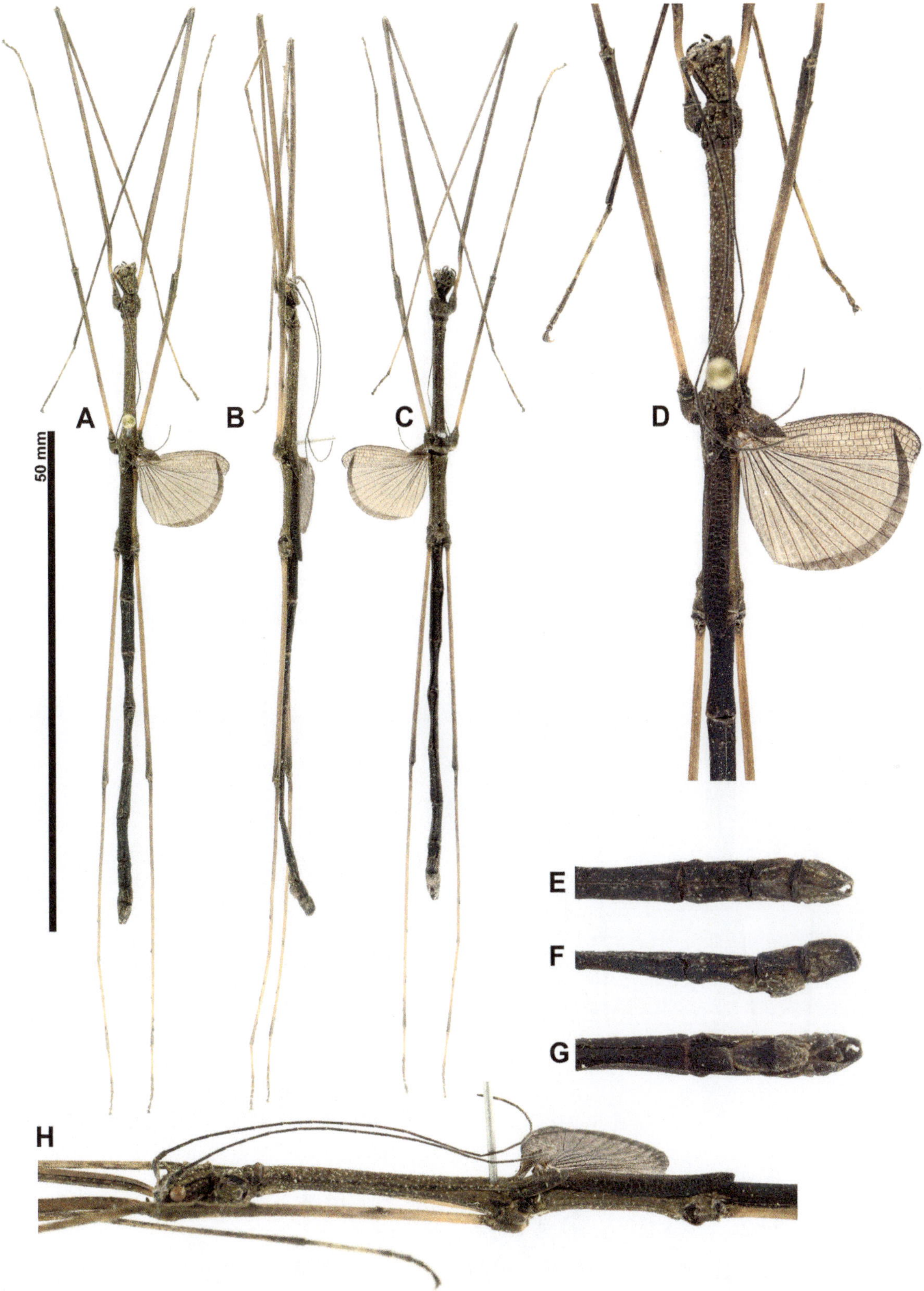

Fig. 6. *Lobofemora scheirei* sp. nov. ♂, holotype. **A**. Habitus, dorsal view. **B**. Habitus, lateral view. **C**. Habitus, ventral view. **D**. Anterior part of body, dorsal view. **E**. Apex of abdomen, dorsal view. **F**. Apex of abdomen, lateral view. **G**. Apex of abdomen, ventral view. **H**. Anterior part of body, lateral view. D–H = not to scale.

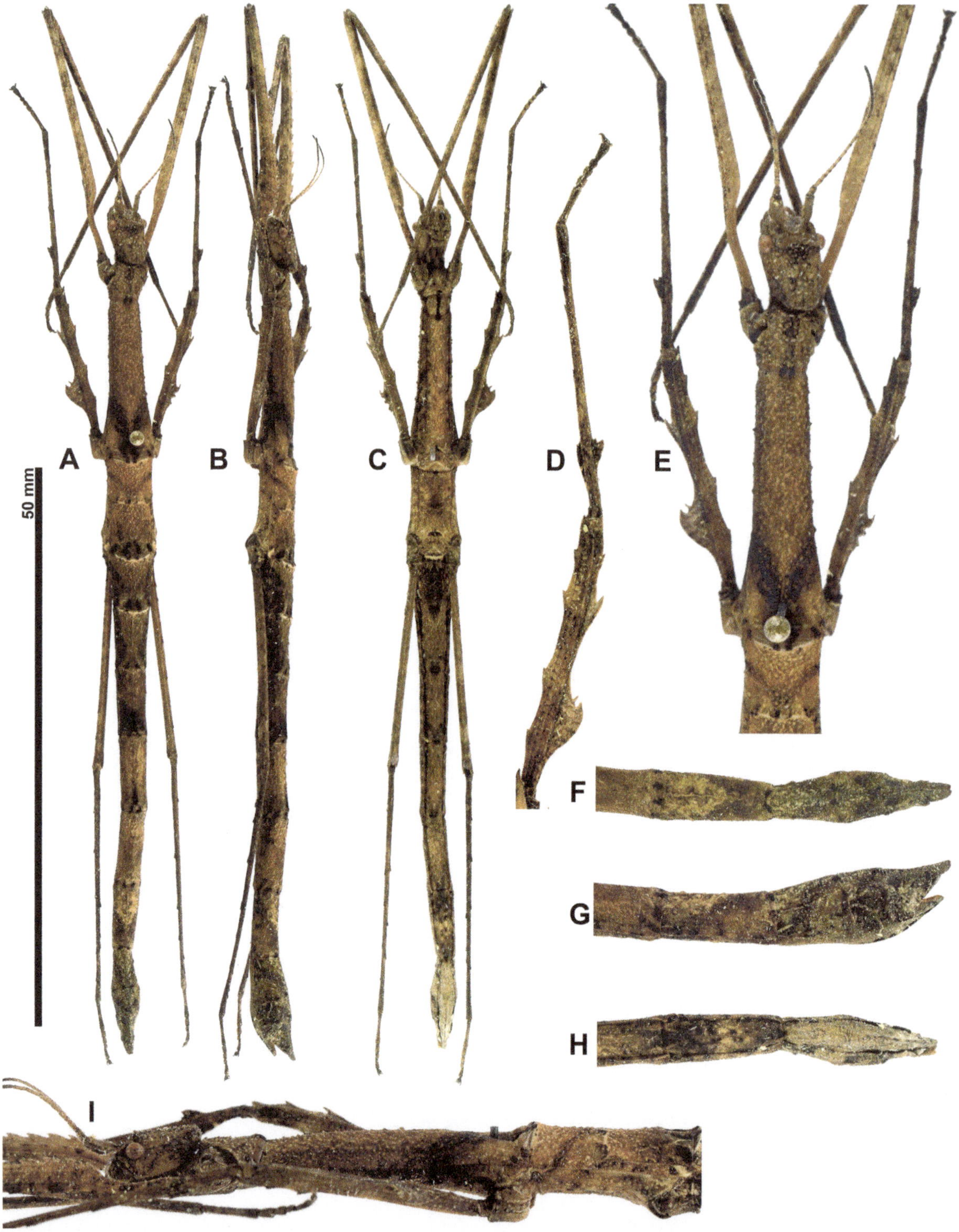

Fig. 7. *Lobofemora scheirei* sp. nov. ♀, paratype. **A**. Habitus, dorsal view. **B**. Habitus, lateral view. **C**. Habitus, ventral view. **D**. Right median leg, detail. **E**. Anterior part of body, dorsal view. **F**. Apex of abdomen, dorsal view. **G**. Apex of abdomen, lateral view. **H**. Apex of abdomen, ventral view. **I**. Anterior part of body, lateral view. D–I = not to scale.

Lobofemora bachmaensis sp. nov.

urn:lsid:zoobank.org:act:87AC14B9-A8A9-48E8-98C4-69A8A8162F67

Figs 1, 2, 3, 10A–C

Etymology

The species epithet refers to the type-locality, Bach Ma National Park in central Vietnam.

Type material

Holotype

VIETNAM: ♂: "Coll. I.R.Sc.N.B., C Vietnam, Bach Ma N.P., 16°12'N 107°52'E, 12-17.vii.2011, leg. J. Constant & J. Bresseel, I.G.: 31.933" (RBINS).

Paratypes (1 ♂, 3 ♀♀, 1 ♀ nymph)

VIETNAM: same collection data as for holotype (RBINS; 1 ♂, 1 ♀: VNMN).

Description

Male (Fig. 2)

MEASUREMENTS. See Table 1.

BODY. Complete body more or less uniformly brown with blackish markings, sometimes with vague longitudinal darker line. Basitarsi and curved anterior part of profemora paler. Prosternum with conspicuous black, longitudinal median line.

HEAD. Longer than wide, slightly tapering posteriorly and granulose. Vertex elongated, projecting over anterior margin of pronotum. Dorsal portion of head flattened, with two definite spines projecting dorsally at posterior margin of eyes. Eyes circular and strongly projecting hemispherically. Antennae projecting over posterior margin of median segment, with 23 segments. Scapus strongly flattened dorso-ventrally and laterally rounded. Pedicellus cylindrical.

THORAX. Pronotum almost rectangular and granulose, shorter than head and slightly widening posteriorly. Anterior margin incurved and raised, followed by faint, median longitudinal impression, not reaching posterior margin. Central transverse impression not reaching lateral edges of pronotum. Mesonotum about five times as long as pronotum and strongly granulose, widening posteriorly; three larger, spinose tubercles on each side, with largest one in middle; fine median longitudinal line. Tegmina with inner portion blackish and outer portion pale cream coloured, short, flattened and just projecting over anterior margin of metanotum; apically rounded. Alae absent. Metanotum with faint median line, shorter than median segment.

ABDOMEN. Abdominal segments II–V gradually increasing in length; V–VI about the same length, slightly granulose and with fine median, longitudinal line. Granules more concentrated along median line. Tergum VII shorter than tergum II. Tergum VIII about half as long as tergum II, slightly widening posteriorly. Tergum IX shorter than VIII and tectiform. Tergum X about as long as VIII and with median longitudinal carina; apically split into two semi-tergites. Inner apical rim of semi-tergite armed with several spines; apex obliquely pointing downwards. Posterolateral angles rounded. Poculum rounded, slightly projecting over apex of abdominal tergum IX. Apex of poculum narrower and rounded. Cerci short, not reaching apex of tergum X, incurving, circular in cross-section, with apices rounded.

LEGS. Profemora slightly longer than head and thorax combined; compressed and curved basally. Anterodorsal carina with few, small serrations. Serrations mostly anteriorly. Other carinae unarmed. Mesofemora slightly shorter than head, pro- and mesonotum combined. Posterodorsal carina with two enlarged triangular spines, one central and one subapical. Anteroventral carina with single triangular subapical spine. Metafemora longer than mesofemora, but shorter than profemora. Protibiae longer than profemora and unarmed. Mesotibiae slightly longer than mesofemora. Posterodorsal carina with triangular spine in basal half. Anterodorsal carina with minute subapical spine. Ventral carinae unarmed. Metatibiae with minute saw-like spines in apical half on the outer carinae.

Female (Figs 3, 10A–C)
MEASUREMENTS. See Table 1.

BODY. Complete body more or less uniformly brown with blackish markings, sometimes with vague longitudinal darker line. Basitarsi and curved anterior part of profemora paler. Prosternum with conspicuous black, longitudinal median line.

HEAD. Longer than wide, slightly tapering posteriorly and strongly granulose. Vertex elongated, almost reaching halfway pronotum. Vertex with enlarged granules apically and asymmetrical: vertex split by shallow longitudinal impression, right hump slightly more pronounced than left one. Dorsal portion of head flattened, with transverse, laterally acute ridge at posterior margin of eyes. Eyes circular and projecting hemispherically. Antennae short, with 25 segments. Scapus strongly flattened dorso-ventrally and rounded laterally. Pedicellus slightly flattened dorso-ventrally. Apical half with very short segments, apical segment about as long as three preceding segments together. Genae granulose.

THORAX. Pronotum trapezoidal, widening posteriorly; shorter than head and slightly granulose. Anterior margin incurved and raised, followed by median longitudinal impression, not reaching posterior margin. Central transverse impression not reaching lateral edges of pronotum. Mesonotum about four times as long as pronotum and strongly granulose, with fine median longitudinal line. Lateral edges with some larger blunt tubercles. Metanotum with anterior margin weakly incurved and raised, longer than wide and strongly granulose. Median segment 1.5 times longer than wide and granulose.

ABDOMEN. Abdominal segments II–VI gradually increasing in length, granulose and with fine median, longitudinal line. Tergum VII about as long as tergum II; sometimes with stalked, irregularly shaped tubercles. Tergum VIII shorter than tergum II, slightly widening posteriorly. Posterior margin weakly incurved. Tergum IX with anterior margin slightly convex, posterior margin slightly narrower and tectiform. Tergum X about as long as IX and with median, longitudinal carina; notched posteriorly and with posterolateral angles rounded. Epiproct elongated, slightly shorter than abdominal tergum X and with median, longitudinal carina; tectiform and apically rounded. Subgenital plate elongated, slightly projecting over apex of tergum X, with median longitudinal carina; apex acute. Cerci short, not reaching apex of tergum X. Gonapophyses strongly elongated, reaching apex of epiproct.

LEGS. Profemora longer than head, pronotum and mesonotum combined; compressed and curved basally. Anterodorsal carina with small serrations, more prominent anteriorly. Posterodorsal carina indistinct with sometimes (two out of three specimens) an enlarged triangular lobe. Ventral carina present and unarmed. Mesofemora with postero-dorsal carina with two lobes; anterior one most prominent. Anterior lobe resulting in different spines; other lobe with one apical spine. Anteroventral carina with three lobes. Basal lobe less prominent and inconspicuous; median and subapical ones with several spines. Posteroventral carina with base more prominent and one subapical lobe. Metafemora with posterodorsal carina sparsely serrated with triangular spines. Posteroventral carina with small triangular lobe basally and with two spines subapically; the latter strongly elongated and "wire"-like (absent in one of the females). Protibiae longer than head and thorax combined. Dorsal carina with small, inconspicuous triangular lobes. Mesotibiae about as long as mesofemora. Posterodorsal carina with conspicuous lobe basally; anterodorsal carina with minute triangular spine near apex. Medioventral carina with conspicuous lobe mirroring the one on posterodorsal carina, and with minute black serrations. Metatibiae shorter than

Table 1. Measurements (mm) of *Lobofemora bachmaensis* sp. nov. HT = holotype; PT = paratype(s).

	HT ♂	PT ♀♀	PT ♂
Body	74.0	87.3–95.5	69.2
Head	4.1	6.5–7.5	3.6
Pronotum	2.4	3.6–4.2	2.8
Mesonotum	14.0	15.3–16.6	12.8
Metanotum	3.7	5.5–6.0	3.7
Median segment	6.2	5.8–6.2	6.0
Profemora	34.8	31.6–32.1	29.6
Mesofemora	17.1	16.2–17.0	16.3
Metafemora	24.5	22.9–25.1	22.3
Protibiae	39.7	35.5–36.7	30.9
Mesotibiae	18.4	16.7–17.6	19.2
Metatibiae	29.4	27.3–28.2	27.0
Tegmina	0.8	–	

profemora; posteroventral carina with few triangular serrations; posterodorsal carina with subapical triangular spine.

Egg
Unknown.

Distribution

So far only known from Bach Ma National Park in central Vietnam (Fig. 1).

Lobofemora bidoupensis sp. nov.

urn:lsid:zoobank.org:act:BA44BAFF-F44A-4476-ABF5-536CADC27B0E

Figs 1, 4, 5, 8, 10D–G

Etymology

The species epithet refers to the type-locality, Bidoup-Nui Ba National Park in central Vietnam.

Type material

Holotype
VIETNAM: ♂: "Coll. I.R.Sc.N.B., Vietnam, Lam Dong prov., Bidoup-Nui Ba N.P., 12°26'N 108°30'E, 21-25.vii.2014, night coll., Leg. J. Constant & J. Bresseel, GTI Project, I.G.: 32.779" (RBINS).

Paratypes (10 ♂♂, 4 ♀♀)
VIETNAM: 6 ♂♂, 1 ♀: same collection data as for holotype (RBINS); 4 ♂♂, 3 ♀♀: "Coll. I.R.Sc.N.B., ex breeding J. Bresseel 2015, origin: Vietnam, Lam Dong prov., Bidoup-Nui Ba N.P., 12°26'N 108°30'E, 21-25.vii.2014, J. Constant & J. Bresseel, GTI Project, I.G.: 32.779" (RBINS; 1 ♂, 1 ♀: VNMN).

Description

Male (Figs 4, 8F–G, 10D)
Measurements. See Table 2.

Body. Base of femora pale brown, darkening towards apex. Tibiae mottled black and pale brown. Head with dorsal surface pale brown. Body darker brown with some black markings.

HEAD. Longer than wide, slightly tapering posteriorly and granulose. Vertex slightly elongated. Dorsal portion of head flattened with two spines projecting dorsally at posterior margin of eyes. Eyes circular and strongly projecting hemispherically. Antennae with 21 segments. Scapus strongly flattened dorso-ventrally and rounded laterally. Pedicellus cylindrical. Upper part of genae granulose.

THORAX. Pronotum longer than wide, almost rectangular and granulose, shorter than head; anterior margin incurved, followed by median longitudinal impression not reaching posterior margin; short, central, transverse impression not reaching lateral edges of pronotum. Mesonotum about six times as long as pronotum and strongly granulose, widening posteriorly, with fine median longitudinal line. Tegmina brown, short and slightly tapering posteriorly, not projecting over metanotum. Alae present as small scales, brown.

ABDOMEN. Abdominal segments II–V gradually increasing in length, V–VI about the same length, slightly granulose. Tergum VII shorter than tergum III. Tergum VIII about half as long as tergum II, slightly widening posteriorly. Tergum IX shorter than VIII and tectiform. Tergum X about as long as VIII and with median longitudinal carina; apically split into two semi-tergites. Inner apical rim of semi-tergite armed with several spines. Apex of semi-tergites obliquely pointing downwards; posterolateral angles rounded. Poculum rounded, slightly projecting over apex of abdominal tergum IX. Apex of poculum narrower and rounded. Cerci short, not reaching apex of tergum X, cylindrical in cross-section, with apices rounded.

LEGS. Profemora slightly longer than head and thorax combined; compressed and curved basally; dorsal carinae with few, minute spines; ventral carinae unarmed. Mesofemora longer than pro- and mesonotum combined; posterodorsal carina with two triangular spines and minute spine subapically; other carinae with few minute spines. Metafemora longer than head, pro-, meso- and metanotum combined; posterodorsal carina with four minute spines. Protibiae longer than profemora and unarmed. Mesotibiae slightly longer than mesofemora; medioventral carina with conspicuous, small triangular lobe. Metatibiae with minute saw-like spines in apical half on all carinae; medioventral carina with conspicuous triangular spine in anterior half.

Female (Figs 5, 8A–E, 10E–G)
MEASUREMENTS. See Table 2.

BODY. Complete body ranging from uniformly brown to mottled brown-black with some green and white markings. Prosternum with conspicuous black, longitudinal line.

HEAD. Longer than wide, slightly tapering posteriorly and granulose. Vertex slightly elongated and slightly asymmetrical: vertex split by shallow longitudinal impression; right portion slightly more pronounced than left one. Dorsal portion of head flattened, with transverse, laterally acute ridge at posterior margin of eyes; ridge split centrally by longitudinal depression. Eyes circular and projecting hemispherically. Antennae short with 24–26 segments. Scapus strongly flattened dorso-ventrally and rounded laterally, with outer lobe more prominent. Pedicellus slightly cylindrical; first antennomere about as long as the following two segments combined. Apical half with very short segments; apical segment about as long as four preceding segments combined. Upper part of genae tuberculate.

THORAX. Pronotum trapezoidal, widening posteriorly; shorter than head and granulose. Anterior margin incurved and raised, followed by median longitudinal impression not reaching posterior margin. Central transverse impression not reaching lateral edges of pronotum. Mesonotum about four times as long as pronotum and strongly granulose, slightly widening posteriorly, with fine median longitudinal line. Metanotum subquadrate with anterior margin weakly incurved and raised; strongly granulose and with median tubercle subapically. Median segment slightly longer than wide, anterior portion slightly

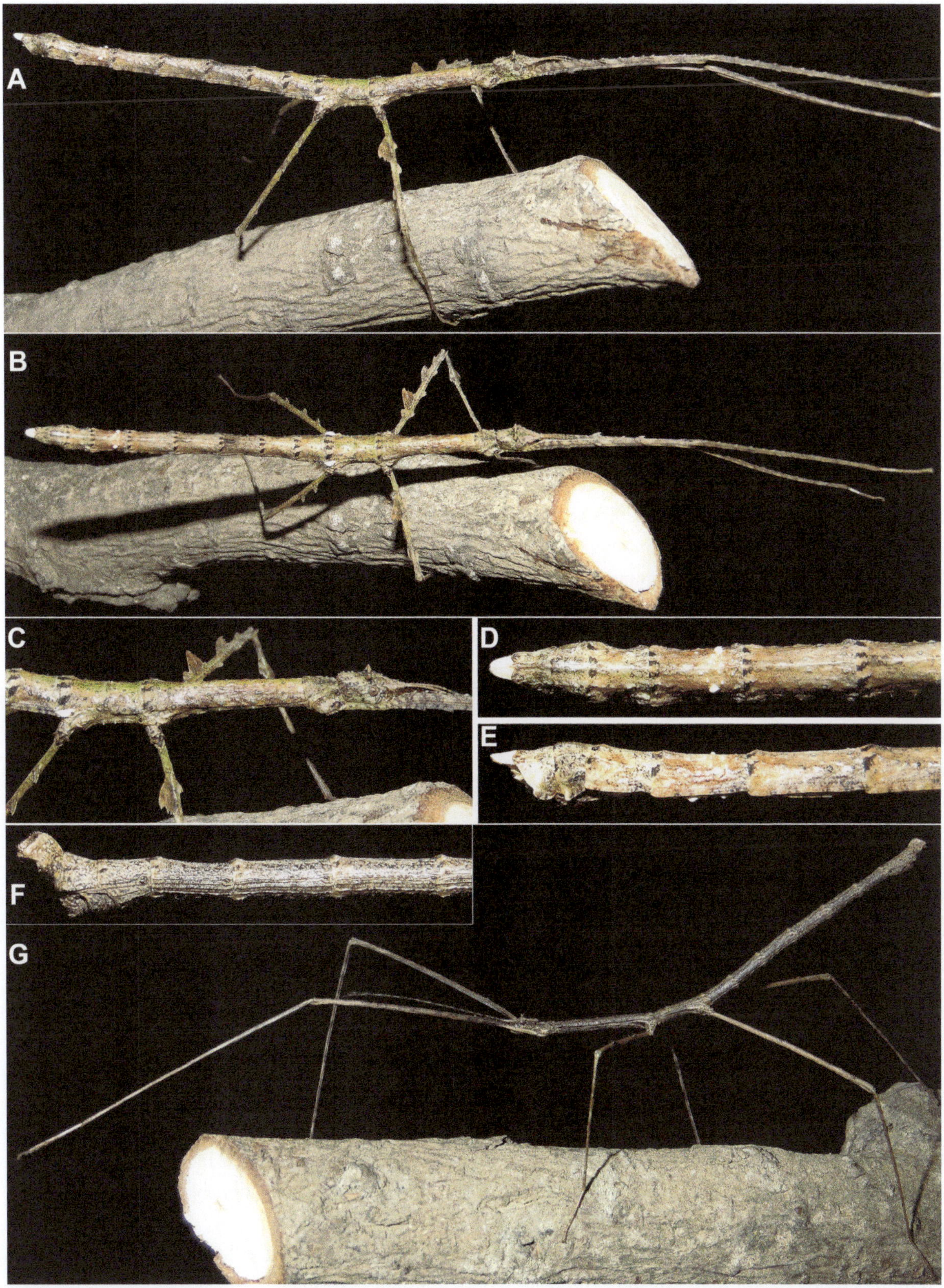

Fig. 8. *Lobofemora bidoupensis* sp. nov., captive reared specimens (photographs by the authors). **A–E**. ♀. **A**. Lateral view. **B**. Dorsal view. **C**. Anterior part of body, lateral view. **D**. Apex of abdomen, dorsal view. **E**. Apex of abdomen, lateral view. **F–G**. ♂. **F**. Apex of abdomen, lateral view. **G**. Lateral view.

granulose. Mesosternum with posteriorly, between legs, Y-shaped impression resulting into two small pseudo-foramina. Metasternum with posteriorly, between legs, X-shaped impression resulting in two small pseudo-foramina anteriorly.

Abdomen. Abdominal segments II–V gradually increasing in length, granulose and with fine median, longitudinal line; larger granules along median line. Terga V–VI nearly equal in length. Tergum VII shorter than VI. Tergum VIII shorter than tergum VI, slightly widening posteriorly; posterior margin weakly incurved. Tergum IX with anterior margin slightly convex; posterior margin slightly narrower. Tergum X slightly longer than IX and with median, longitudinal carina; strongly notched posteriorly; posterolateral angles rounded. Terga IX–X with lateral edges more pronounced and lobe-like. Epiproct elongated; about ⅔ as long as abdominal tergum X and with very shallow apical notch. Subgenital plate elongated, slightly projecting over apex of tergum X; apical half with median longitudinal carina; apex rounded. Cerci short, not reaching of apex tergum X, cylindrical in cross-section, with apices rounded. Gonapophyses strongly elongated, reaching apex of epiproct.

Legs. Profemora longer than head, pro-, meso- and metanotum combined; compressed and curved basally; anterodorsal carina with small serrations, more prominent anteriorly; posterodorsal carina armed with two enlarged triangular bispinose teeth; posteroventral carina sometimes with few serrations anteriorly. Mesofemora with postero-dorsal carina with three definite lobes; anterior one most prominent and gradually becoming smaller; anterior and central lobes resulting in different spines; other lobe with one apical spine; anterodorsal carina with few, small triangular lobes resulting in different minute spines; anteroventral carina with three small lobes resulting in minute spines. Posteroventral carina with one small lobe in apical half. Metafemora with posterodorsal carina armed with four triangular spines resulting in minute small spines; anterodorsal carina sometimes with some minute spines; posteroventral carina with few small spines. Protibiae about as long as head and thorax combined; posterodorsal carina with few inconspicuous triangular lobes; anterodorsal carina with small serrations over its complete length; posteroventral carina with few minute spines. Mesotibiae slightly longer than mesofemora; posterodorsal carina with conspicuous lobe anteriorly and few small serrations posteriorly; medioventral carina with conspicuous lobe mirroring the one on posterodorsal carina and with minute black serrations; other carinae with few, small serrations. Metatibiae shorter than profemora; dorsal carinae and postero- and anteroventrally with minute serrations; medioventral carina armed with two small triangular lobes basally.

Egg (Fig. 5J–O)

Measurements (in mm). Length: 3.0–3.5; width: 1.2; height: 1.5.

Elongate-oval, general colouration, light brown with black patches. Capsule oval in cross section, with the surface granulose. Anteriorly with short brown fringes along the anterior margin of the capsule. Dorsal side slightly concave. Micropylar plate more or less inverted heart shaped, with a median impression posteriorly. Anteriorly triangular, then widening, posterolateral angles only slightly projecting over micropylar cup. Micropylar cup definite, followed by relatively long median line. Micropylar plate and median line coloured as capsule. Ventral side slightly convex. Polar area with shallow indention. Operculum slightly darker than capsule, with central dark, circular marking.

Distribution

So far only known from Bidoup-Nui Ba National Park in central Vietnam (Fig. 1).

Table 2. Measurements (mm) of *Lobofemora bidoupensis* sp. nov. HT = holotype; PT = paratype(s).

	HT ♂	PT ♀♀	PT ♂
Body	52.5	71.2–77.8	55.0
Head	2.7	4.5–4.7	2.7
Pronotum	1.8	2.6–2.7	2.0
Mesonotum	11.2	13.6–16.4	11.7
Metanotum	1.8	3.6–3.9	2.5
Median segment	4.7	4.3–4.4	4.4
Profemora	26.3	25.1–29.2	28.2
Mesofemora	15.8	15.8–15.5	16.9
Metafemora	20.3	21.3–24.5	21.5
Protibiae	31.1	30.1–33.1	33.8
Mesotibiae	17.8	16.6–18.4	19.3
Metatibiae	25.0	25.3–27.0	27.1
Tegmina	1.4	–	

Lobofemora scheirei sp. nov.
urn:lsid:zoobank.org:act:9CE86F8D-E644-4A37-A278-DB4D99971378
Figs 6, 7, 9, 10H–J

Etymology

The species epithet refers to the Flemish TV show presenter Mr Lieven Scheire. We dedicate this new species to Lieven both in recognition of his efforts in raising public awareness of science (including taxonomy), especially for young people, in his TV shows, and as a present for his recent wedding.

Type material

Holotype

VIETNAM: ♂: "Coll. I.R.Sc.N.B., Vietnam, Dong Nai Biosphere Res. 11°18'N 107°06'E, 25.vi-6.vii.2012, night collecting, Leg. J. Constant & J. Bresseel, I.G.: 32.161" (RBINS).

Paratypes (39 ♂♂, 28 ♀♀, 17 eggs)

VIETNAM: 12 ♂♂, 12 ♀♀: same collection data as for holotype (RBINS; 1 ♂, 1 ♀: VNMN); 5 ♂♂, 5 ♀♀: "Coll. I.R.Sc.N.B., Vietnam, Cat Tien N.P., 11°26'N 107°26'E, 6-16.vii.2012, Leg. J. Constant & J. Bresseel, I.G.: 32.161" (RBINS; 1 ♂, 1 ♀: VNMN); 12 eggs: "Coll. I.R.Sc.N.B., ex breeding J. Bresseel i.2015, origin: Vietnam, Cat Tien N.P., 11°26'N 107°26'E, 6-16.vii.2012, J. Constant & J. Bresseel" (RBINS; 2: VNMN); 22 ♂♂, 11 ♀♀, 5 eggs: "Ex culture F. Hennemann 2015, Origin: S-Vietnam, Lam Dong Prov., Cat Tien N.P., 11°26'N 107°26'E, leg. J. Constant & J. Bresseel 6.-16.VII.2012" (3 ♂♂, 1 ♀: MNHU; 3 ♂♂, 1 ♀: ZMUH; 4 ♂♂, 1 ♀: ZSMC; 3 ♂♂, 1 ♀: RMNH; 4 ♂♂, 2 ♀♀: BMNH; 5 ♂♂, 5 ♀♀, 5 eggs: FHC).

Additional material

VIETNAM: 25 ♂♂, 23 ♀♀, 45 eggs: "Ex culture F. Hennemann 2015, Origin: S-Vietnam, Lam Dong Prov., Cat Tien N.P., 11°26'N 107°26'E, leg. J. Constant & J. Bresseel 6-16.VII.2012" (FHC).

Description

Male (Figs 6, 9I–M, 10H)

Measurements. See Table 3.

Body. Complete body blackish-brown. Profemora dark brown, paler basally. Protibiae light brown. Other femora pale brown with darker apex. Corresponding tibiae pale brown.

Head. Longer than wide, slightly tapering posteriorly and granulose. Vertex slightly elongated. Dorsal portion of head flattened, with two spines projecting dorsally at posterior margin of eyes. Eyes circular and strongly projecting hemispherically. Antennae almost reaching posterior margin of median segment, with 23 segments. Scapus strongly flattened dorso-ventrally and rounded laterally. Pedicellus cylindrical. Upper part of genae granulose.

Thorax. Pronotum trapezoidal and granulose, widening posteriorly; shorter than head; anterior margin incurved and raised, followed by median longitudinal impression not reaching posterior margin. Central transverse impression not reaching lateral edges of pronotum. Mesonotum about six to seven times as long as pronotum and strongly granulose, widening posteriorly; with fine median longitudinal line. Tegmina blackish, short and with definite hump; slightly tapering posteriorly and with apex rounded; external margin densely covered with minute tubercles. Alae projecting over apex of median segment; costal area coloured as elytra, anal area greyish brown; subcostal and radial veins densely covered with minute tubercles.

Abdomen. Abdominal segments II–V gradually increasing in length, V–VI about equal in length, slightly granulose and with fine median, longitudinal line; granules more concentrated along median line. Tergum VII shorter than tergum III. Tergum VIII about half as long as tergum II, slightly widening posteriorly. Tergum IX shorter than VIII and tectiform. Tergum X about as long as VIII and with median longitudinal carina; apically split into two semi-tergites. Inner apical rim of semi-tergite armed with several spines; apex obliquely pointing downwards; posterolateral angles rounded. Poculum rounded, slightly projecting over apex of abdominal tergum IX. Apex of poculum narrower and rounded. Cerci short, not reaching apex of tergum X, cylindrical in cross-section, with apices rounded.

Legs. Profemora slightly longer than head and thorax combined; compressed and curved basally; anterodorsal carina with few, small serrations anteriorly; other carinae unarmed. Mesofemora about as long as pro- and mesonotum combined; posterodorsal carina variably with few triangular spines. Metafemora longer than head, pro-, meso- and metanotum combined and unarmed. Protibiae longer than profemora and unarmed. Mesotibiae slightly longer than mesofemora; anterodorsal carina with minute subapical spine; anteroventral carina with minute spine situated more basally than spine on anterodorsal carina; posteroventral carina with minute spine more basally than spine on anteroventral carina. Metatibiae with minute saw-like spines in posterior half on all carinae.

Female (Figs 7, 9A–H, 10I–J)
Measurements. See Table 3.

Body. Colouration variable. Most specimens more or less uniformly brown with some black markings; some specimens with head, legs and apex of abdomen red-brown and very pale fawn body. Prosternum with conspicuous black, longitudinal median line.

Head. Longer than wide, slightly tapering posteriorly and granulose. Vertex slightly elongated and asymmetrical: vertex split by shallow longitudinal impression, right hump slightly more pronounced than left one. Dorsal portion of head flattened, with transverse, laterally acute ridge at posterior margin of eyes. Eyes circular and projecting hemispherically. Antennae short, with 23–25 segments, most females possess 25 segments. Scapus strongly flattened dorso-ventrally and rounded laterally, with outer lobe more prominent. Pedicellus slightly flattened dorso-ventrally. First antennomere about as long as following two ones combined. Apical half of antenna with very short segments; apical segment about as long as four preceding segments combined. Upper part of genae granulose.

Fig. 9. *Lobofemora scheirei* sp. nov., captive reared specimens (photographs by Dr B. Kneubühler). **A–H**. ♀. **A**. Dorsal view. **B**. Ventral view. **C**. Lateral view. **D**. Dorsolateral view, morph whitish with head, apex of abdomen and legs chestnut brown. **E**. Head, lateral view. **F**. Apex of abdomen, dorsal view. **G**. Apex of abdomen, lateral view. **H**. Apex of abdomen, ventral view. **I–M**. ♂. **I**. Dorsal. **J**. Wing spread, lateral view. **K**. Head and thorax, lateral view. **L**. Apex of abdomen, lateral view. **M**. Apex of abdomen, dorsal view. **N**. Nymph. **O**. Eggs. **P–S**. Egg. **P**. Detail of micropylar plate. **Q**. Ventral view. **R**. Detail of operculum. **S**. Detail of polar area. **T**. Eggs laid in a slice of sponge.

Fig. 10. *Lobofemora* gen. nov. spp., photographs *in natura* by the authors. **A–C**. *L. bachmaensis* sp. nov., Bach Ma National Park. **A–B**. ♀, 11 Jul. 2011. **C**. ♀, 14 Jul. 2011. **D–G**. *L. bidoupensis* sp. nov., Bidoup-Nui Ba National Park. **D**. ♂, 22 Jul. 2014. **E–F**. ♀, 25 Jul. 2014. **G**. ♀, 23 Jul. 2014. **H–J**. *L. scheirei* sp. nov. **H–I**. Dong Nai Biosphere Reserve, 30 Jun. 2012. **H**. ♂. **I**. ♀. **J**. ♀, Cat Tien National Park, 9 Jul. 2012.

THORAX. Pronotum trapezoidal, widening posteriorly; shorter than head and granulose; anterior margin incurved and raised, followed by median longitudinal impression not reaching posterior margin. Central transverse impression not reaching lateral edges of pronotum. Posterior margin with small posteromedian tubercle. Mesonotum about 3.7 times as long as pronotum and strongly granulose, widening posteriorly; fine median longitudinal line; tiny vestigial wing remnants posterolaterally. Metanotum with anterior margin weakly incurved and raised, subquadrate and strongly granulose. Median segment slightly longer than wide; anterior portion strongly granulose. Meso- and metasternum with posteriorly, between legs, Y-shaped impression resulting in two small pseudo-foramina.

ABDOMEN. Abdominal segments II–VI gradually increasing in length, granulose and with fine median, longitudinal line; larger granules along median line. Tergum VII about as long as tergum III; sometimes with stalked, irregularly shaped tubercles. Tergum VIII shorter than tergum II, slightly widening posteriorly; posterior margin weakly incurved. Tergum IX with anterior margin slightly convex; posterior margin slightly narrower. Tergum X about as long as IX and with median, longitudinal carina; strongly notched posteriorly; posterolateral angles rounded. Terga IX–X with lateral edges more pronounced and lobe-like. Epiproct elongated, about ⅔ as long as abdominal tergum X, and with median, longitudinal carina; tectiform and with shallow apical notch. Subgenital plate elongated, slightly projecting over apex of tergum X; apical half with median longitudinal carina; apex rounded. Cerci short, not reaching apex of tergum X, cylindrical in cross-section, with apices rounded. Gonapophyses strongly elongated, reaching apex of epiproct.

LEGS. Profemora longer than head, pronotum and mesonotum combined; compressed and curved basally; anterodorsal carina with small serrations, more prominent anteriorly; posterodorsal carina indistinct; posteroventral carina sometimes with few minute serrations. Mesofemora with three lobes on posterodorsal carina; anterior one most prominent and gradually becoming smaller; anterior lobe resulting in different spines; two other lobes with one apical spine; antero-ventral carina with three small lobe-like spines; posteroventral carina with one small lobe-like spine in apical half. Metafemora with anterodorsal carina sparsely serrated; other carinae unarmed. Protibiae longer than head and thorax combined; posterodorsal carina with inconspicuous triangular lobe positioned at about basal ¼ of length; anterodorsal carina with few minute serrations. Mesotibiae slightly longer than mesofemora; posterodorsal carina with conspicuous lobe at about basal ¼ of length; dorsal carinae with minute black spine near apex; medioventral carina with conspicuous lobe mirroring the one on posterodorsal carina and with minute black serrations; postero- and anteroventral carinae with minute black serrations. Metatibiae shorter than profemora; all ventral carinae with minute black serrations; dorsal carinae sparsely armed with minute black serrations, more concentrated subapically.

Nymph (Fig. 9N)

Newly hatched nymphs are about 17 mm long and have comparatively long legs. Antennae short, slightly longer than head. Body entirely mottled green, brown and black.

Egg (Fig. 9O–S)

MEASUREMENTS (in mm). Length: 4.0; width: 1.2; height: 1.5.

Elongate-oval, general colouration light brown with black markings. Capsule oval in cross section, with the surface granulose and with a net-like sculpturing. Anteriorly with short brown fringes along the anterior margin of the capsule. Dorsal side slightly concave. Micropylar plate more or less bi-lobate with a median impression anteriorly and posteriorly. Anteriorly triangular, then widening; lateral expansions axe-shaped, with posterolateral angles projecting over micropylar cup. Micropylar cup definite, blackish and followed by relatively long median line. Micropylar plate and median line coloured as capsule.

Table 3. Measurements (mm) of *Lobofemora scheirei* sp. nov. HT = holotype; PT = paratypes.

	HT ♂	**PT ♀♀**	**PT ♂♂**
Body	66.5	72.3–79.9	56.3–73.4
Head	3.3	5.4–5.5	2.5–3.4
Pronotum	2.1	3.7–3.8	1.9–2.1
Mesonotum	14.3	13.2–15.7	11.1–14.8
Metanotum	Covered by wings	3.9–4.1	Covered by wings
Median segment	Covered by wings	4.2–4.9	Covered by wings
Profemora	30.0	24.6–26.9	25.2–31.2
Mesofemora	17.1	15.0–16.0	14.1–17.1
Metafemora	23.5	19.3–21.4	18.8–22.2
Protibiae	36.0	30.9–31.6	30.0–37.0
Mesotibiae	19.5	15.7–16.0	16.0–19.2
Metatibiae	27.5	24.3–24.9	23.2–27.6
Tegmina	2.8	–	1.9–2.9
Alae	9.4	–	8.5–9.5

Ventral side slightly convex. Polar area with shallow indention. Operculum slightly darker than capsule, with deep central impression.

Distribution

So far only known from Cat Tien National Park and Dong Nai Biosphere Reserve in south Vietnam (Fig. 1).

Biological observations

Several eggs of this species hatched and nymphs were reared in captivity using ferns (Dryopteridacea), which seem to be their natural foodplant. Substitute plants like bramble (*Rubus* L. spp., Rosaceae), beech (*Fagus sylvatica* L., Fagaceae) and *Hypericum* L. spp. (Hypericaceae) are also well accepted in captivity in Europe. This species is easy to maintain in captivity. Females try to stick their eggs into cracks and holes. Sponges cut into slices are suitable as substrate for this (Fig. 9T). Only few eggs will lay on the cage floor (pers. comm. B. Kneubühler, Jan. 2015). To collect eggs from wild caught females, paper tissue was used as substrate for egg laying.

Males are capable of stridulating, but in order to do so, they have to completely open their wings. When wings are open, the alae are not flat as in most stick insects, but have the outer margin curved downwards (Fig. 9J). By rubbing the outer margins of the tegmina against the subcostal and radial veins of the alae they are able to produce sound. The stridulation occurs when the specimens are disturbed.

Discussion

The currently known distribution of the genus *Lobofemora* gen. nov. is restricted to the southern half of Vietnam. The three new species were found at altitudes ranging from close to sea level (in Bach Ma National Park) up to 1600 m (in Bidoup-Nui Ba National Park).

From our observations in the field and in culture, it appears that the colouration in males is very constant, while females show more variable patterns and colours.

When identifying specimens of *Lobofemora* gen. nov., one must be aware that regenerated legs do not show the specific features of the "normal" specimens: lobes and spines are less developed and can even be absent.

The tribal placement of *Lobofemora* gen. nov. suggested herein is based on a set of characters which key out to the tribe Clitumnini Brunner von Wattenwyl, 1893. This due to having shortened antennae, that are not longer than the profemora, with a considerably broadened scapus; profemora that are triangular in cross-section with the anterodorsal carina strongly raised and the medioventral carina conspicuously displaced towards the anteroventral carina; females with strongly elongated lower gonapophyses (gonapophyses VIII) which clearly project over the apex of the subgenital plate; eggs are elongate and bullet-shaped, with an open internal micropylar plate and a well developed median line and lacking a central capitulum. Furthermore, eggs have a hairy rim on the border of the capitulum. Males lack an external vomer.

However, as mentioned above the definition of the Clitumnini had to be adapted to include this genus. The genus actually shares and violates several morphological characters with both Clitumnini and Medaurini Hennemann & Conle, 2008. Therefore, the genus has a fairly isolated position within the tribe.

The beak-like ovipositor is shared with several other genera in Clitumnini, e.g., *Entoria* Stål, 1875, *Metentoria* Brunner von Wattenwyl, 1907 and *Rhamphophasma* Brunner von Wattenwyl, 1893.

The aberrant and remarkable characters exhibited by *Lobofemora* gen. nov., with especially the length of the median segment and presence of tegmina and vestigial or shortened alae in the males, are unique within the entire tribe Clitumnini and warrant further investigation.

As usual for stick insects that have wings in one or both sexes, the median segment is fairly long in relation to the metanotum, an additional feature which distinguishes *Lobofemora* gen. nov. from all currently known genera of the tribe Clitumnini.

Traditionally, the presence of wings has *a priori* been regarded as a plesimorphic character, but Whiting *et al.* (2003) have shown that wings within the Phasmatodea can not only be lost but also recovered during the evolutionary process of certain clades, a hypothesis which is supported by several examples and has already received acceptance by certain subsequent authors (e.g. Hennemann & Conle 2008). The decision on whether the presence of wings in the males of *Lobofemora* gen. nov. is plesiomorphic or apomorphic for the tribe Clitumnini is difficult to make at present and certainly warrants a detailed phylogenetic analysis of this and potentially closely related taxa. If it is regarded as an autapomorphy of *Lobofemora* gen. nov., this would be a further and very striking example for the recovery of wings in stick insects and support the hypothesis of Whiting *et al.* (2003), since all other known taxa of the very diverse and speciose tribe Clitumnini and also the related Medaurini are exceptionally wingless. Another fact that makes the presence of wings in *Lobofemora* gen. nov. even more remarkable is the capability of the winged males of *L. scheirei* sp. nov. to stridulate. This striking specialization is also not found in the winged males of the supposedly closely related tribe Pharnaciini (regarded as a subordinate taxon of the subfamily Clitumninae by Hennemann & Conle 2008) and hence is unique in that entire subfamily. The sounds are produced by rubbing the tegmina against the radial and subcostal veins of the alae, which bear numerous very minute tooth-like structures. Morphologically this is an analogy to the wing structures, e.g., seen in certain winged taxa of the predominantly New Guinean genus *Dimorphodes* Westwood, 1859 (Phasmatidae: Xeroderinae: Xeroderini), members of the Mascarean tribe Achriopterini or the New World genus *Pterinoxylus* Audinet-Serville, 1838 (Cladomorphinae: Hesperophasmatini). It is hoped that future phylogenetic studies will deal with the interesting new genus described herein to clarify its taxonomic and phylogenetic position.

However, one must be aware that most genera in Clitumnini need to be redefined and several appear to be polyphyletic in their present configuration. For that reason, it is of utmost importance that DNA-based phylogenetic studies contain, for each genus, at least the type species to be well supported.

Acknowledgments

We thank Dr Hong Thai Pham (VNMN), Mr Vu Tru Hoang and Mr Thieu Du Tran (Institute of Ecology and Biological Resources, Hanoi, Vietnam) for all their help and friendship during collecting trips, Dr Bruno Kneubühler (Luzern, Switzerland) for his excellent pictures and all his help in breeding the species we study, Mr Tim Bollens (Herselt, Belgium) for providing live specimens, Mr Ludovic Nicolas for mounting numerous specimens we brought back from the field and the authorities of the different national parks and reserves for their support to our studies. All the authors' collecting trips were supported through grants issued by the capacity building Programme of the Belgian Global Taxonomic Initiative National Focal Point that runs with financial support from the Belgian Directorate-General for Development Cooperation. Dr Patrick Grootaert, Dr Yves Samyn and Dr Marie-Lucie Susini (RBINS) are also thanked for their permanent support to our projects in Vietnam. We warmly thank Mr Frank Hennemann (Hohenecken, Germany) for his valuable comments, which greatly improved our work.

References

Brock P.D. 2015. Phasmida Species File Online. Version 5.0/5.0.

Clark Sellick J.T. 1997. Descriptive terminology of the phasmid egg capsule, with an extended key to the phasmid genera based on egg structure. *Systematic Entomology* 22 (2): 97–122.

Clark Sellick J.T. 1998. The micropylar plate of the eggs of Phasmida, with a survey of the range of plate form within the order. *Systematic Entomology* 23 (3): 203–228.

Hennemann F.H. & Conle O.V. 2008. Revision of Oriental Phasmatodea: The tribe Pharnaciini Günther, 1953, including the description of the world's longest insect, and a survey of the family Phasmatidae Gray, 1835 with keys to the subfamilies and tribes (Phasmatodea: "Anareolatae": Phasmatidae). *Zootaxa* 1906: 1–316.

Whiting M.F., Bradler S. & Maxwell T. 2003. Loss and recovery of wings in stick insects. *Nature* 421: 264–267.

8

Revalidation of *Acanthinozodium* Denis, 1966 with description of three new species and discovery of a remarkable male palpal character (Araneae, Zodariidae)

Rudy JOCQUÉ [1] & Arnaud HENRARD [2]

[1, 2] Royal Museum for Central Africa, B-3080 Tervuren, Belgium.
[2] Earth and Life Institute, Biodiversity Research Center, UCL-17.07.04, Bâtiment Carnoy, Croix du Sud, 5, B-1348, Louvain-la-Neuve, Belgium.

Abstract. The genus name *Acanthinozodium* Denis, 1966 is considered available and valid, with *A. spinulosum* Denis, 1966 as its type species. The genus is characterized by the presence of ventral rows of long setae on the femora, large anterior median eyes and a large, dorsal, crater-like pit on the cymbium. *Zodariellum* Andreeva & Tyschenko, 1968 is removed from its synonymy and now only contains its type species *Z. surprisum* Andreeva & Tyschenko, 1968. Three new *Acanthinozodium* species are described: *A. crateriferum* sp. nov. (♂♀) from Ethiopia, *A. sahelense* sp. nov. (♂♀) from a wide range in the Sahel region and *A. quercicola* sp. nov. (♂) from Morocco. The genus appears to have a large distribution in and around the Sahara. The possible function of the cymbial pit is discussed.

Keywords. Name availability, mating plug, Sahel region, *Zodarion*, *Zodariellum*.

Introduction

The spider family Zodariidae is fairly well analyzed and its phylogeny at genus level well resolved (Jocqué 1991; Ramirez *et al.* 2014). However, a few areas in the phylogeny remain to be studied. One of them is the huge genus *Zodarion* Walckenaer, 1826 and its closest relatives *Zodariellum* Andreeva & Tyshchenko, 1968, *Tropizodium* Jocqué & Churchill, 2005 and *Parazodarion* Ovtchinnikov, Ahmad & Gurko, 2009. Their relationship cannot be solved with the characters that were mentioned in the original descriptions because some crucial characters, e.g. the femoral organ, were not yet known at that time. Among them, the genus *Zodariellum* Andreeva & Tyshchenko, 1968 was only superficially described. It was synonymized with *Acanthinozodium* Denis, 1966 by Jocqué (1991) but it remained doubtful whether this synonymy was well founded.

In the present paper we address the nomenclatural problem of *Acanthinozodium* and reanalyze the presumed synonymy with *Zodariellum* on the base of classical and new morphological characters. Three new species are described.

Material and methods

Specimens were observed, drawn and measured with a WILD M 10 stereo microscope. Details of the female genitalia and male palps were observed with a Zeiss Stemi 2000 stereo microscope. Measurements and photographs of the habitus, details of mouthparts, detached male palps and female genitalia were taken with a Leica MZ16 using the LAS automontage software (ver. 3.8). The female genitalia were dissected and digested with pancreatin, then immersed in 75% ethanol.

For SEM photos, specimens were dried in Hexamethyldisilazane (36h), gold coated and examined and photographed with a JEOL 6480 LV scanning electron microscope. All types are deposited in the Royal Museum for Central Africa, Tervuren, Belgium (MRAC).

All measurements are in mm. All palp illustration are from left palps.

Abbreviations

ALE = anterior lateral eyes
AME = anterior median eyes
ALS = anterior spinnerets
CRB = collection Robert Bosmans
F = femur
MA = median apophysis
MOQ = median ocular quadrangle
MRAC = Musée Royal de l'Afrique Centrale (Tervuren, Belgium)
Mt = metatarsus
P = patella
PLE = posterior lateral eyes
PLS = posterior lateral spinnerets
PME = posterior median eyes
PMS = posterior median spinnerets
RTA = retrolateral tibial apophysis
t = tarsus
T = tibia

Results

Class Arachnida Cuvier, 1812
Order Araneae Clerck, 1757
Family Zodariidae Thorell, 1881

Genus ***Acanthinozodium*** Denis, 1966

Diagnosis

Members of the genus *Acanthinozodium* are recognized by the wide cephalothorax (L/W < 1.3), large AME (Fig. 4F), and all femora without flattened setae but with a double row of long ventral setae (Fig. 4C). Males are characterized by the deep dorsal pit at the base of the cymbium (Figs 9A–B, 10C, 12B–D); females by the wide epigyne with a pair of large atria and the large spermathecae with a double chamber (Fig. 9F–G).

Type species

Acanthinozodium spinulosum Denis, 1966.

Remark

The types of *A. spinulosum* could not be traced but the description mentions two of the diagnostic characters of the genus: large AME and a double row of long ventral setae on the femora. Large AME are widely distributed among the taxa in the femoral organ clade but are important to differentiate the genus from *Zodariellum* that supposedly is a synonym (Jocqué 1991). The presence of dorsal spines on the femora is not diagnostic for the genus and also occurs in *Zodarion nitidum* (Audouin, 1826), the type species of *Zodarion* (Ono & Jocqué 1986; Levy 1992).

Description

Medium size spiders (2.8–5.7 mm) with smooth teguments. Carapace wide (L/W < 1.3) almost hairless apart from a few longer hairs on clypeus; widest at level of coxae II–III, narrowed to about 0.68 times maximum width in males and 0.55 times maximum width in females (cephalic width measured on posterior tangent of PME). Cervical grooves poorly indicated. Profile fairly flat, often with a slight dip between eyes and fovea.

Colour. Carapace orange to shades of yellow, suffused with dark pattern; chelicerae, legs, mouthparts and sternum orange to yellow; abdomen dorsum uniform dark, venter uniform grey, darker towards sides.

Eyes. In two rows: anterior row straight (eye centres), strongly procurved as seen from in front, posterior row strongly procurved. AME dark, much larger (two to four times) than remainder which are pale; AME less than half their diameter apart and as far from ALE; PME oval, three to five times their diameter apart and 1.5 to 3 times their diameter from PLE. MOQ as wide in back as in front and slightly wider than long. Clypeus convex, height about 1.5 times diameter of ALE, with some long dispersed setae.

Chilum. Absent. Chelicerae fused in the middle; intercheliceral triangle well developed; paturon dorsally covered with stout hairs; with distal cusp with one tooth; fang shorter than wide at base. Labium triangular, narrowed base. Endites roughly triangular, strongly converging. Sternum shield-shaped with rounded sides, provided with triangular extensions corresponding with coxal concavities. Slightly longer than wide.

Legs. Slender. Formula 4123 or 4132. Spination reduced but a few dorsal and sometimes prolateral spines on femora and ventral spines on tibiae. Femora with conspicuous double row of long ventral setae. No incised hairs on femora. Femoral organ with barbed hairs on legs I–III (Figs 9E, 12E).

Female palp. With pectinated claw turned inwards over nearly 90°, with distal patch of chemosensitive setae, modified prolateral, pectinated setae reduced or absent.

Abdomen. Oval; tracheal spiracle just in front of spinnerets; fairly wide, around 0.25 times maximum width of abdomen) preceded by row of modified flat setae (Figs 5F, 12E–F). Males with two spinnerets, females with six. ALS large, conical, biarticulate on common base. PLS and PMS very small. Colulus represented by haired field.

Male palp. Tibia with small prolateral apophysis apart from the well developed RTA. Cymbium with basolateral flange; with distal and prolateral smooth spines and dorsal patch of chemosensitive setae; prolateral modified pectinated setae reduced; base with deep crater-like pit; bulbus with hook-like median apophysis situated near distal tip; embolus flat, distal part with flange.

Distribution

Acanthinozodium is found in the semi-arid areas in and around the Sahara Desert from Senegal to Yemen.

Acanthinozodium crateriferum sp. nov.
urn:lsid:zoobank.org:act:D55F4819-EEB4-489B-9AD5-A52FA259E42B
Figs 1–3, 13

Diagnosis

The male of this species is recognized by the shape of the RTA, which is straight and strongly tapered towards the sharp tip (Figs 2B–C, 3A–B); the female has a wide epigyne with two triangular atria divided by a thin septum (Figs 1E, 3C), very different from that of *A. zavattarii* Caporiacco, 1941 in which the atria are covered and the septum invisible from the outside.

Etymology

The name refers to the large crater-like pit at the base of the cymbium.

Type material

Holotype

ETHIOPIA: ♂, Near Ras Hotel, Awash National Park, 8°59' N 40°10' E, 29 May 1986, Russell-Smith A. (MRAC 220817).

Fig. 1. *Acanthinozodium crateriferum* sp. nov. **A**. Holotype ♂. **B–E**. Paratype ♀. **A**. Male habitus, dorsal view. **B**. Female habitus, dorsal view. **C**. Idem, ventral view. **D**. Idem, lateral view. **E**. Epigyne, ventral view. Scale bars: A–D = 1 mm; E = 0.2 mm.

Paratypes

1 ♂, 1 ♀, same data as holotype (MRAC 243548).

Other material examined

1 ♀, 1 Oct. 1986, under piles of grass, further as holotype (MRAC 220696); 1 ♂, 1 ♀, 9 Nov. 1988, in grass and gravel, further as previous (MRAC 224602).

Description

Male

TOTAL LENGTH. 3.00.

CARAPACE. (Fig. 1A). 1.56 long and 1.21 wide.

COLOUR. (Fig. 1A). Carapace and chelicerae yellow; cephalic part with dense, brown suffusion forming butterfly pattern in front of dark red fovea; sternum and mouthparts pale yellow; legs yellow, anterior pairs with darker distal half of femora. Abdomen: Dorsum dark sepia with white spot in front of pale yellow spinnerets, the latter surrounded by ill-defined dark ring; sides turning into pale along pale cream venter.

CLYPEUS. 0.21 high.

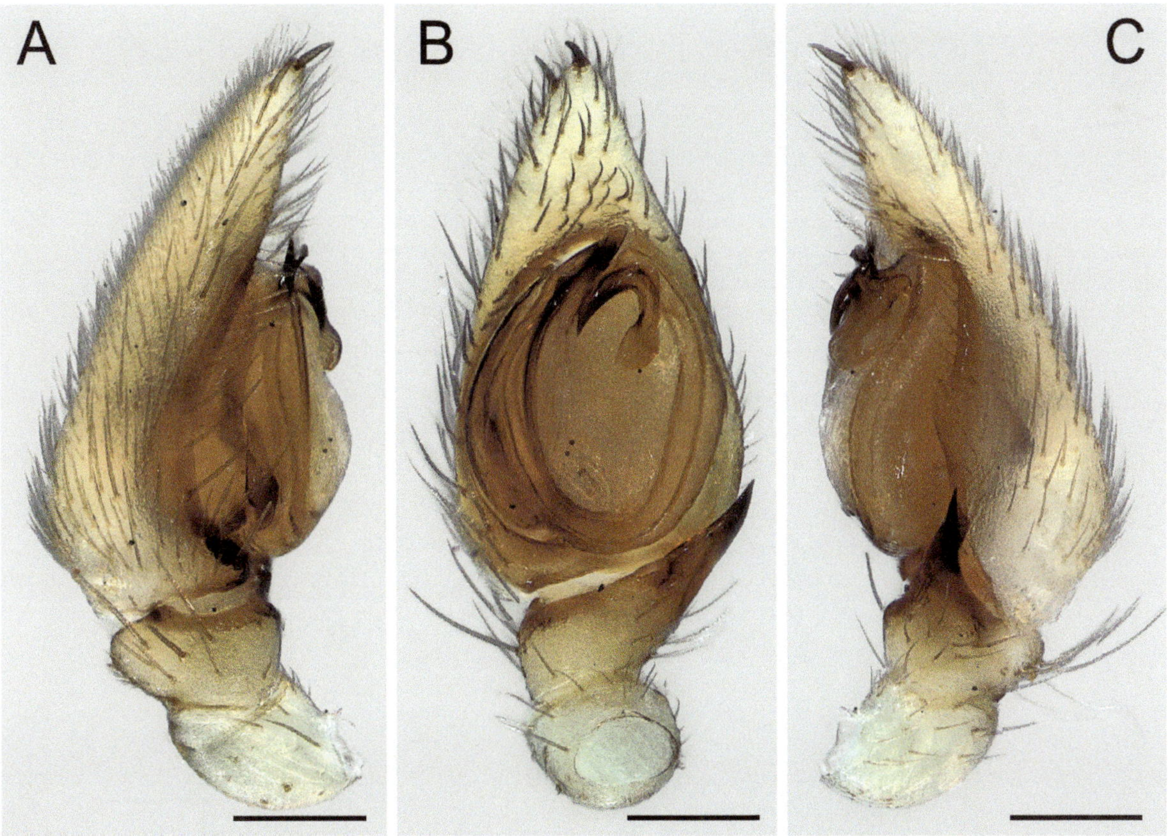

Fig. 2. *Acanthinozodium crateriferum* sp. nov., holotype ♂, palp. **A**. Prolateral view. **B**. Ventral view. **C**. Retrolateral view. Scale bars = 0.2 mm.

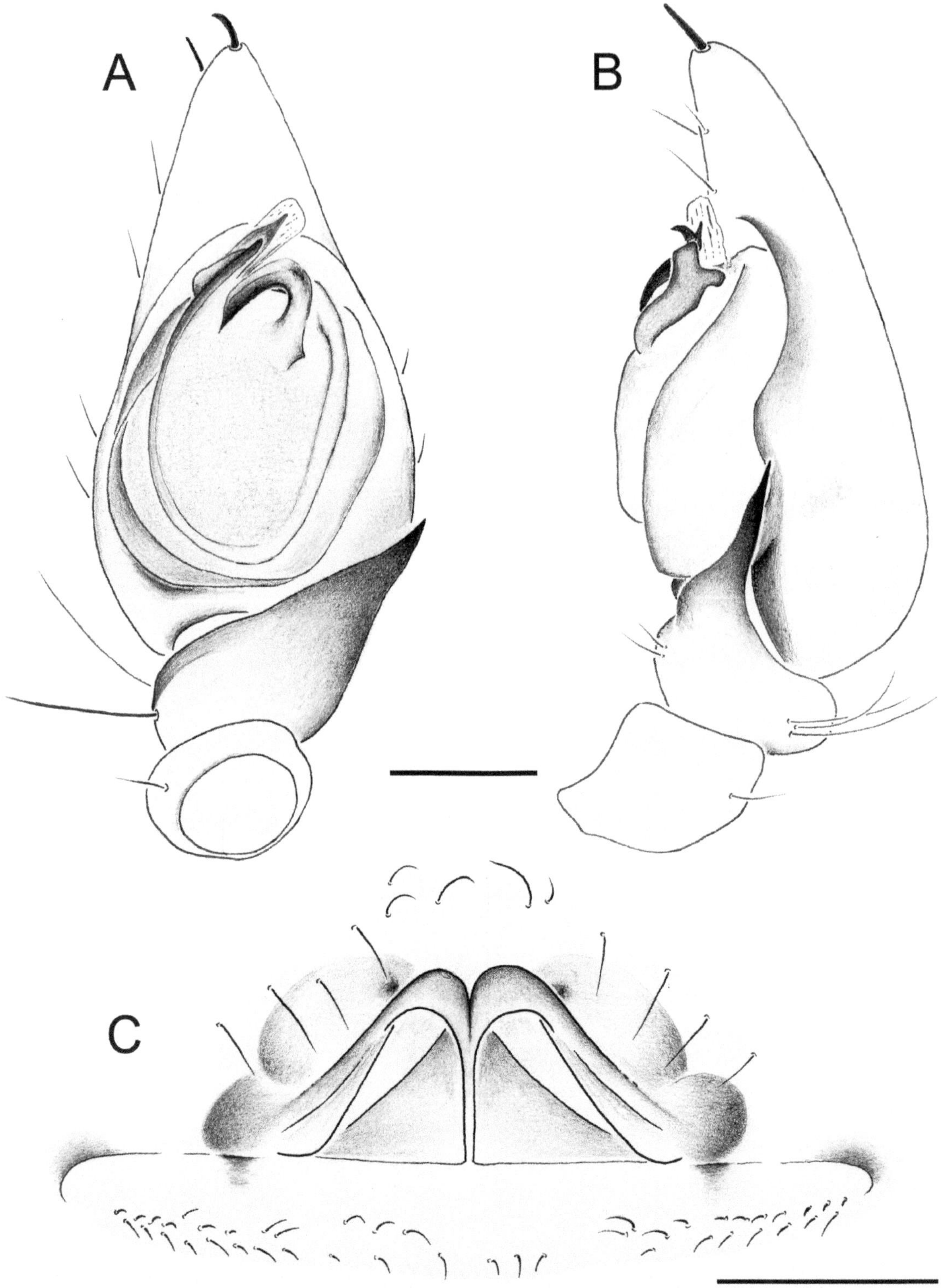

Fig. 3. *Acanthinozodium crateriferum* sp. nov. **A–B**. Holotype ♂. **C**. Paratype ♀. **A**. Male palp, ventral view. **B**. Idem, retrolateral view. **C**. Epigyne, ventral view. Scale bars = 0.2 mm.

EYES. AME large, round; anterior eye row straight, posterior row procurved; AME: 0.15; ALE: 0.10; PME: 0.08; PLE: 0.10; AME-AME: 0.08; AME-ALE: 0.02; PME-PME: 0.21; PME-PLE: 0.03. MOQ: width 0.39 in front, 0.34 at the back; 0.33 long.

STERNUM. 0.85 long, 0.78 wide.

LEGS. Spination: all femora with two dorsal spines and two rows of long ventral setae.

ABDOMEN. With row of thin modified setae in front of spinnerets.

LEG MEASUREMENTS.

Leg	F	P	T	Mt	t	Total
I	1.75	0.57	1.49	1.78	1.11	6.65
II	1.65	0.57	1.28	1.63	0.99	6.10
III	1.56	0.57	1.28	1.78	0.92	6.11
IV	2.13	0.50	1.85	2.34	0.99	7.81

PALP. (Figs 2, 3A–B). Tibia protruding dorsally; RTA straight, tapered to sharp tip; cymbium with large plug-pit just in front of tibial bulge; embolus reaching distal margin of bulbus with flat distal part; MA sickle-shaped with distal bulge; conductor rounded, membranous.

Female

TOTAL LENGTH. 3.98.

CARAPACE. (Fig. 1B). 1.85 long and 1.42 wide.

COLOUR. (Fig. 1B–D). Similar to male, but all femora distally darkened.

CLYPEUS. 0.25 high.

EYES. AME: 0.16; ALE: 0.10; PME: 0.08; PLE: 0.10; AME-AME: 0.08; AME-ALE: 0.02; PME-PME: 0.23; PME-PLE: 0.03. MOQ: width 0.44 in front, 0.38 at the back; 0.34 long.

STERNUM. 0.89 long, 0.82 wide.

LEG MEASUREMENTS.

Leg	F	P	T	Mt	t	Total
I	1.63	0.71	0.99	1.56	1.07	5.96
II	1.70	0.64	1.14	1.49	0.92	5.89
III	1.56	0.64	1.07	1.35	0.78	5.40
IV	2.06	0.64	1.70	2.13	0.85	7.38

EPIGYNE. (Figs 1E, 3C). Much wider than long; with two triangular atria divided by thin median septum; spermathecae visible in transparency along distal side of triangle.

Variation

Total length of paratype male 2.8. Carapace colour may vary from bright orange to yellowish orange.

Distribution

Known only from the type locality (Fig. 13).

Acanthinozodium sahelense sp. nov.
urn:lsid:zoobank.org:act:01A0026D-5A5E-4CCD-A6CE-241708236D18
Figs 4–9, 13

Diagnosis

The male of this species is recognized by the shape of the RTA, which is about as long as the length of the tibia, slightly curved down and dorsoventrally flattened; it has a typical large ventral bulge; extremity of the MA is slightly indented (Fig. 8A). Females are characterized by the epigyne (Figs 5G–H, 6C) in which the three sections of the frontal margin are strongly curved.

Etymology

The name refers to the wide area of distribution in the Sahel zone of western Africa.

Type material

Holotype

CAMEROON: ♂, Waza, 11°25' N 14°34' E, 28 Aug. 1972, Geraert E. (MRAC 144790).

Fig. 4. *Acanthinozodium sahelense* sp. nov., holotype ♂. **A**. Habitus, dorsal view. **B**. Idem, ventral view. **C**. Idem, frontal view. **D**. Idem, lateral view. **E**. Idem, dorsolateral view. **F**. Carapace, eye region, dorsal view. Scale bars: A–E = 1 mm; F = 0.5 mm.

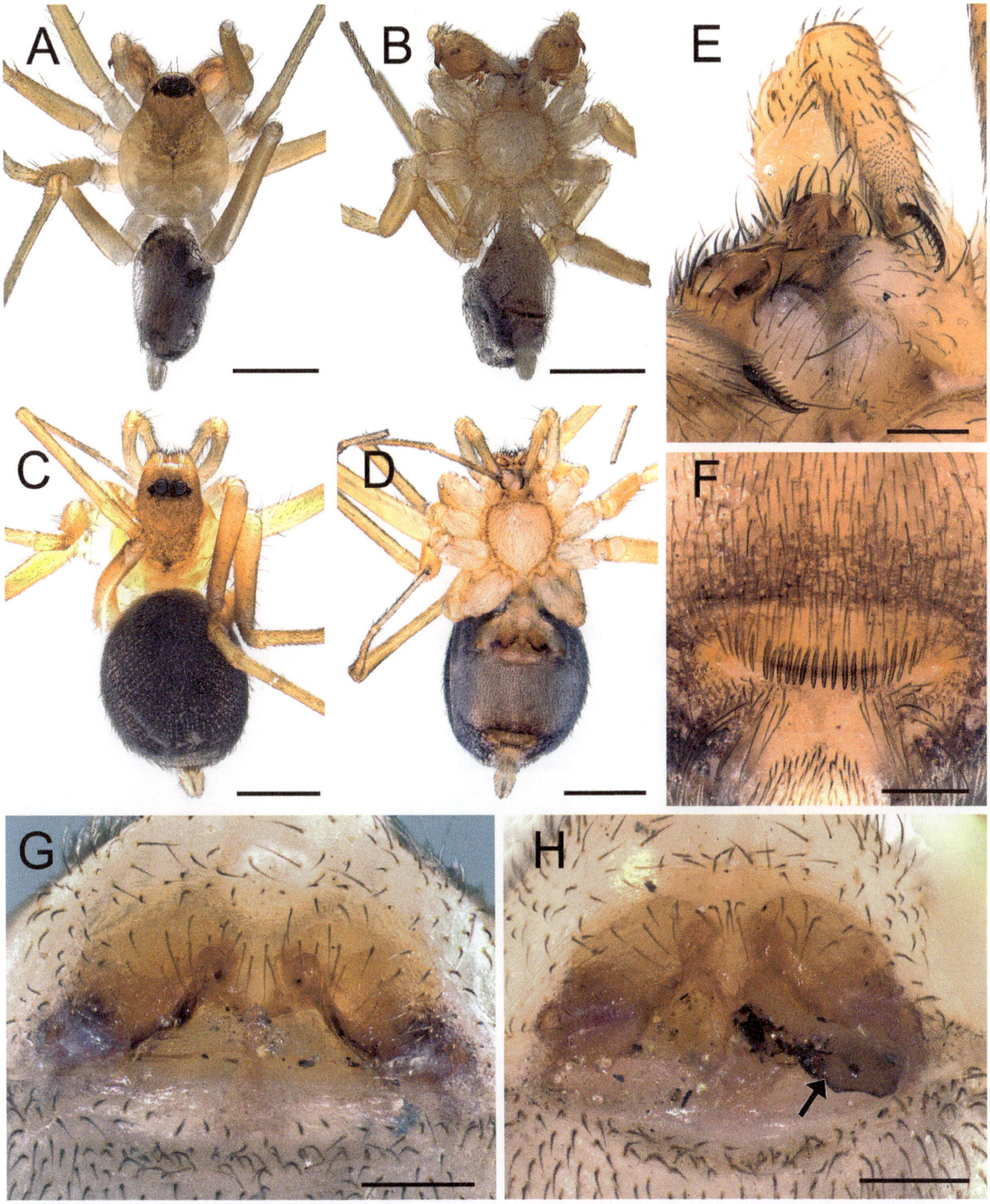

Fig. 5. *Acanthinozodium sahelense* sp. nov. **A–B**. ♂ from Faro Game Reserve (MRAC 221205). **C–H**. ♀♀ from same sample. **A**. Male habitus, dorsal view. **B**. Idem, ventral view. **C**. Female habitus, dorsal view. **D**. Idem, ventral view. **E**. Female, mouthparts and palps, ventral view. **F**. Female, abdomen, modified setae in front of tracheal spiracle, ventral view. **G**. Epigyne, ventral view. **H**. Epigyne, other specimen; arrow indicates plug. Scale bars: A–D = 1 mm; E–F = 0.2 mm.

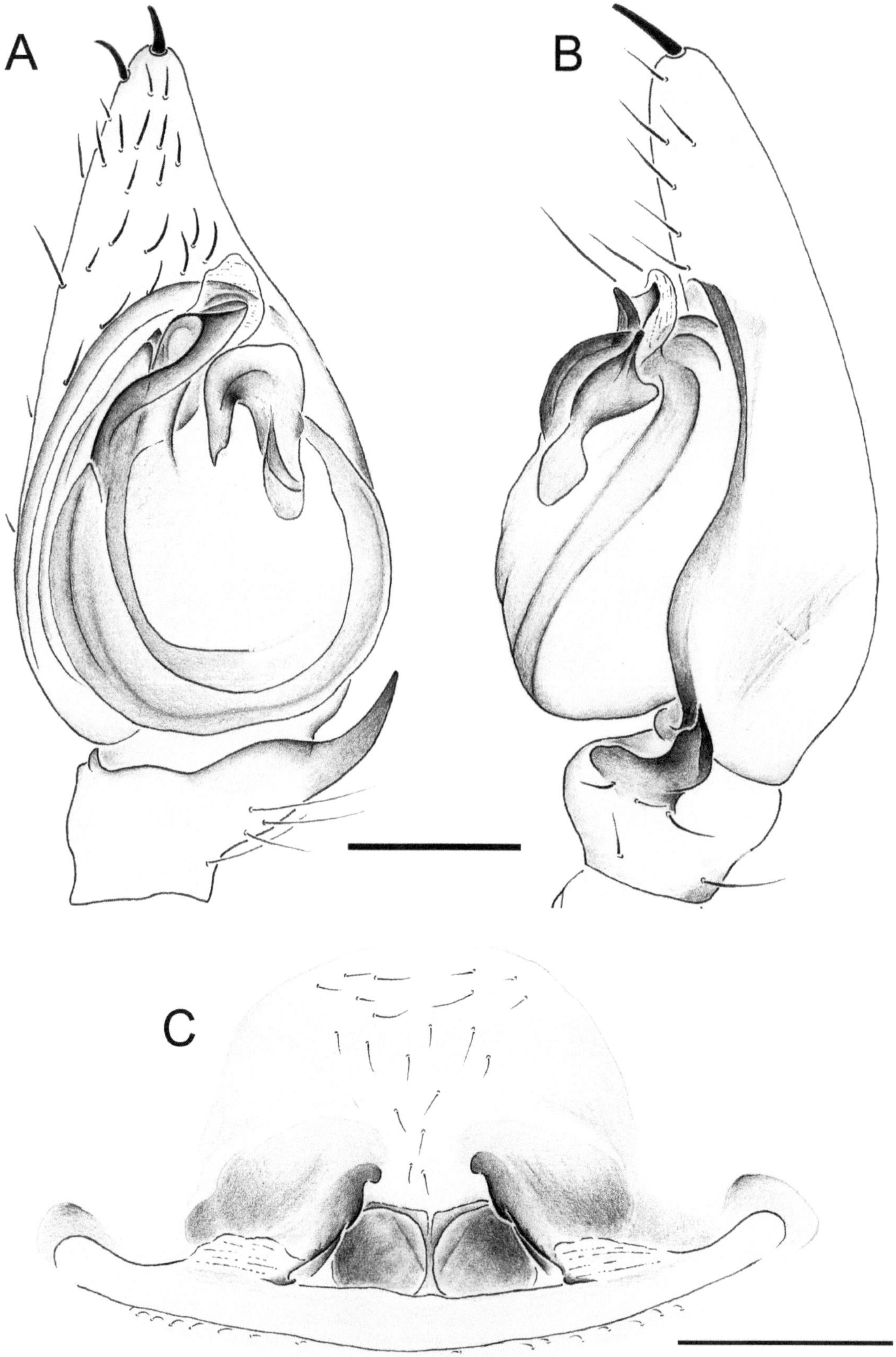

Fig. 6. *Acanthinozodium sahelense* sp. nov. **A–B**. Holotype ♂. **A**. Palp, ventral view. **B**. Idem, retrolateral view. **C**. ♀ from Faro Game Reserve (MRAC 221205), epigyne, ventral view. Scale bars = 0.2 mm.

Other material examined

CAMEROON: 3 ♂♂, 4 ♀♀: Faro Game Reserve, 8°24' N 12°49'E, 4 May 2007, wooded savanna, pitfall, Jocqué R., Loosveldt K., Baert L., Alderweireldt M. (MRAC 221205); 2 ♀♀: as previous, 1 May 2007 (MRAC 221147); 1 ♀: as previous, 5 May 2007, gallery forest, pitfall (MRAC 221183). IVORY COAST: 1 ♂, 1 ♀: N of Korhogo, Bandama river, 9°27' N 5°38' W, 4 Feb. 1980, edge of riverine forest, Everts J. (MRAC 174271); 1 ♂: Bouaké, Mbé, WARDA The Africa Rice Centre, 7°52' N 5°06' W, 22 Aug. 1992, pitfall trap, hydromorphic rice, Russell-Smith A. (MRAC 235937). BURKINA FASO: 1 ♂: Sissamba, 13°30' N 2°28' W, 19 Oct. 1993, pitfalls, treated C1, De Visscher M.-N. & Balança G. (MRAC 207576). SENEGAL: 1 ♀: S. of Richard Toll, 16°20' N 15°30' W, 28 Aug. 1989, pitfall, savanna, Everts J. (MRAC 170712); 1 ♀: as previous, 20 Aug. 1989 (MRAC 170716); 2 ♂♂: as previous, 2 Sep. 1989 (MRAC 170719).

Description

Male

TOTAL LENGTH 3.55.

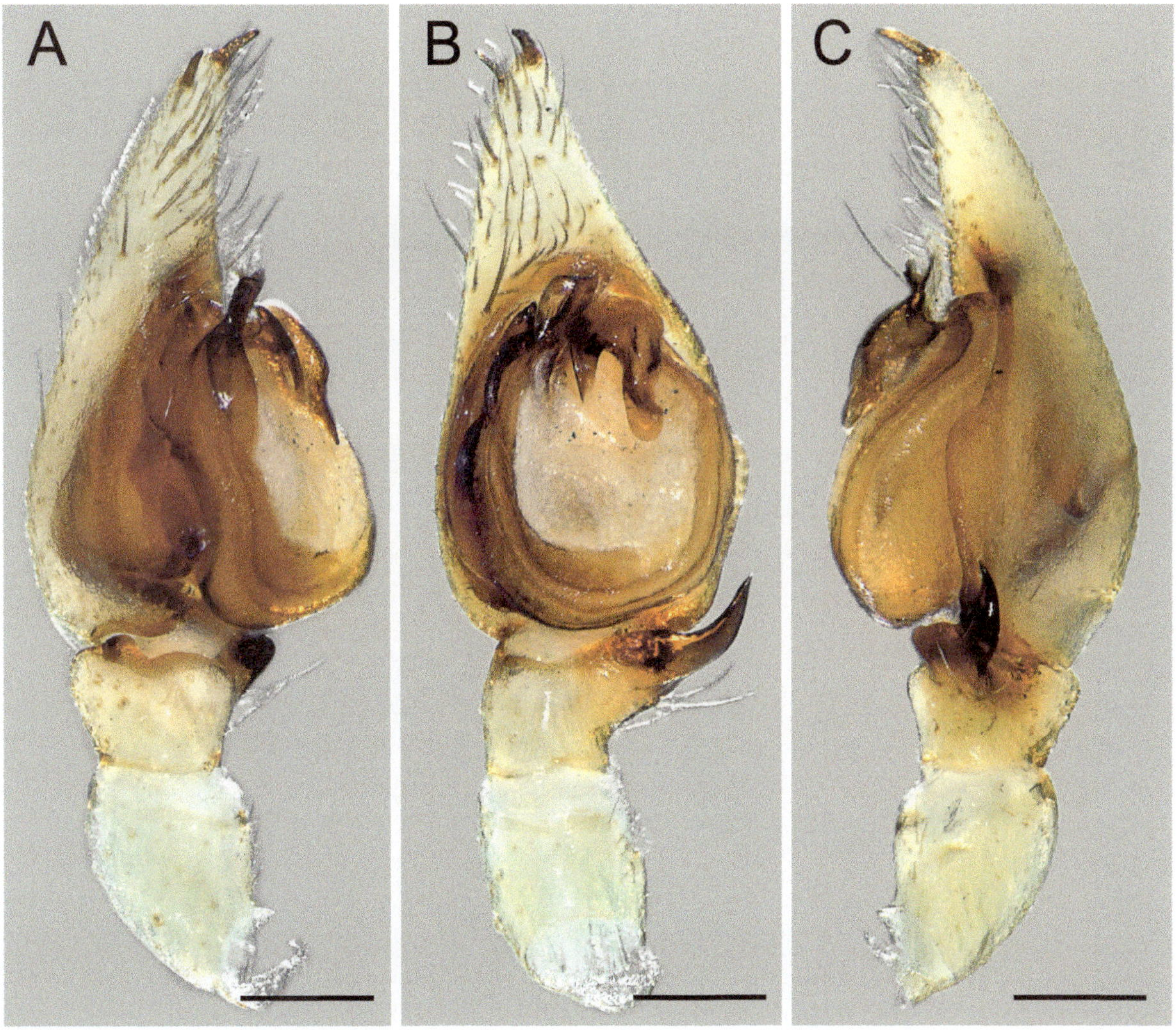

Fig. 7. *Acanthinozodium sahelense* sp. nov., holotype ♂, palp. **A**. Prolateral view. **B**. Ventral view. **C**. Retrolateral view. Scale bars = 0.2 mm.

Carapace. (Figs 4A, C–E, 5A–B). 1.85 long and 1.49 wide.

Colour. (Figs 4, 5A–B). Carapace and chelicerae yellow; brown suffusion on cervical grooves; fovea dark reddish brown; sternum and mouthparts pale yellow; legs uniform yellow. Abdomen: dorsum dark sepia with large white spot in front of pale yellow spinnerets, the latter surrounded by ill-defined dark ring; sides turning into pale along pale cream venter.

Clypeus. 0.31 high.

Eyes. (Fig. 4F). AME large, round; anterior eye row straight, posterior row strongly procurved; AME: 0.20; ALE: 0.10; PME: 0.10; PLE: 0.08; AME-AME: 0.10; AME-ALE: 0.03; PME-PME: 0.20; PME-PLE: 0.07. MOQ: width 0.49 in front, 0.44 at the back; 0.39 long.

Sternum. 1.42 long, 0.99 wide.

Legs. Spination: all femora with one dorsal spine and two rows of long ventral setae.

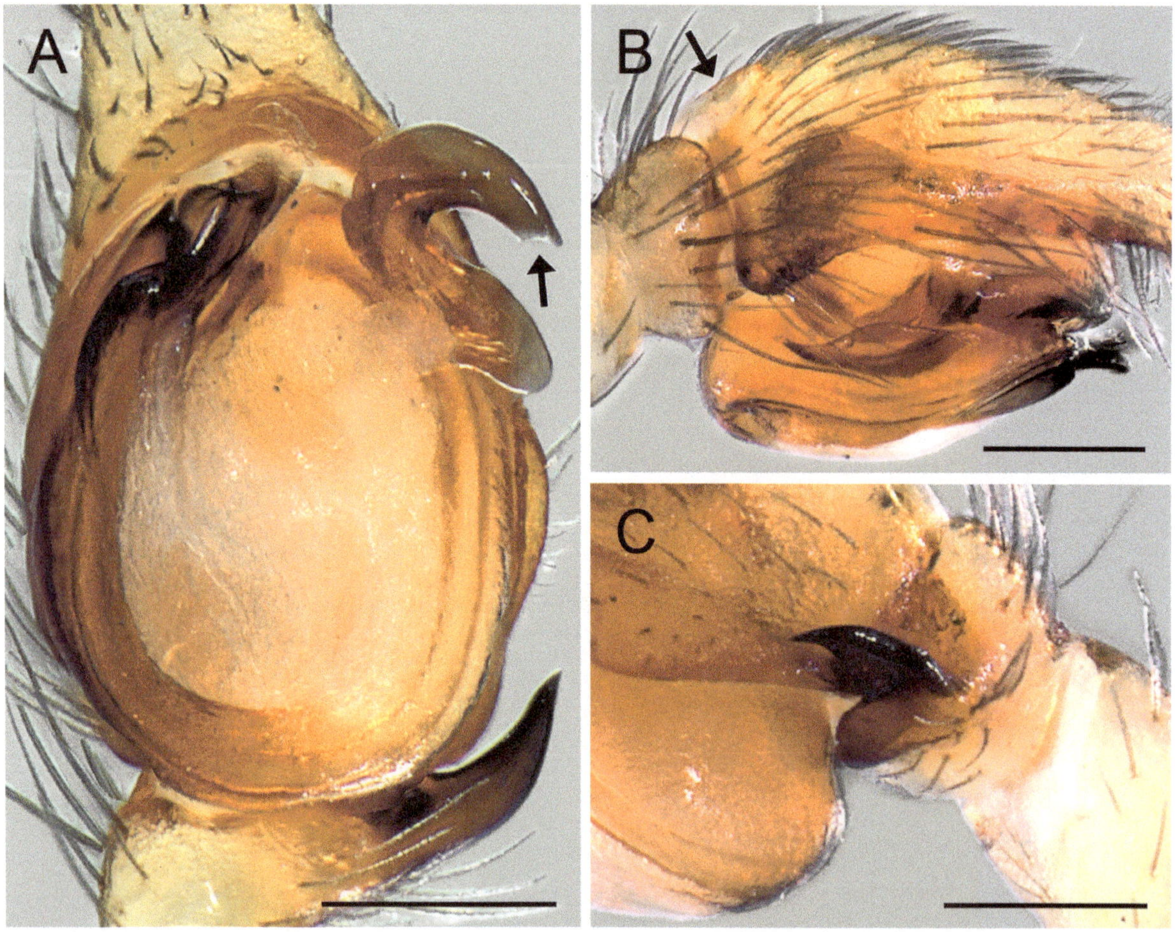

Fig. 8. *Acanthinozodium sahelense* sp. nov., ♂ from Faro Game Reserve (MRAC 221205), palp. **A**. Detail of bulbus with MA in expanded position, ventral view; arrow indicates the indentation of the MA tip. **B**. Tibia and base of bulbus, prolateral view; arrow indicates position of the crater-like plug pit. **C**. Retrolateral view. Scale bars = 0.2 mm.

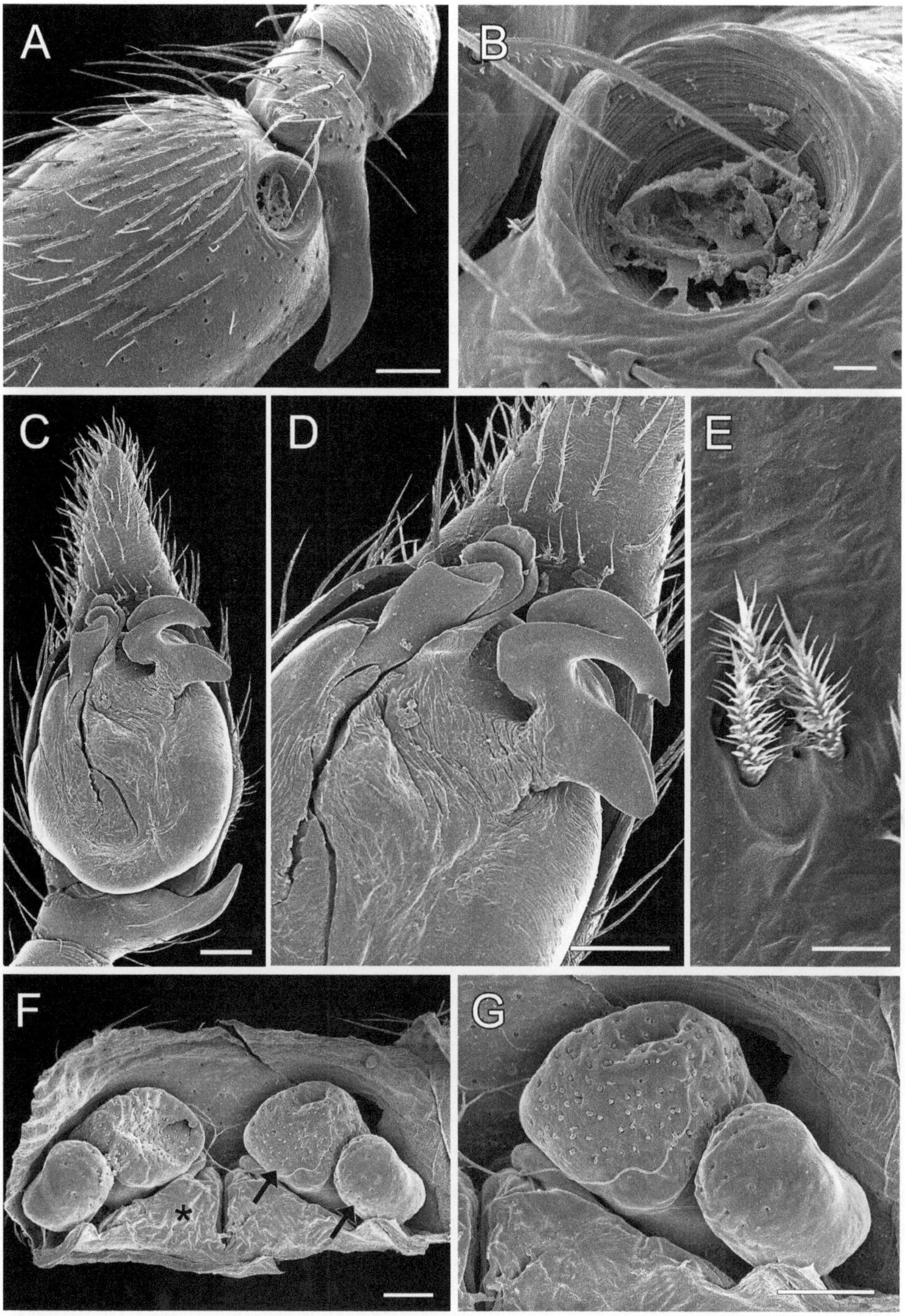

Fig. 9. *Acanthinozodium sahelense* sp. nov. SEM photos of ♂ and ♀ from Faro Game Reserve (MRAC 221183). **A**. Male palp, base of cymbium, dorsal view. **B**. Detail of previous showing "plug pit", dorsal view. **C**. Male palp, ventral view. **D**. Detail of previous, showing MA in expanded position. **E**. Male, femoral organ. **F**. Epigyne, after digestion, dorsal view; arrows indicate the two spermatheca chambers; star indicates the atria. **G**. Detail of previous showing spermatheca with two chambers. Scale bars: A, C–D, F–G = 100 μm; B, E = 10 μm.

ABDOMEN. With row of thin modified setae in front of spinnerets.

LEG MEASUREMENTS.

Leg	F	P	T	Mt	t	Total
I	2.66	0.63	2.17	2.10	1.47	9.03
II	2.10	0.70	1.54	2.17	1.33	7.84
III	2.03	0.63	1.40	2.10	1.26	7.42
IV	2.10	0.84	2.17	2.80	1.96	9.87

PALP. (Figs 6A–B, 7–8, 9A–D). Tibia protruding dorsally; RTA about as long as length of tibia, slightly curved down and dorso-ventrally flattened; with conspicuous ventral bulge; cymbium with large plug-pit just in front of tibial bulge; embolus curved, flat, distal part with dorsal and ventral lamellae; MA strongly curved backward, distal extremity slightly indented; conductor rounded, membranous.

Female

TOTAL LENGTH. 5.04.

CARAPACE. (Fig. 5C) 2.34 long and 1.85 wide.

COLOUR. (Fig. 5C–F). Similar to male, but cephalic area more strongly suffused with dark brown.

CLYPEUS. 0.26 high.

EYES. AME: 0.20; ALE: 0.10; PME: 0.08; PLE: 0.10; AME-AME: 0.10; AME-ALE: 0.03; PME-PME: 0.33; PME-PLE: 0.08. MOQ: width 0.46 in front, 0.43 at the back; 0.39 long.

STERNUM. 1.28 long, 1.07 wide.

LEG MEASUREMENTS.

Leg	F	P	T	Mt	t	Total
I	2.52	0.70	2.03	2.80	1.54	9.59
II	2.45	0.70	1.89	2.66	1.19	8.89
III	1.68	0.70	1.82	2.87	0.98	8.05
IV	2.94	0.77	2.52	2.80	1.19	10.22

EPIGYNE. (Figs 5G–H, 6C). Wider than long; with two atria divided by thin median septum; anterior atria margin with three sections, one central, two lateral; all curved; spermathecae visible in transparency at extremities of central section.

Variation

Intensity of suffusion in cephalic area is variable. Total length of females varies between 3.4 and 5.1; that of males between 3.3 and 3.5.

Distribution

Very wide, from northern Cameroon in the east, to Senegal in the west (Fig. 13).

Acanthinozodium quercicola sp. nov.

urn:lsid:zoobank.org:act:917D7A4F-2400-416F-8BC2-08013EB70E44

Figs 10–11, 13

Diagnosis

The male of this species is characterized by the shape of the sinuous, by dorsolaterally flattened RTA and by the MA with dorsal bulge.

Etymology

The name refers to the habitat where the types were found: *Quercus suber* woodland.

Material examined

Holotype

MOROCCO: ♂, Oulmès, 33°26" N 6°1'0" W, 4 Nov. 2011, sieving leaf litter of *Quercus suber*, Brustel H. (MRAC 243549).

Paratype

1 ♂: same data as holotype (MRAC 243549).

Description

Male

TOTAL LENGTH. 3.91.

CARAPACE. (Fig. 10A). 1.92 long and 1.49 wide.

COLOUR. (Fig. 10A–B). Carapace and chelicerae brownish yellow; clypeus, around eyes and cephalic area with dark brown suffusion forming butterfly pattern in front of black fovea; thoracic area with darker striae; mouthparts pale yellow; sternum yellow with dark margin; legs brownish yellow. Abdomen: dorsum dark sepia, no pale spot in front of pale cream spinnerets; sides and venter pale sepia.

CLYPEUS. 0.20 high.

EYES. AME large, round; anterior eye row straight, posterior row strongly procurved; AME: 0.15: ALE: 0.10; PME: 0.08; PLE: 0.08; AME-AME: 0.08; AME-ALE: 0.02; PME-PME: 0.20; PME-PLE: 0.05. MOQ: width 0.34 in front, 0.46 at the back; 0.31 long.

STERNUM. 0.99 long, 0.85 wide.

LEGS. Spination: femora: I with 3 dorsal 1 prolateral, II with 2, III and IV with 1 dorsal spine; two rows of long ventral setae.

ABDOMEN. With row of modified flat setae in front of spinnerets preceded by dense cover of flattened setae.

LEG MEASUREMENTS:

Leg	F	P	T	Mt	t	Total
I	1.89	0.70	1.75	2.03	1.19	7.56
II	1.89	0.70	1.61	1.75	1.05	7.00
III	1.61	0.70	1.26	1.96	0.84	6.37
IV	2.24	0.77	2.10	2.31	0.91	8.33

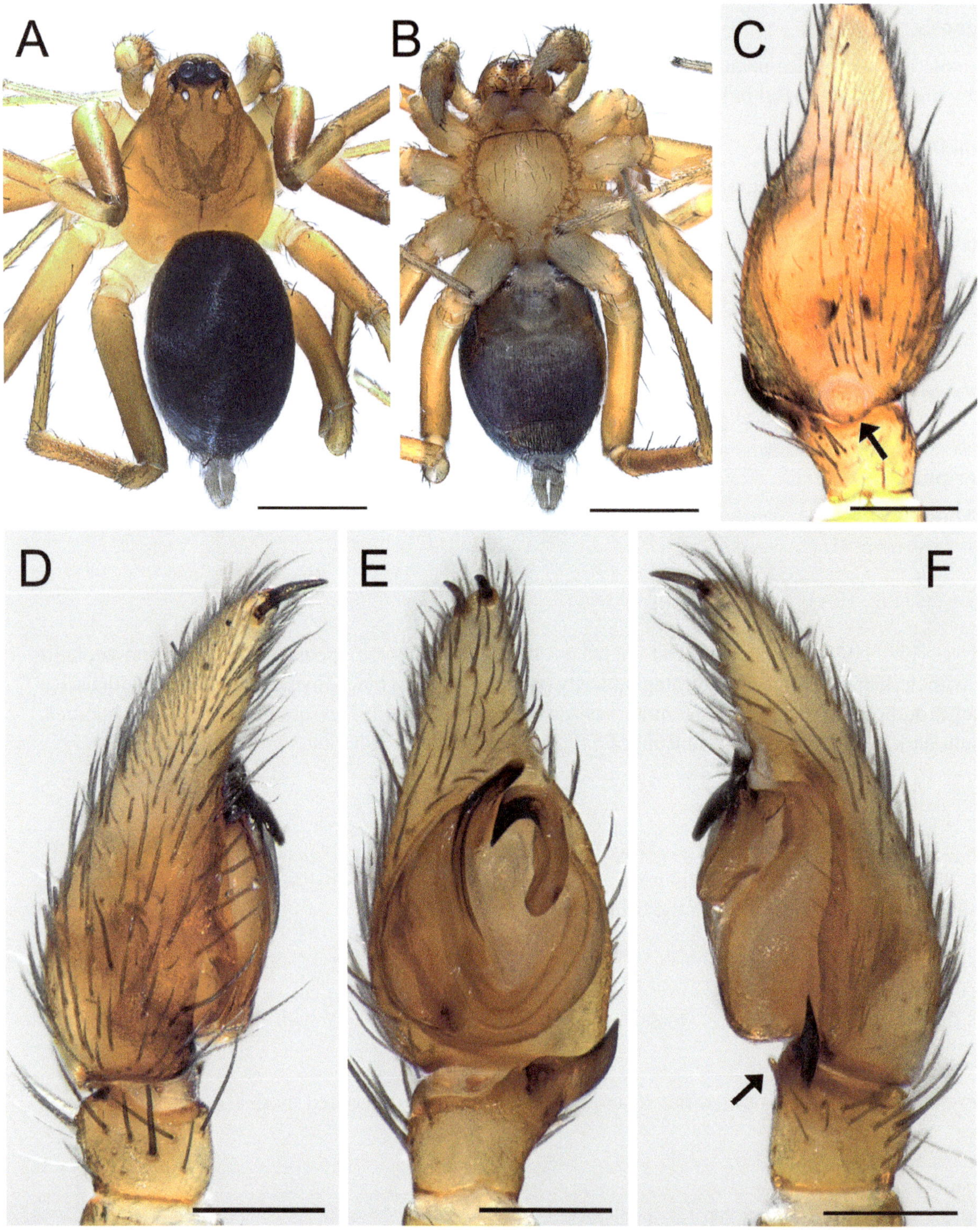

Fig. 10. *Acanthinozodium quercicola* sp. nov., holotype ♂. **A**. Habitus, dorsal view. **B**. Idem, ventral view. **C**. Palp, dorsal view; arrow indicates the crater-like plug pit. **D**. Idem, prolateral view. **E**. Idem, ventral view. **F**. Idem, retrolateral view; arrow indicates the tooth at base of the RTA. Scale bars: A–B = 1 mm; C–F = 0.2 mm.

Palp. (Figs 10C–F, 11). Tibia slightly produced forward dorsally; RTA short, dorsoventrally flattened, sinuous in ventral view; cymbium with dorsal plug pit in front of tibial protrusion; embolus broad at base, tapered to rounded tip with dorsal and ventral lamellae; MA fairly long, sickle shaped, with strong dorsal boss; conductor membranous, rounded.

Female

Unknown.

Distribution

Only known from the type locality in Morocco (Fig. 13).

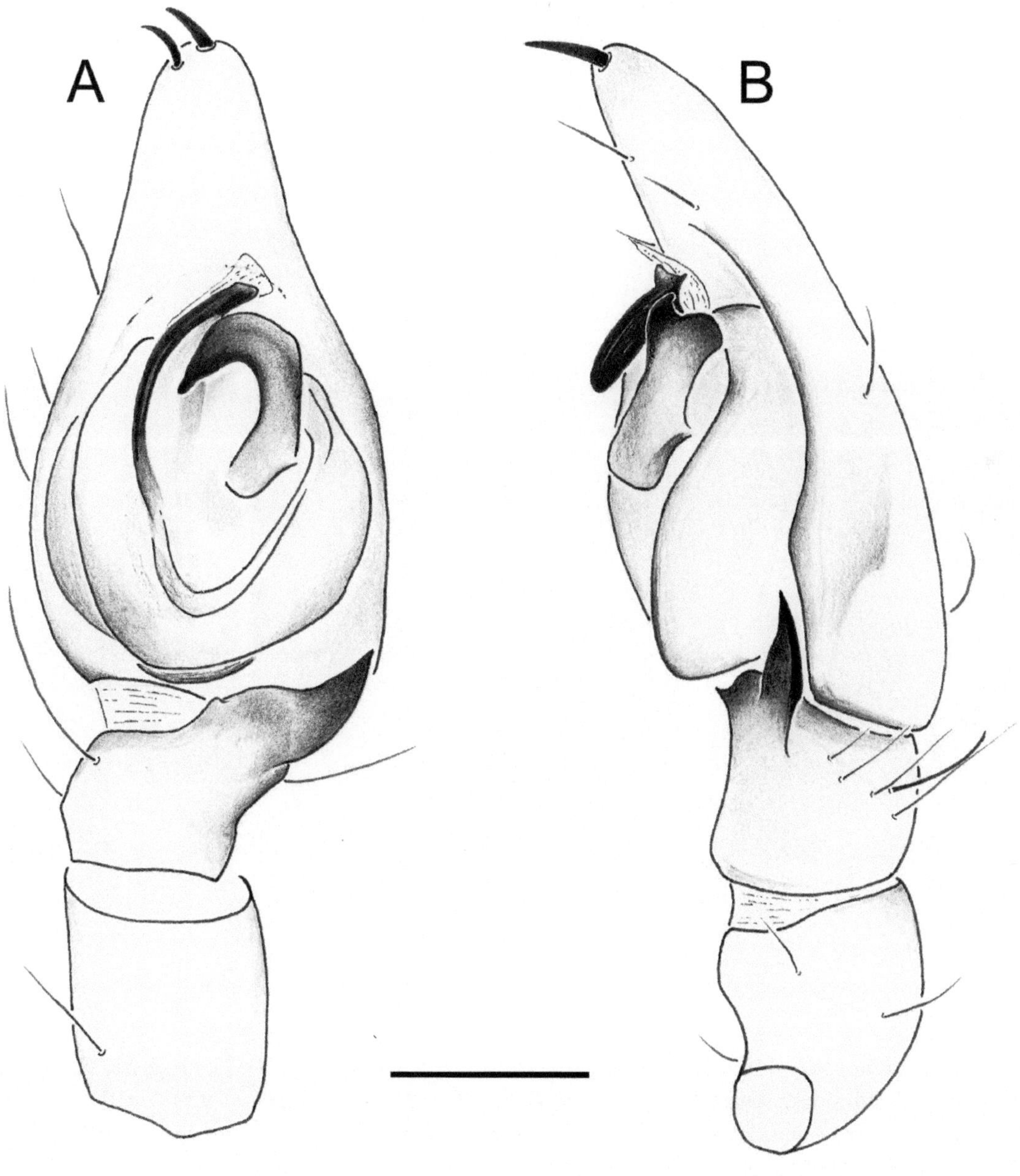

Fig. 11. *Acanthinozodium quercicola* sp. nov., holotype ♂, palp. **A**. Ventral view. **B**. Retrolateral view. Scale bar = 0.2 mm.

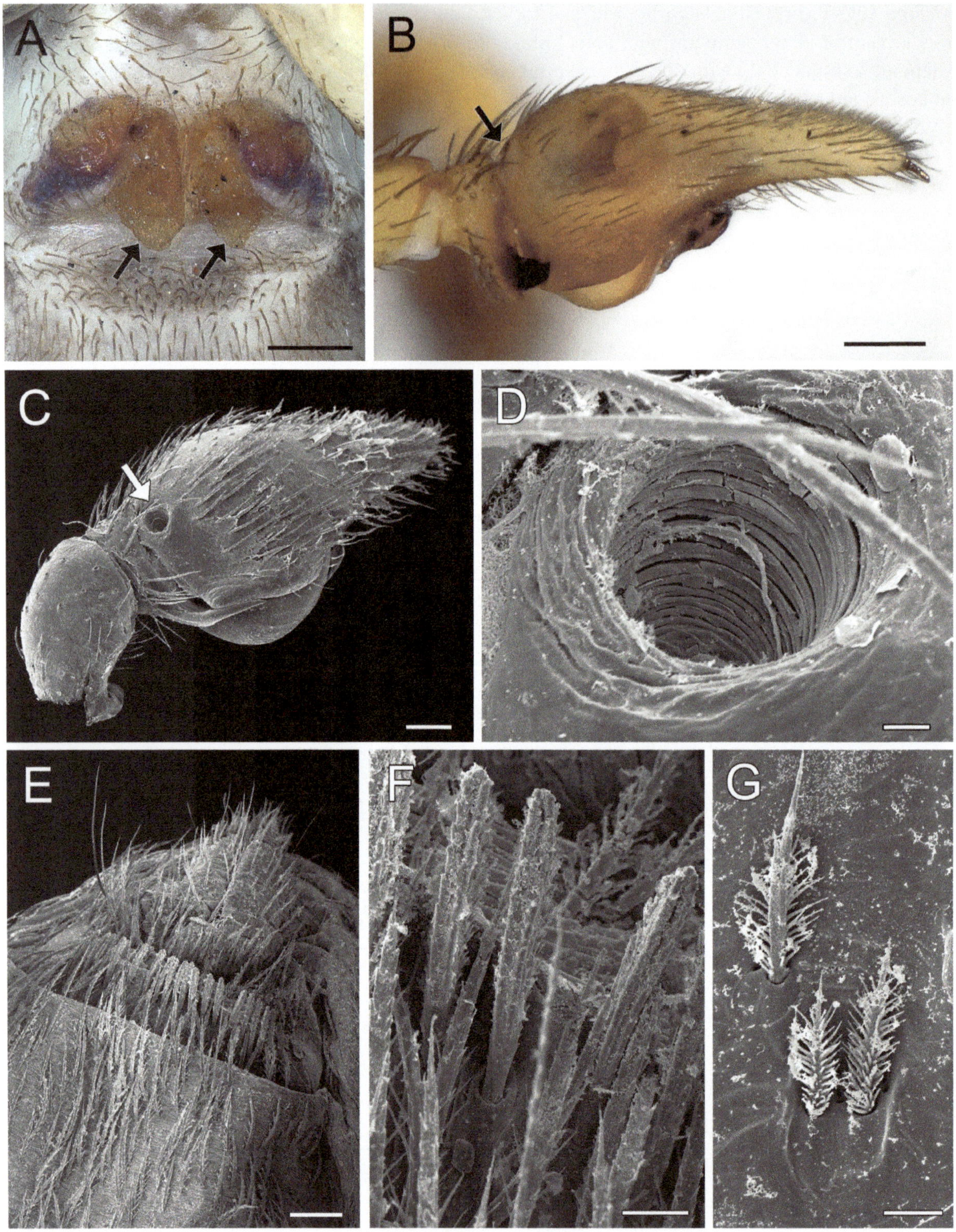

Fig. 12. A. *Acanthinozodium tibesti* Jocqué, 1991, holotype ♀ (MRAC 133011), epigyne, ventral view; arrows indicate both sides plugged. **B**. *Acanthinozodium sahariense* Denis, 1959 (CRB), male palp, dorsolateral view; arrow indicates the crater-like plug pit. **C**. *Acanthinozodium sericeum* Denis, 1956 (CRB), male palp, dorsolateral view; arrow shows the crater-like plug pit. **D**. Idem, detail of previous. **E**. Same specimen, abdomen, ventral view. **F**. Idem, detail of previous. **G**. Same specimen, femoral organ on leg II. Scale bars: A–B = 0.2 mm; C, E = 100 μm; F = 20 μm D, G = 10 μm.

Discussion

The nomenclatural problem

In 1952 J. Denis described (in French) three species in a genus "*Acanthinozodium*" without designating a type species. He did not do this because the publication in which he wanted to describe the genus was greatly delayed and only appeared in 1966. In the latter paper he described *Acanthinozodium spinulosum* and mentioned that this should have been the type species: "Le genre *Acanthinozodium* est intermédiaire entre les *Diores* et les *Zodarium*: son type aurait dû être *A. spinulosum* : mais par suite du retard apporté à la publication de cette étude, j'ai été amené à en donner une description très sommaire (in the 1952 paper)" (our remark in brackets).

In his catalogue, Brignoli (1983) regarded the name *Acanthinozodium* as available, with the type species *A. spinulosum*, because the paper of Denis (1966) could be recognized as the first "revision" of the genus thereby establishing its availability. In a revision of the Zodariidae, Jocqué (1991) also considered *Acanthinozodium* as available and valid, and designated the first species described in the 1952 paper (*A. cirrisulcatum* Denis, 1952) as the type species. In 2001 Marusik & Koponen considered *Acanthinozodium* unavailable because no type species was designated at the first publication of the name. As a result, all species regarded as congeneric with *Acanthinozodium* were transferred to *Zodariellum* Andreeva & Tyshchenko, 1968 (Platnick 2002, see also the World Spider Catalog 2015). We here use "Tyshchenko" spelled as in this late author's most recent publications (see Marusik & Koponen 2001).

However, before the intended type species *A. spinulosum* was described by Denis, 1966, no other authors used the name, but Denis published three other papers with *Acanthinozodium* before that year (Denis 1952, 1956 and 1959). The first one contained two species and one subspecies of the genus, the second only one: *A. sericeum* Denis, 1956, the third one three. The case is complicated because, if Denis

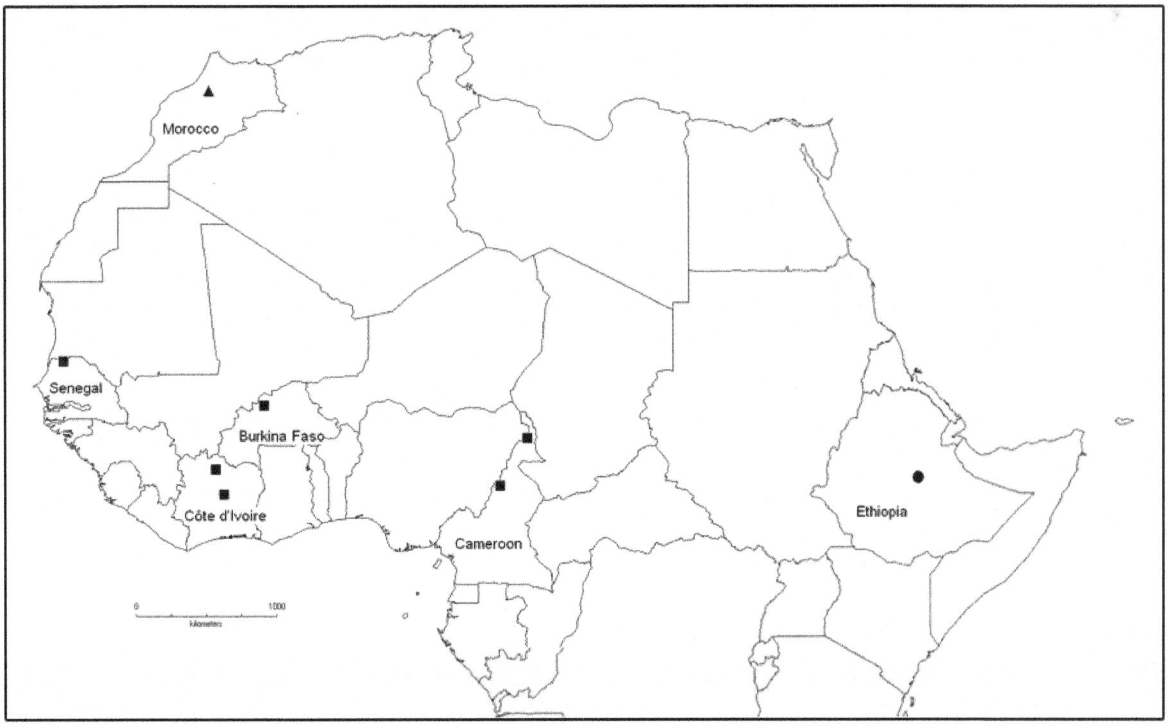

Fig. 13. Distribution map of *A. crateriferum* sp. nov. (●), *A. sahelense* sp. nov. (■) and *A. quercicola* sp. nov. (▲).

(1956) had used a reference to the new genus in 1956, *A. sericeum* would have been the type species by monotypy and combined description. But he did so in the 1966 publication in which the description of the intended type species *A. spinulosum* was finally published. By that action *Acanthinozodium* becomes available and valid, and its type species is *A. spinulosum* Denis, 1966. This is confirmed in the catalogue by Brignoli (1983) who also stipulated that the 1966 paper of Denis makes the name available.

The vast majority of the comments (*in litteris*) about this case agree that *Acanthinozodium* is a valid genus. However, there is no complete unanimity as to whether the type species should be the first species described by Denis in 1952 (*A. cirrisulcatum*), as argued by Jocqué (1991), or the species Denis selected and confirmed as his choice in 1966 (*A. spinulosum*). The main argument to prioritize the former is that the selection of *A. spinulosum* would jeopardize the stability of the name *Acanthinozodium*. Since the name was not used by any other author before 1966 and possible synonyms were only created thereafter, we regard this argument as invalid. A second argument is "priority", but rules in this connection only concern synonymies. We therefore consider *A. spinulosum* the type species with the following rationale:

As no type species was proposed in 1952, the genus name is not available from Denis, 1952, according to the *International Code of Zooological Nomenclature* (International Commission on Zoological Nomenclature 1999, Article 13.3). The first time a type species was proposed – in 1966 – Denis designated *A. spinulosum* as the type species. Assuming that Denis included a statement that would satisfy Article 13.1 (ICZN 1999), then the correct name is *Acanthinozodium* Denis, 1966. There is no requirement for Denis to have nominated one of the originally included species as the type species. Also, Article 11.9.3.1 guarantees that any species described in *Acanthinozodium* prior to 1966 is validly formed, despite the genus-group name being not available until Denis (1966).

A further complication of this case is the synonymy with *Zodariellum* that was established by Jocqué (1991). The paper by Andreeva & Tyshchenko (1968), in which the latter genus was described, is troubling as it contains several inconsistencies (see also Marusik & Koponen 2001). The authorship of the type species of *Zodariellum* changes from Tyshchenko in the text to Andreeva & Tyshchenko in the abstract. The heading of the new genus is "*Lodariellum*" instead of *Zodariellum*. These mistakes have little bearing on the use of the new genus name and its possible synonymy with *Acanthinozodium*, but the following has. *Zodarion continentalis* (Andreeva & Tyschchenko, 1968) (which should be "*continentale*") was mentioned in the English abstract as belonging to "the three new *Zodarion*" (there mentioned as "*Zodarium*") species in the paper but described in between the *Zodarion* species as *Zodariellum* in the text, although the single illustration for that species (fig. 5) is attributed to "*Z. continentalis*" among the other *Zodarion* species.

The core of the paper, the particular characters of the type species of *Zodariellum*, its unusual small AME (fig.7 in Andreeva & Tyshchenko 1968), the size and position of the eyes in the posterior row and the flat carapace, apparently escaped the attention both of Jocqué (1991) and Marusik & Koponen (2001). The confusion in the Russian (Cyrillic) paper and the very superficial description of the genus with only a lateral view of the habitus and a dorsal view of the male palpal tibia, may explain these mistakes. *Acanthinozodium* can indeed not be a synonym of *Zodariellum* because one of the main characters of the former is exactly the large size of the AME, but also the profile of the carapace is different. Marusik & Koponen (2001) based the attribution of their new species from Mongolia to *Zodariellum* on the similarity of the male palps with that in the crucial fig. 5 of Andreeva & Tyshchenko (1968), which, in their description, was erroneously mentioned as belonging to "*Zodariellum*" *continentalis*.

The outcome of this discussion is as follows. Apart from those described in the present paper, *Acanthinozodium* now contains the following species: *A. cirrisulcatum* Denis, 1952, *A. sahariense* Denis, 1959*, *A. sericeum* Denis, 1956*, *A. spinulosum* Denis, 1966, *A. subclavatum* Denis, 1952,

A. tibesti Jocqué, 1991* and *A. zavattarii* (Caporiacco, 1941)*. Only specimens of the species marked with * have been examined by the present authors. The genus *Zodariellum* now only contains its type species (*Z. surprisum* Andreeva & Tyschchenko, 1968), as it is apparently the only species in the femoral organs clade with AME that have the same size as the other eyes. Other species that have been placed there, need to be transferred to *Zodarion*. However, it is worth to note that at least some species of the latter genus, i.e. species of the *Z. thoni* group appear to have small AME, the diagnostic character of *Zodariellum*. These observations show a need for revisions of these very diverse groups.

The cymbial pit

Species of *Acanthinozodium* have a conspicuous circular pit at the base of the cymbium. This hole is not always easily observed as it is sometimes filled with a colourless gel-like substance. However, when it is empty, it appears as a large crater-like depression with a ringed inner wall (Figs 9A–B, 10C, 12B–D). The most plausible hypothesis so far is that it represents the outlet of a gland that produces a sticky matter used to seal the epigyne after copulation. The surface of the amorphous matter in the pit is irregular and sometimes has long thin smears (Jocqué & Harten, in preparation), just like glue that has been pressed out of a tube. Epigynes of *Acanthinozodium* have large atria that are often filled with a large sticky plug that protrudes above the surrounding tegument (Figs 1A, 5H, 12A). Epigyne plugs are not uncommon in spiders (Uhl *et al.* 2010). The most commonly mentioned plugs are parts of the palp, more precisely of the embolus. The consequence of this plugging system is that males, leaving the emboli behind as plugs (Fromhage & Schneider 2006), loose the possibility to mate another time. In other species, male spiders produce a sticky matter that is meant to be applied to the epigyne either to impede subsequent matings by other males or to prevent backflow of sperm, but remain polygamous themselves. There are apparently several types of sticky plugs: glands that produce a sticky matter within the male palp, released through a separate opening at the base of the embolus base as in some *Amaurobius* C.L. Koch, 1837 species (Suhm *et al.* 1996) or ejected through the embolus itself. In the latter case the plug has to be produced and introduced into the epigyne copulatory tract after sperm transfer (Uhl *et al.* 2014). This type of glands was well documented by Uhl *et al.* (2014) for *Oedothorax retusus* (Westring, 1851). Its efficiency is fairly high (Kunz *et al.* 2014), but depends on the size and the time it takes to harden. Since the canal through which the matter is ejected is very narrow, it is not surprising that the amount of sticky matter that can be transferred is restricted and should also have a low viscosity. These prerequisites are presumably less stringent in the second type of sticky plugs that are produced by a gland that ejects the matter through a wide canal in the cymbium. Such glands are supposed to be present in Storenomorphinae (Jocqué 1991, figs 39–40) where the cymbium is provided with hollow setae and epigynes are very often completely covered by a rubbery matter. The *Acanthinozodium* crater-like plug-pit can also be assumed to be an efficient outlet for sufficient viscous sticky matter to block the large atria of the epigyne (Figs 5H, 12A).

Acknowledgements

We are indebted to Mark Harvey, Norman Platnick, Thomas Pape, Stephen Thorpe and Laurent Raty for their interesting and useful comments in connection with the nomenclatural problem presented here. We thank Alain Reygel for the drawings and Robert Bosmans and Sylvain Déjean for the loan of specimens. We are grateful to Gabriele Uhl for critical comments on the plug-pit hypothesis. This paper is publication BRC 325 of the Biodiversity Research Center (Université Catholique de Louvain). This publication is financed thanks to the FRIA (Fonds pour la Formation à la Recherche dans l'Industrie et dans l'Agriculture).

References

Andreeva E.M. & Tyschchenko V.P. 1968. Materials on the fauna of spiders (Aranei) of Tadjikistan. II. Zodariidae. *Zoologicheskiĭ Zhurnal* 47: 684–689.

Brignoli P.M. 1983. *A catalogue of the Araneae described between 1940 and 1981*. Manchester University Press, Manchester.

Denis J. 1952. Zodariides recueillis au Maroc et en Mauritanie par M. L. Berland. *Bulletin de la Société d'Histoire naturelle d'Afrique du Nord* 41: 58–63.

Denis J. 1956. Notes d'aranéologie marocaine.-VI. Bibliographie des araignées du Maroc et addition d'espèces nouvelles. *Bulletin de la Société des Sciences naturelles du Maroc* 35: 179–207.

Denis J. 1959. Description de deux Zodariides (Aranéides) nouveaux du Sahara. *Bulletin du Muséum national d'Histoire naturelle* 31 (3): 238–241.

Denis J. 1966. Les araignées du Fezzân. *Bulletin de la Société d'Histoire Naturelle de l'Afrique du Nord* 55: 103–144.

Fromhage L. & Schneider J.M. 2006. Emasculation to plug up females: the significance of pedipalp damage in *Nephila fenestrata*. *Behavioral Ecology* 17 (3): 353–357.

International Commission on Zoological Nomenclature 1999. *International Code of Zoological Nomenclature*. 4th Edition. International Trust for Zoological Nomenclature, London.

Jocqué R. 1991. A generic revision of the spider family Zodariidae (Araneae). Bulletin of the American Museum of Natural History 201, American Museum of Natural History, New York.

Levy G. 1992. The spider genera Palaestina, Trygetus, Zodarion and Ranops (Araneae, Zodariidae) in Israel with annotations on species of the Middle East. *Israel Journal of Zoology* 38 (2): 67–110.

Marusik Y.M. & Koponen S. 2001. Spiders of the family Zodariidae from Mongolia (Arachnida: Araneae). *Reichenbachia* 34: 39–48.

Kunz K., Witthuhn M. & Uhl G. 2014. Do the size and age of mating plugs alter their efficacy in protecting paternity? *Behavioral Ecology and Sociobiology* 68 (8): 1321–1328

Ono H. & Jocqué R. 1986. Two new species of the family Zodariidae from Saudi Arabia (Arachnida: Araneae). *Fauna Saudi Arabia* 8: 6–11.

Platnick N.I. 2002. The world spider catalog, version 2.5. American Museum of Natural History [online].

Ramírez M.J., Grismado C.J., Labarque F.M., Izquierdo M.A., Ledford J. M., Miller J.A., Haddad C.R., Griswold C.E. 2014. The morphology and relationships of the walking mud spiders of the genus *Cryptothele* (Araneae: Zodariidae). *Zoologischer Anzeiger* 253 (5): 382–393.

Suhm M., Thaler K & Alberti G. 1996. Glands in the male palpal organ and the origin of the mating plug in *Amaurobius* species (Araneae: Amaurobiidae). *Zoologischer Anzeiger* 234 (3):191–199.

Uhl G., Nessler S.H. & Schneider J.M. 2010. Securing paternity in spiders? A review on occurrence and effects of mating plugs and male genital mutilation. *Genetica* 138 (1): 75–104.

Uhl G., Kunz K., Vöcking O. & Lipke E. 2014. A spider mating plug: origin and constraints of production. *Biological Journal of the Linnean Society* 113 (2): 345–354.

World Spider Catalog 2015. World Spider Catalog, version 16. Natural History Museum Bern.

9

Taxonomy of *Bradysia* Winnertz (Diptera, Sciaridae) in the Northern Holarctic, with the description of four new species

Kai HELLER [1], Heikki HIPPA [2] & Pekka VILKAMAA [3]

[1] Erwin-Salomon-Str. 25, D-25451 Quickborn, Germany.

[2] Gribbylunds allé 2, SE-183 65 Täby, Sweden.

[3] Finnish Museum of Natural History, Zoology Unit, P.O. Box 17, FI-00014 University of Helsinki, Finland.

Abstract. Four species of *Bradysia* Winnertz (Diptera, Sciaridae) from the Northern Holarctic are described and illustrated for the first time: *Bradysia bigeminata* sp. nov. (Finland, Canada), *B. falciceps* sp. nov. (Finland, Canada), *B. oelandica* sp. nov. (Sweden) and *B. plusiospina* sp. nov. (Finland). A few *Bradysia* species, described previously and now found in Northern Europe, are also redefined and illustrated.

Key words. Sciaroidea, Holarctic region, biodiversity, systematics, new taxa

Introduction

Within the framework of the research project 'Diversity of the Fennoscandian black-winged fungus gnats (Diptera, Sciaridae),' financed by the Swedish Taxonomy Initiative, and the research programme 'Deficiently known and threatened forest species' (PUTTE, see Juslén *et al.* 2008), financed by the Ministry of the Environment, Finland, a number of new species in various genera (Vilkamaa *et al.* 2013a, b, c), including species of the genus *Bradysia* Winnertz, 1867, have been found in Northern Europe, while others are new geographical records. Furthermore, conspecific specimens originating from saproxylic studies ongoing in Canada have indicated that some of these species are Holarctic in distribution.

The main aim of this paper is to further expand our knowledge of northern Sciaridae by describing species of *Bradysia* that are new to science and redescribing others that are new records.

Material and methods

All specimens of the new species were detected and picked out from unsorted sciarid or general insect material stored in ethanol. Most specimens originated from Malaise trap samples, but some had been caught by sweeping or with an aspirator, a photoeclector or pan traps. The specimens were mounted on microscope slides in Euparal, after dehydrating them in absolute ethanol. The terminology and methods of measuring and illustrating morphological structures follow Hippa & Vilkamaa (1991, 1994) and Hippa *et al.* (2010).

The material is deposited in the following collections:

MZH	=	Zoological Museum, Finnish Museum of Natural History, Helsinki, Finland
PKHH	=	Private collection of Kai Heller, Heikendorf, Germany
PRDM	=	Private collection of Rob Deady, Montréal, Canada
PWMP	=	Private collection of Werner Mohrig, Poseritz, Germany
SMNH	=	Swedish Museum of Natural History, Stockholm, Sweden
ZFMK	=	Zoological Museum Bonn, Germany

Results

Class Insecta Linnaeus, 1758
Order Diptera Linnaeus, 1804
Superfamily Sciaroidea Billberg, 1820
Family Sciaridae Billberg, 1820

Genus ***Bradysia*** Winnertz, 1867

Bradysia Winnertz, 1867: 180.

Type-species: *Bradysia angustipennis* Winnertz, 1867, by subsequent designation of Enderlein (1911: 127, 183).

Bradysia ascenda Rudzinski, 1994
Fig. 1A–D

Material examined

SWEDEN: Torne Lappmark, Kiruna, Abisko, 68°20' N, 18°52' E, north slope, 440 m, edge of birch forest, white pan trap, 13–17 Sep. 1991, v. Tschirnhaus, 3 ♂♂, PWMP.

Redescription (male)

HEAD. Dark brown, antenna unicolorous pale brown; maxillary palpus very pale brown. Eye bridge 2–3 facets wide. Face with 10–12 scattered longer and shorter setae. Clypeus with 1–4 setae. Maxillary palpus with 3 segments; segment 3 longer than segment 1, segment 2 shortest; segment 1 with 3–4 setae, with dorsal pit of sensilla; antennal flagellomeres with slightly rough surface, body of flagellomere 4 2.15–2.4× as long as wide, neck shorter than broad, longest setae longer than width of flagellomere.

THORAX. Dark brown, setae pale. Anterior pronotum with 6–12 setae. Episternum 1 with 13–15 setae. Scutum with long dorsocentrals, with some longer and shorter laterals, scutellum with 2 longer and some short setae.

WING. Fumose. Length 1.8–1.9 mm. Width/length 0.40. Veins distinct. R1/R 0.65. c/w 0.70–0.75. r-m shorter than bM, r-m with 1 seta, bM non-setose.

LEGS. Pale brown. Coxal setae pale. Front tibial organ with pale vestiture forming a row. Front tibial spur longer than tibial width. Claws without teeth.

Abdomen. Pale brown, setae pale.

Hypopygium (Fig. 1A–D). Brown, concolorous with abdomen.

Gonocoxa. Narrow, longer than gonostylus, mesial margin with normal setosity.

Gonostylus. Narrow, rather straight, slightly thickened at middle, slightly narrowed towards apex, with mesial side not impressed; with dense apical setosity, without apical tooth, with 8–9 apical-subapical megasetae; megasetae subequal in size, slightly curved and sharp. Tegmen conical, apically truncate, with straight lateral sides, with indistinct basolateral corners, laterally sclerotized, with area of aedeagal teeth. Aedeagal apodeme long.

Remarks

Bradysia ascenda was described on the basis of three males from the Austrian Alps (Rudzinski 1994). In having its setae pale and in having a distinct sensory pit in the maxillary palpus, it belongs to the *Bradysia tilicola* group of Menzel & Mohrig (2000).

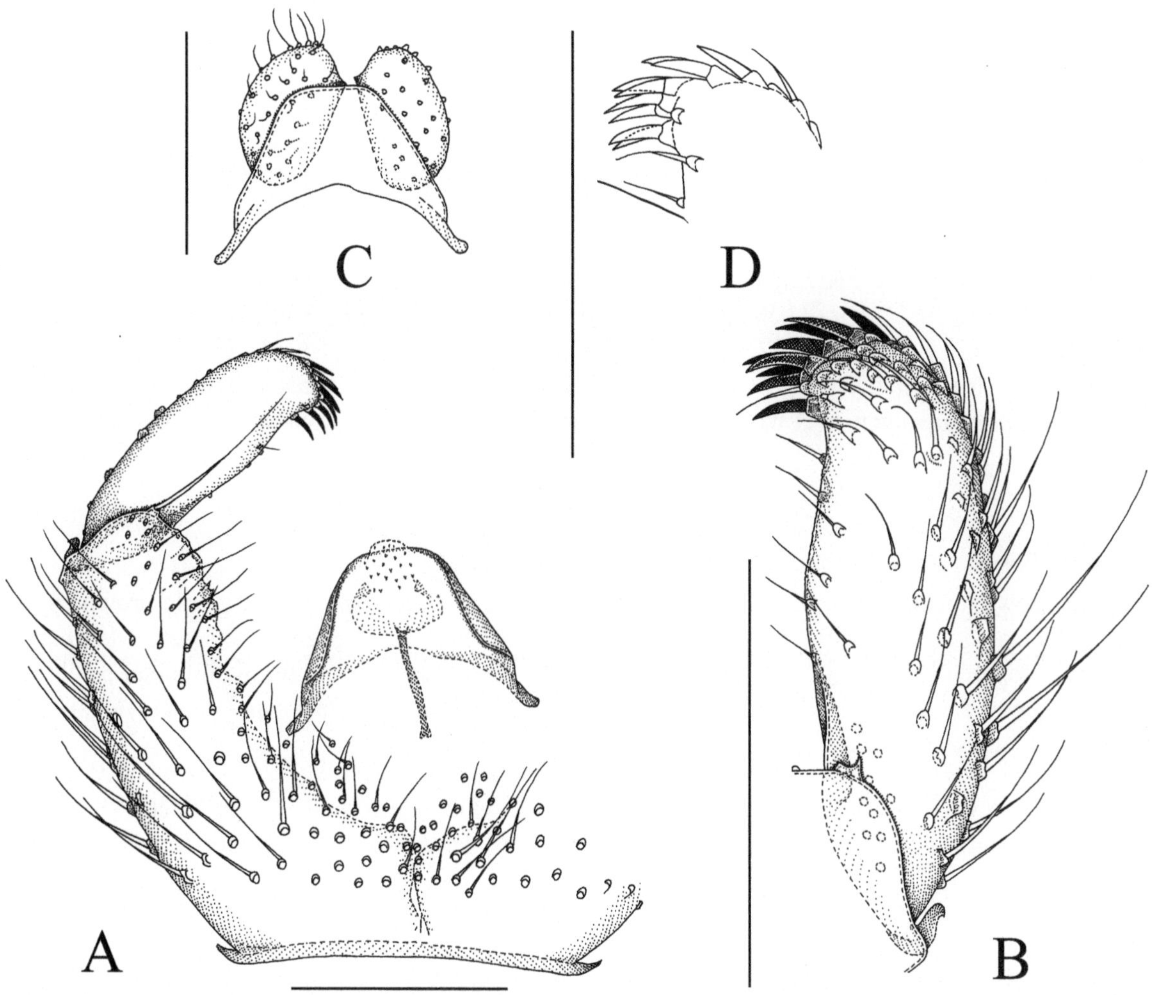

Fig. 1. *Bradysia ascenda* Rudzinski, 1994 (from Sweden). **A**. Part of hypopygium, ventral view. **B**. Gonostylus, ventral view. **C**. Tegmen and cerci, ventral view. **D**. Apical part of gonostylus, ventral view. Scales = 0.1 mm.

Bradysia bigeminata sp. nov.
urn:lsid:zoobank.org:act:2B3EB8AC-0D40-48B8-8453-AC9CA78B6316
Fig. 2A–B

Diagnosis

Small-sized *Bradysia*. Dark brown. Maxillary palpus with 2 segments. Wing length 1.5 mm, anal lobe strong. Gonostylus narrow, without apical tooth, with 4 subequal megasetae.

Etymology

The name is derived from the Latin words *bis-* (two) and *geminus* (twin), referring to the two pairs of gonostylar megasetae of the species.

Material examined

Holotype

FINLAND: ♂, Ostrobothnia borealis, Pudasjärvi, Murto-oja (Grid 7267183:3525642), near a brook, Malaise trap, 1 May–3 Jul. 2006, J. Ilmonen & J. Salmela, MZH.

Paratypes

FINLAND: Karelia borealis, Kuhmo, Elimyssalo, bred from *Fomitopsis rosea* collected 23–28 May 1998, A. Komonen, 1 ♂, MZH.

CANADA, Ontario, Sudbury Co., Chapleau, Nimitz Forest, from wind thrown *Pinus banksiana* trees, photoeclector, 10–24 Jul. 2012, R. Deady & T. Work, 14 ♂♂, 1 [no. 8622] in PKHH, 3 [ZMFK-TIS-2527999, ZMFK-TIS-2527999, ZMFK-TIS-2528275] in ZMFK, 1 in MZH, 9 in PRDM; same data but 16–29 May 2012, 4 ♂♂, PRDM, 1 ♂, MZH; same data but 28 May–12 Jun. 2012, 2 ♂♂, PRDM, 1 ♂, MZH; same data but 25 Apr.–11 Jul. 2012, 2 ♂♂, PRDM; same data but 23 Jul.–8 Aug. 2012, 5 ♂♂, PRDM; same data but 7–20 Aug. 2012, 1 ♂, PRDM; Ontario, Sudbury Co., Chapleau, Superior Forest, mature (99 years old) stand composed of 90% *Pinus banksiana* and 10% *P. mariana* with an understory shrub layer of *Vaccinium*, very wet with considerable bryophyte coverage, pine log, photoeclector, 21 Jul.–6 Aug. 2013, R. Deady & T. Work, 18 ♂♂, PRDM; same data but 5–19 Aug. 2013, 15 ♂♂, PRDM; same data but 7–22 Jul. 2013, 2 ♂♂, PRDM, 1 ♂, MZH.

Description (male)

Head. Dark brown, antennal flagellomeres unicolorous brown; maxillary palpus very pale brown. Eye bridge 2–3 facets wide. Face with 23 scattered longer and shorter setae. Clypeus with 1 seta. Maxillary palpus with 2 segments; segment 1 longer than segment 2, segment 1 with 1 seta, with dorsal patch of sensilla; surface of antennal flagellomeres smooth, body of flagellomere 4 2.65× as long as wide, neck shorter than broad, longest setae longer than width of flagellomere.

Thorax. Dark brown, setae pale. Anterior pronotum with 3 setae. Episternum 1 with 7 setae. Scutum with long dorsocentrals, with some longer and shorter laterals, scutellum with 4 longer and some short setae.

Wing. Hyalinous. Length 1.5 mm. Width/length 0.50. Anal lobe strong. Veins distinct. R1/R 0.55. c/w 0.65. r-m and bM subequal in length, both non-setose.

Legs. Yellow. Coxal setae pale. Front tibial organ with pale vestiture forming a short row. Front tibial spur as long as tibial width. Tibial spurs of middle and hind tibia subequal. Tarsal setosity normal. Claws without teeth.

Abdomen. Brown, setae pale.

Hypopygium (Fig. 2A–B). Brown, concolorous with abdomen.

Gonocoxa. Basally broad, longer than gonostylus, mesial margin with sparse setosity.

Gonostylus. Narrow, rather evenly broad and straight, tumid, with mesial side impressed at apical half; with dense apical setosity, without apical tooth, with two pairs of sharp subapical megasetae. Tegmen truncate, with straight lateral sides, laterally and apically sclerotized, with small area of solitary aedeagal teeth. Aedeagal apodeme rather long.

Remarks

Bradysia bigeminata sp. nov. is here formally assigned to the genus *Bradysia* because of the presence of a comb-like row of elements in its tibial organ and because of its subequal middle and hind tibial spurs, but it cannot be placed in one of the currently defined species groups. In the structure of its gonostylus it also resembles the species of the *Corynoptera vagula* group (see Vilkamaa & Hippa 2006). By its elongated gonostylus with four apical megasetae, and by its short front tibial comb placed in a non-setose area at the apical part of the tibia, *Bradysia bigeminata* sp. nov. resembles *Scatopsciara* (*Xenopygina*) *vinea* Rudzinski & Baumann, 2013, but differs in having a simpler tegmen and a different placement of the gonostylar megasetae. In a molecular phylogenetic analysis, the *Corynoptera vagula* group of species appeared close to species of *Scatopsciara* Edwards (Shin *et al*. 2013). In this analysis, no species of *Scatopsciara* (*Xenopygina*) Frey, 1942 was included. It is obvious that a more inclusive analysis is needed to clarify the relationships of *B. bigeminata* sp. nov., the *Corynoptera vagula* group and the species currently placed in *Scatopsciara* (*Xenopygina*).

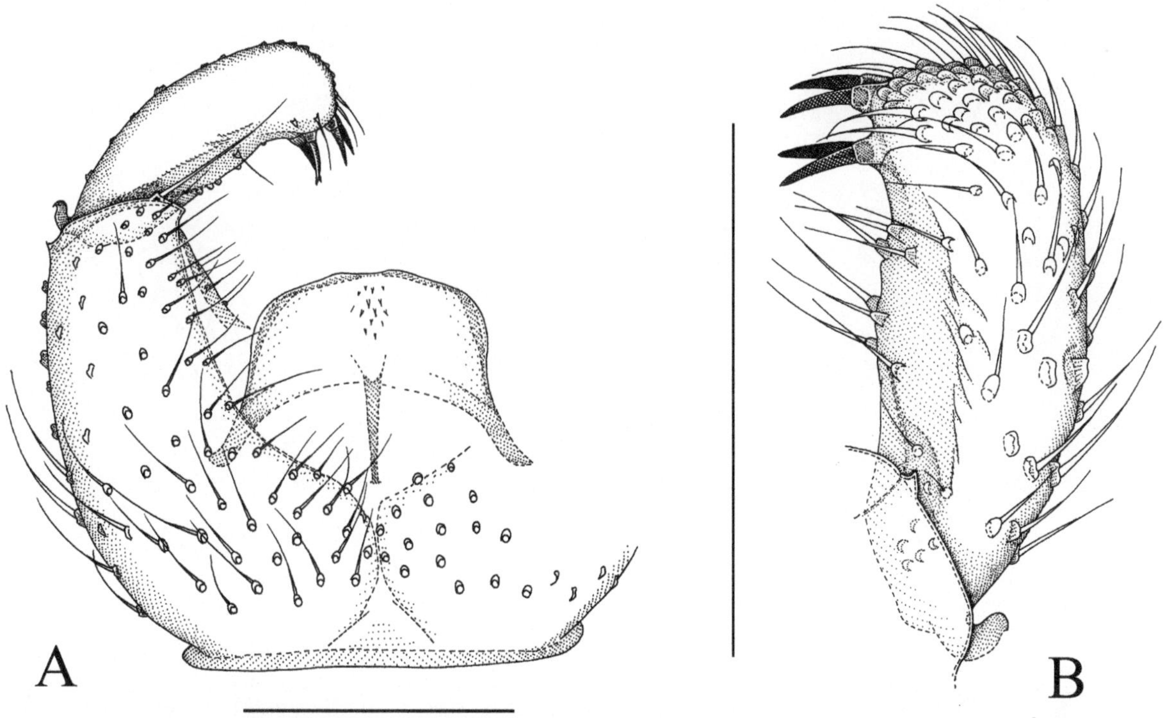

Fig. 2. *Bradysia bigeminata* sp. nov. (holotype). **A**. Part of hypopygium, ventral view. **B**. Gonostylus, ventral view. Scales = 0.1 mm.

Ecology and distribution

Bradysia bigeminata sp. nov. was found to be very numerous in deadwood sampled in Canada. This may suggest deadwood to be the natural habitat of the species. The COI sequence of the collected specimens was allocated to the BIN ACA4746 on BOLD (www.boldsystems.org), and, interestingly, the results showed that the species is widely distributed in Eastern Canada.

Bradysia chlorocornea Mohrig & Menzel, 1992
Fig. 3A–B

Material examined

FINLAND: Regio kuusamoensis, Kuusamo, Kuohusuo-Kalliovaara 30 km S of Kuusamo, 65.43° N, 29.05° E, 270 m, spruce/birch swamp forest, Malaise trap, 2–15 Aug. 2004, M. & C. Jaschhof, 4 ♂♂, MZH.

Redescription (male)

HEAD. Brown, antennal flagellomeres 1–2 yellow, scapus, pedicellus and flagellomere 3–14 pale brown; maxillary palpus very pale brown. Eye bridge 3 facets wide. Face with 10–15 scattered longer and shorter setae. Clypeus with 1–3 setae. Maxillary palpus with 3 segments; segment 1 longer than segment 3, segment 2 shortest; segment 1 with 3–4 setae, with dorsal patch of sensilla; surface of antennal flagellomeres smooth, body of flagellomere 4 2.3–2.7× as long as wide, neck shorter than broad, longest setae as long as width of flagellomere.

THORAX. Brown, setae pale. Anterior pronotum with 3–8 setae. Episternum 1 with 7–15 setae. Scutum with long dorsocentrals, with some longer and shorter laterals, scutellum with 2 longer and some short setae.

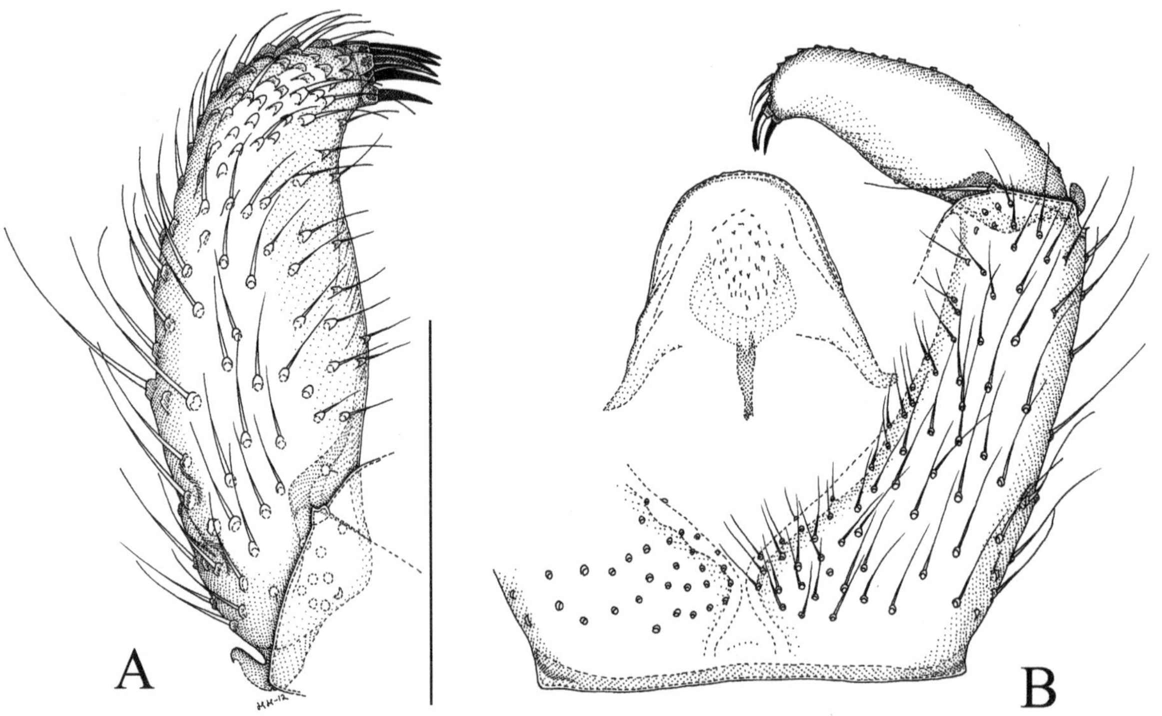

Fig. 3. *Bradysia chlorocornea* Mohrig & Menzel, 1992 (from Finland). **A**. Gonostylus, ventral view. **B**. Part of hypopygium, ventral view. Scales = 0.1 mm.

Wing. Hyalinous. Length 1.7–1.9 mm. Width/length 0.40. Veins distinct. R1/R 0.70–0.85. c/w 0.70–0.80. r-m and bM subequal in length. bM non-setose, r-m non-setose or with 1 seta.

Legs. Yellow. Coxal setae pale. Front tibial organ with pale vestiture forming a short row. Front tibial spur slightly longer than tibial width. Claws with teeth.

Abdomen. Pale brown, setae pale.

Hypopygium (Fig. 3A–B). Brown, concolorous with abdomen.

Gonocoxa. Narrow, longer than gonostylus, mesial margin with rather dense setosity.

Gonostylus. Narrow, broadest at middle, narrowed towards apex, with mesial side weakly impressed at apical half; with dense apical setosity, without apical tooth, with 5–6 apical megasetae whip-lash in form. Tegmen conical, apically roundish, with straight lateral sides, weakly sclerotized, with large area of aedeagal teeth, laterally straight. Aedeagal apodeme short.

Remarks

Bradysia chlorocornea was described from three males from Japan (Mohrig *et al.* 1992). By having bi-coloured antennal flagellomeres and the tarsal claws with teeth, it belongs to the *Bradysia fungicola* group of Menzel & Mohrig (2000). By its slender gonostylus with an apical group of slender, subequal megasetae it resembles *B. affinis* (Zetterstedt, 1838), but it can be distinguished by having the megasetae more slender and longer.

Bradysia falciceps sp. nov.

urn:lsid:zoobank.org:act:60FA7B62-2FFA-43FE-A5A2-5AC02CF29B13

Fig. 4A–B

Diagnosis

Small-sized *Bradysia*. Brown. Maxillary palpus with 3 segments, segment 1 with shallow sensory area. Wing length 1.5–1.7 mm, veins indistinct. Gonostylus narrow, mesially impressed, with apical tooth and 5–6 megasetae in one group.

Etymology

The name is derived from the Latin words *falx* (scythe) and *-ceps* (of the head), referring to the curved apical tooth of the gonostylus of the species.

Material examined

Holotype

FINLAND: ♂, Lapponia inariensis, Utsjoki, tundra, Malaise trap, 14 Aug. 2000, Fatouros, MZH.

Paratypes

FINLAND: Same data as holotype, 1 ♂, PKHH 3368; Regio aboensis, Turku, Pomponrahka (Grid 6719377:3240217), open bog, Malaise trap, 1 Aug.–28 Sep. 2011, J. Salmela, 1 ♂, MZH.

CANADA: Ontario, Sudbury Co., Chapleau, Superior Forest, mature (99 years old) stand composed of 90% *Pinus banksiana* and 10% *P. mariana* with an understory shrub layer of *Vaccinium*, very wet, with considerable bryophyte coverage, photoeclector on pine log, 23 Jun.–8 Jul. 2013, R. Deady & T. Work, 3 ♂♂, PRDM; same data but 7–22 Jul. 2013, 1 ♂, MZH.

Description (male)

Head. Brown, antennal flagellomeres 1–2 yellow, scapus, pedicellus and flagellomeres 3–14 unicolorous pale brown; maxillary palpus very pale brown. Eye bridge 3 facets wide. Face with 15–21 scattered longer and shorter setae. Clypeus with 1–2 setae. Maxillary palpus with 3 segments; segment 3 longer than segment 1, segment 2 shortest; segment 1 with 1–2 setae, with small shallow dorsal sensory area; surface of antennal flagellomeres smooth, body of flagellomere 4 1.85–2.3× as long as wide, neck shorter than broad, longest setae as long as width of flagellomere.

Thorax. Brown, setae pale. Anterior pronotum with 4–7 setae. Episternum 1 with 8–9 setae. Scutum with long dorsocentrals, with some longer and shorter laterals, scutellum with 4 longer and some short setae.

Wing. Hyalinous. Length 1.5–1.7 mm. Width/length 0.45–0.50. Veins indistinct. R1/R 0.60–0.65. c/w 0.75–0.85. r-m and bM subequal in length. r-m with 1–2 setae, bM non-setose.

Legs. Yellow. Coxal setae pale. Front tibial organ with pale vestiture forming short row. Front tibial spur longer than tibial width. Claws without teeth.

Abdomen. Pale brown, setae pale and fine.

Hypopygium (Fig. 4A–B). Brown, concolorous with abdomen.

Gonocoxa. Narrow, longer than gonostylus, mesial margin with normal setosity.

Gonostylus. Narrow, evenly straight, with shallow lateral notch at apical fourth, with mesial side impressed at all its length; with dense apical setosity, with strong curved apical tooth-like megaseta,

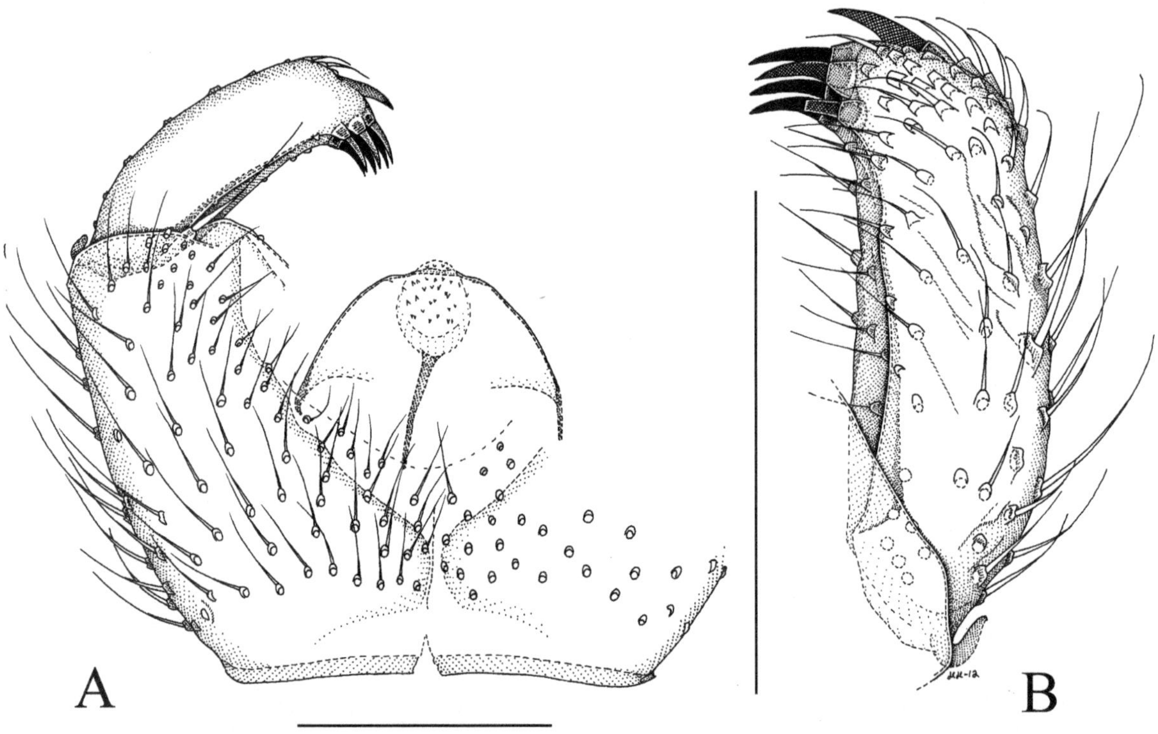

Fig. 4. *Bradysia falciceps* sp. nov. (holotype). **A**. Part of hypopygium, ventral view. **B**. Gonostylus, ventral view. Scales = 0.1 mm.

with 5–6 subapical megasetae, megasetae sharp, in single group. Tegmen conical, weakly sclerotized, with area of aedeagal teeth. Aedeagal apodeme long.

Remarks

Bradysia falciceps sp. nov. belongs to the *Bradysia praecox* group and is most similar to *B. nitidicollis* (Meigen, 1818), *B. iridipennis* (Zetterstedt, 1938) and *B. quercina* Menzel & Köhler, 2014 by sharing a strong apical tooth at the dorsoapical part of the gonostylus (see Köhler *et al.* 2014). *Bradysia falciceps* sp. nov. can be distinguished from all other species of the *B. praecox* group in having pale basal antennal flagellomeres with dark necks.

Bradysia oelandica sp. nov.

urn:lsid:zoobank.org:act:D1390D14-F74E-4595-A927-A287B8A331DF

Fig. 5A–C

Diagnosis

Large-sized *Bradysia*. Brown. Maxillary palpus with 3 segments, segment 1 with a small shallow sensory patch. Wing length 2.3 mm, anal lobe small. Setose intercoxal lobe of hypopopygium present. Gonocoxa and gonostylus broad; gonocoxa with elongated ventromesial setae; gonostylus bulged mesiad, with a short apical tooth.

Etymology

The name is derived from the Latinized name of the type locality (the island of Öland, Sweden) of the species.

Material examined

Holotype

SWEDEN: ♂, Öland, Mörbylonga, Gamla Skogsby (Kalkstad), 56°37.002' N, 16°30.457' E, meadow with bushes, Malaise trap (Trap ID 22), 20 May–28 Jun. 2006, Swedish Malaise Trap Project, SMNH 2652.

Description (male)

HEAD. Dark brown, antennal flagellomeres unicolorous, paler brown; maxillary palpus very pale brown. Eye bridge 3 facets wide. Face with 15 scattered longer and shorter setae. Clypeus with 3 setae. Maxillary palpus with 3 segments; segment 1 longer than segment 3, segment 2 shortest; segment 1 with 3 setae, with small dorsal patch of sensilla in shallow depression; surface of antennal flagellomeres rough, body of flagellomere 4 2.9× as long as wide, neck shorter than broad, longest setae as long as width of flagellomere.

THORAX. Brown, setae dark. Anterior pronotum with 7 setae. Episternum 1 with 6 setae. Scutum with long dorsocentrals, with some longer and shorter laterals, scutellum with short setae.

WING. Fumose. Length 2.3 mm. Width/length 0.45. Anal lobe small. Veins distinct. R1/R 0.75. c/w 0.70. r-m shorter than bM, both non-setose.

LEGS. Yellowish brown. Coxal setae pale. Front tibial organ with pale vestiture forming a row. Front tibial spur longer than tibial width. Claws without teeth.

ABDOMEN. Setae dark and long.

Hypopygium (Fig. 5A–C). Brown, concolorous with abdomen. Intercoxal area with lobe and possessing *ca.* 10 long setae.

Gonocoxa. Broad, longer than gonostylus, ventromesial margin with elongated setae, dorsomesial margin with distinct lobes.

Gonostylus. Globular, medially bulged mesiad, with dense apical setosity, with short narrow apical tooth, with apical (lateral) megaseta, with specialized subapical seta. Tegmen conical, weakly sclerotized, with area of small aedeagal teeth. Aedeagal apodeme short but strong.

Remarks

Bradysia oelandica sp. nov. belongs to the *Bradysia polonica* group of Menzel & Mohrig (2000) in having a broad intercoxal lobe of the hypopygium and an indistinct apical tooth at the end of the gonostylus. The general structure of the gonostylus is fairly similar to the other species of that group, but *B. oelandica* sp. nov. is easy to distinguish by the elongated setae at the apical part of the gonocoxa.

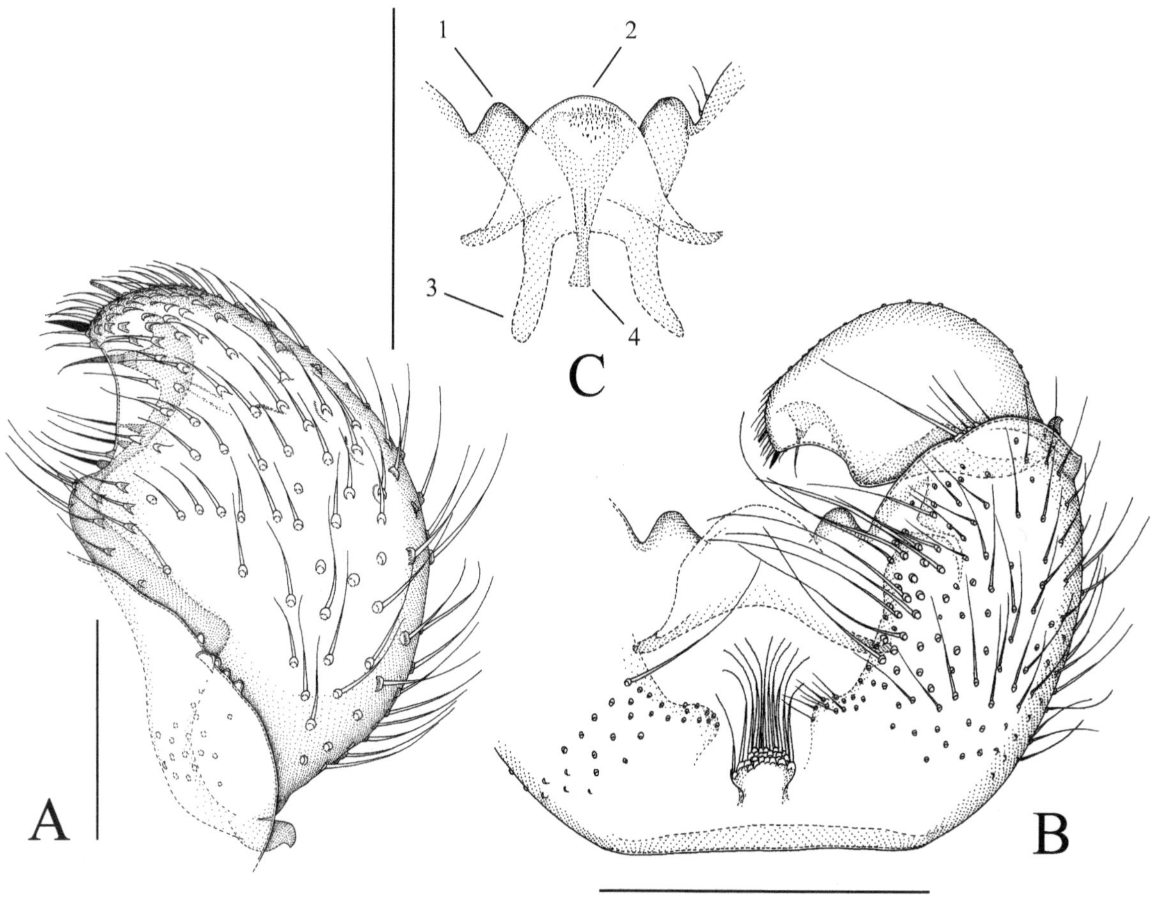

Fig. 5. *Bradysia oelandica* sp. nov. (holotype). **A**. Gonostylus, ventral view. **B**. Part of hypopygium, ventral view. **C**. Medial parts of hypopygium, ventral view. Scale bars: A = 0.1 mm, B–C = 0.3 mm. 1 = lobe at dorsal mesial margin of gonocoxa, 2 = tegmen, 3 = gonocoxal apodeme, 4 = aedeagal apodeme.

Bradysia plusiospina sp. nov.

urn:lsid:zoobank.org:act:9EFEEAC1-6A5A-49FA-881B-1AA96CA76C28

Fig. 6A–B

Diagnosis

Medium-sized *Bradysia*. Brown. Maxillary palpus with 3 segments, segment 1 with a dorsal patch of sensilla. Wing length 1.8–1.9 mm, anal lobe small. Gonocoxa and gonostylus narrow, gonostylus straight, mesially weakly impressed, without apical tooth, with 7 sharp megasetae in two groups.

Etymology

The name is derived from the Latinized Greek word *plousious* (wealthy), and the Latin word *spina* (spine), referring to the many gonostylar megasetae of the species.

Material examined

Holotype

FINLAND: ♂, Ostrobothnia borealis, Tervola, Yrttijänkä (Grid 7346827:340814), near a spring, Malaise trap, 1–28 Jun. 2004, J. Salmela, MZH.

Paratype

FINLAND: same data as holotype, ♂, MZH.

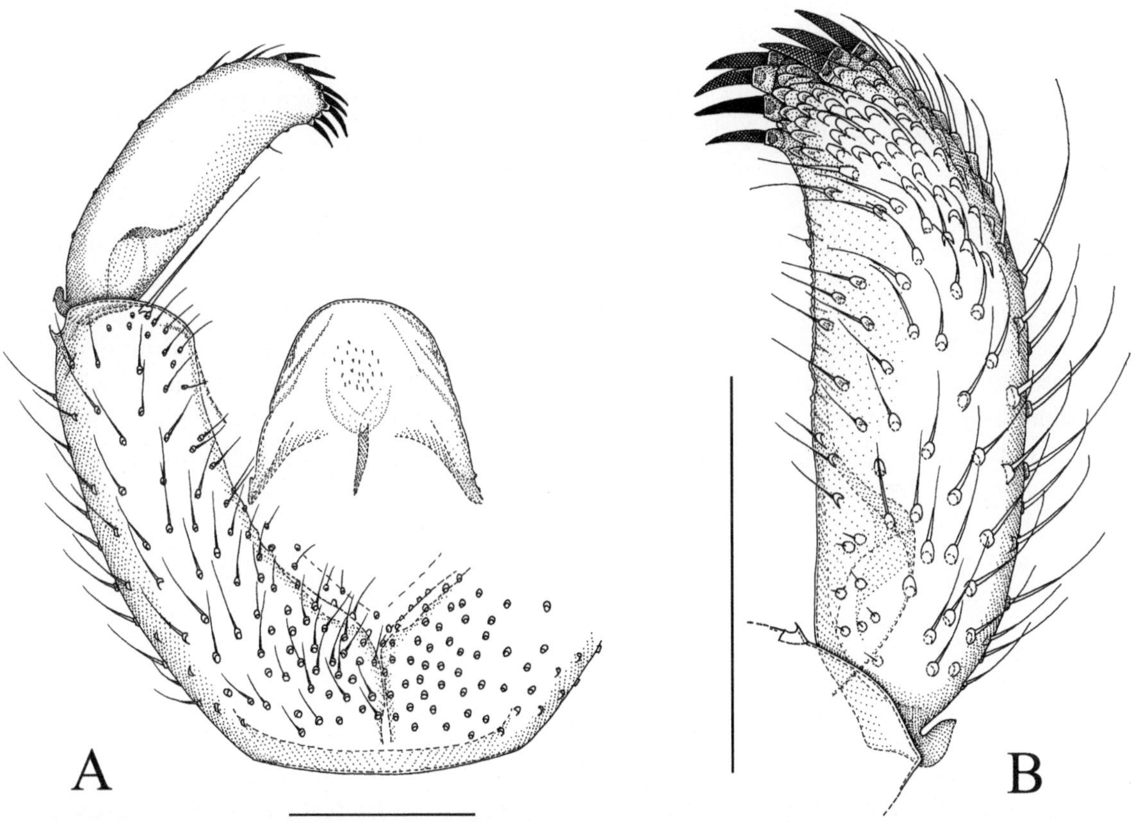

Fig. 6. *Bradysia plusiospina* sp. nov. (holotype). **A**. Part of hypopygium, ventral view. **B**. Gonostylus, ventral view. Scales = 0.1 mm.

Description (male)

HEAD. Brown, antennal flagellomeres unicolorous, paler brown, maxillary palpus very pale brown. Eye bridge 3 facets wide. Face with 31–40 scattered longer and shorter setae. Clypeus with 1 seta. Maxillary palpus with 3 segments; segment 1 longer than segment 3, segment 2 shortest; segment 1 with 4–6 setae, with dorsal patch of sensilla; surface of antennal flagellomeres smooth, body of flagellomere 4 3.1× as long as wide, neck shorter than broad, longest setae slightly longer than the width of flagellomere.

THORAX. Brown, setae pale. Anterior pronotum with 6–7 setae. Episternum 1 with 8–9 setae. Scutum with long dorsocentrals, with some longer and shorter laterals, scutellum with 2 longer and some short setae.

WING. Fumose. Length 1.8–1.9 mm. Width/length 0.40. Anal lobe small. Veins distinct. R1/R 0.80. c/w 0.60–0.70. r-m and bM subequal in length. r-m with 1 seta, bM non-setose.

LEGS. Yellow. Coxal setae pale. Front tibial organ with pale vestiture forming short row. Front tibial spur longer than tibial width. Claws without teeth.

ABDOMEN. Pale brown, setae pale and short.

HYPOPYGIUM (Fig. 6 A–B). Brown, concolorous with abdomen.

GONOCOXA. Narrow, longer than gonostylus, mesial margin with short setosity.

GONOSTYLUS. Narrow, rather straight, slightly narrowed and curved towards apex, with mesial side weakly impressed; with dense apical setosity, without apical tooth, with 7 megasetae, megasetae in apical and subapical groups, subequal in size, sharp. Tegmen truncate, with straight lateral sides, weakly sclerotized, with area of aedeagal teeth. Aedeagal apodeme short.

Remarks

Bradysia plusiospina sp. nov. resembles *B. pseudocampestris* Mohrig, 1978 and *B. angustata* Tuomikoski, 1960 in the general form of the hypopygium and gonostylus, as well as in having the anal lobe of the wing poorly developed. *Bradysia angustata* differs from the others in having an apical tooth and slightly stronger megasetae in its gonostylus, and in having its wing even narrower than the others. *Bradysia plusiospina* sp. nov. can be distinguished from *B. pseudocampestris* in having stronger megasetae, some megasetae in a more lateral position than in the latter species and in having a more setose basal part of its hypopygium (see Mohrig 1978: figs 4, 7). *Bradysia plusiospina* sp. nov. is placed in the *Bradysia fungicola* group of Menzel & Mohrig (2000).

Bradysia quercina Menzel & Köhler, 2014
Fig. 7A–C

Material examined

Paratypes

SWEDEN: Småland, Gränna, Lönnemålen, 58°02.935' N, 14°34.382' E, next to old cellar in Norway Spruce forest, Malaise trap (Trap ID 17), 10–24 Sep. 2003, Swedish Malaise Trap Project, 3 ♂♂, no. 3184, 3185, SMNH, 4103, MZH; Småland, Kalmar län, Bäckebo, Grytsjön Nature Reserve, old aspen forest in boulder terrain, Malaise trap, 15 Jun. 2006, Swedish Malaise Trap Project, 1 ♂, SMNH 6781.

SWITZERLAND: Sihlwald near Zürich, window trap, 5 Jun. 1996, Schiegg, 1 ♂, PKHH 1805.

Remarks

Bradysia quercina Menzel & Köhler, 2014 was described as belonging to the *Bradysia praecox* group on the basis of a number of specimens from Norway, Italy, Sweden and Switzerland (Köhler *et al.* 2014), and its affinities were thoroughly discussed. To compare to the figures of the holotype provided in the original description, we present drawings of the hypopygium and gonostylus of a paratype from Sweden (Fig. 7).

Discussion

Bradysia is by far the most species-rich genus of Sciaridae, including about 400 species worldwide, most in the Palaearctic region. With the present new species included, about 70 species are known from both Finland and Sweden. In the Nearctic region, only 20 species are known (Mohrig *et al.* 2013). This is a strongly biased figure and our current knowledge of the Nearctic fauna is still very incomplete. Even less is known of *Bradysia* from other biogeographical regions (see Vilkamaa *et al.* 2012). The newly described species show that even in the areas for which the sciarid fauna is best known, new species can be discovered. With many of the species morphologically very similar to each other, the genus is probably the most challenging of all Sciaridae regarding its taxonomy. The Holarctic fauna has been divided into sixteen species groups (Menzel & Mohrig 2000), but these have been formed mainly based on general similarity. A recent phylogeny based on molecular characters challenges even the monophyly of the genus (Shin *et al.* 2013).

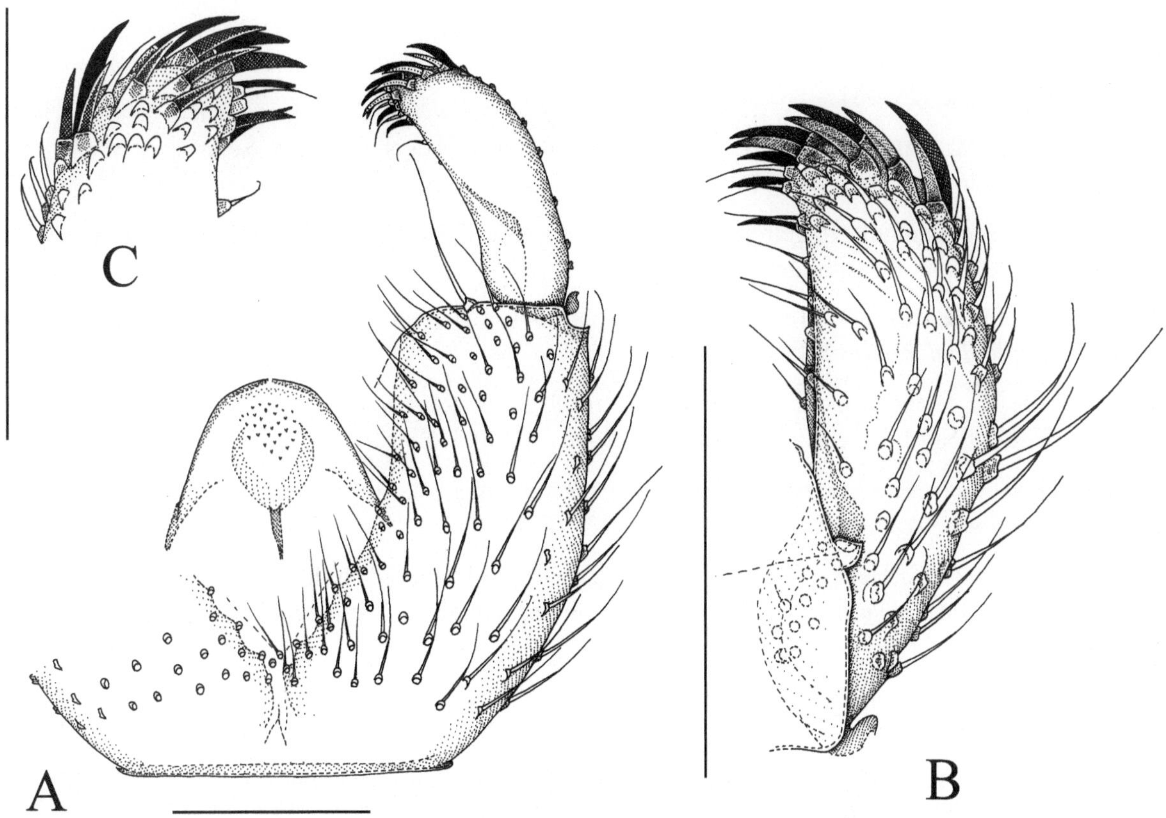

Fig. 7. *Bradysia quercina* Menzel & Köhler, 2014 (from Sweden). **A**. Part of hypopygium, ventral view. **B**. Gonostylus, ventral view. **C**. Apical part of gonostylus, ventral view. Scales = 0.1 mm.

Acknowledgements

We thank Rob Deady (Montréal, Canada), Werner Mohrig (Poseritz, Germany) and Jukka Salmela (Rovaniemi, Finland) for materials and Björn Rulik (Bonn, Germany) for doing the DNA analyses. This work was supported by the Swedish Taxonomy Initiative. Rob Deady kindly checked the English.

References

Enderlein G. 1911. Die phyletischen Beziehungen der Lycoriiden (Sciariden) zu den Fungivoriden (Mycetophiliden) und Itonididen (Cecidomyiiden) und ihre systematische Gliederung. *Archiv für Naturgeschichte* 77, Suppl. 3: 116–201.

Hippa H. & Vilkamaa P. 1991. The genus *Prosciara* Frey (Diptera, Sciaridae). *Entomologica Fennica* 2: 113–155.

Hippa H. & Vilkamaa P. 1994. The genus *Camptochaeta* gen. n. (Diptera, Sciaridae). *Acta Zoologica Fennica* 194: 1–85.

Hippa H., Vilkamaa P. & Heller K. 2010. Review of the Holarctic *Corynoptera* Winnertz, 1867, s. str. (Diptera, Sciaridae). *Zootaxa* 2695: 1–197.

Juslén A., Kuusinen M, Muona J., Siitonen J. & Toivonen H. 2008. Puutteellisesti tunnettujen ja uhanalaisten metsälajien tutkimusohjelma [Research Programme of Deficiently Known and Threatened Forest Species] (PUTTE). *Suomen Ympäristö* 1: 1–146.

Köhler A., Menzel F., Thunes K.H. & Søli G.E.E. 2014. Black fungus gnats (Diptera: Sciaridae) in oak canopies: description of *Bradysia quercina* Menzel & Köhler spec. nov. and new records for Norway. *Studia Dipterologica* 20: 325–331.

Menzel F. & Mohrig W. 2000. Revision der paläarktischen Trauermücken (Diptera: Sciaridae). *Studia Dipterologica*, Suppl. 6 (for 1999): 1–761.

Mohrig W. 1978. Zur Kenntnis flügelreduzierter Dipteren der Bodenstreu. IX. Beitrag: Gattungen *Corynoptera, Bradysia* und *Plastosciara* (Sciaridae). *Zoologischer Anzeiger* 201: 424–432.

Mohrig W., Menzel F. & Kozánek M. 1992. Neue Trauermücken (Diptera, Sciaridae) aus Nord-Korea und Japan. *Dipterological Research* 3: 17–32.

Mohrig W., Heller K., Hippa H., Vilkamaa P. & Menzel F. 2013. Revision of the black fungus gnats (Diptera: Sciaridae) of North America. *Studia Dipterologica* 19: 141–286.

Rudzinski H.-G. 1994. Neue Mitteilungen zur Trauermückenfauna Österreichs (Diptera Nematocera: Sciaridae). *Entomofauna, Zeitschrift für Entomologie* 15: 281–292.

Shin S., Jung S., Menzel F., Heller K., Lee H. & Lee S. 2013. Molecular phylogeny of black fungus gnats (Diptera: Sciaroidea: Sciaridae) and the evolution of larval habitats. *Molecular Phylogenetics and Evolution* 66: 833–846.

Tuomikoski R. 1960. Zur Kenntnis der Sciariden Finnlands. *Annales Zoologici Societatis Zoologiae Botanicae Fennicae ,Vanamo'* 21 (4): 1–164.

Vilkamaa P. & Hippa H. 1996. Review of the genus *Prosciara* Frey (Diptera, Sciaridae) in the Indomalayan region. *Acta Zoologica Fennica* 203: 1–57.

Vilkamaa P. & Hippa H. 2006. *Corynoptera vagula* Tuomikoski and allied species (Diptera: Sciaridae). *Zootaxa* 1150: 31–42.

Vilkamaa P., Hippa H. & Mohrig W. 2012. The genus *Bradysia* Winnertz (Diptera, Sciaridae) in New Caledonia, with the description of thirteen new species. *Zootaxa* 3489: 25–44.

Vilkamaa P., Hippa H. & Heller K. 2013a. Review of the genus *Camptochaeta* Hippa & Vilkamaa (Diptera, Sciaridae), with the description of nine new species. *Zootaxa* 3636: 476–488.

Vilkamaa P., Hippa H. & Heller K. 2013b. Notes on the taxonomy of the Holarctic *Corynoptera* Winnertz *sensu lato* (Diptera, Sciaridae), with description of six new species. *Zootaxa* 3710: 322–332.

Vilkamaa P., Hippa H. & Heller K. 2013c. Taxonomy of the Sciaridae (Diptera) of Northern Europe: description of eight new species. *Studia Dipterologica* 20: 47–58.

Winnertz J. 1867. *Beitrag zu einer Monographie der Sciarinen*. Zoologisch-Botanischen Gesellschaft in Wien, Vienna.

10

New *Spirinia* and *Stygodesmodora* species (Nematoda, Spiriniinae) from the Southwest Pacific, and a revision of the related genera *Spirinia*, *Chromaspirina* and *Perspiria*

Daniel LEDUC[1] & Dominick VERSCHELDE[2]

[1] National Institute of Water and Atmospheric Research, Private Bag 14-901, Wellington, New Zealand.

[2] Ghent University - Zoology Museum, K.L. Ledeganckstraat 35, B-9000 Ghent, Belgium

Abstract. Two new species of the family Desmodoridae are described from the upper continental slope of New Zealand, Southwest Pacific, and the genera *Spirinia, Chromaspirina* and *Perspiria* are revised. *Spirinia verecunda* sp. nov. is characterised by a short, stout body, cuticle covered in minute, hair-like structures, unispiral amphideal aperture and cryptospiral amphideal fovea, buccal cavity with small dorsal tooth and minute subventral teeth, eight oblong glands surrounding anterior portion of pharynx, large sperm cells, spicules with weak capitulum, and the absence of precloacal supplements. *Stygodesmodora confusa* sp. nov. is characterised by a relatively short body, spiral amphids with 1.2–1.4 turns, cephalic setae situated at or slightly posterior to mid-level of amphid, and males with four precloacal supplements consisting of short setae on wide bases. Like other species of the genus, *S. confusa* sp. nov. is characterised by an annulated head region, but in some specimens the cuticle annulations are restricted to the dorsal and ventral sides of the head. *S. confusa* sp. nov. shows similarities with species of other desmodorid genera (i.e., *Echinodesmodora, Bolbonema*), but can be differentiated from them by the presence of an annulated head region and amphideal plates. The genera *Spirinia, Chromaspirina* and *Perspiria* have very similar morphologies and differ mainly in the size and structure of the buccal cavity (*Spirinia vs Chromaspirina*) or the shape of the tail and placement of the amphids relative to the cuticle annulations (*Spirinia vs Perspiria*); their diagnoses are clarified and some nomenclatural changes are proposed to eliminate overlap in the definitions of these genera.

Keywords. Desmodoridae, deep-sea, meiofauna, Chatham Rise, southern Hikurangi Margin

Introduction

Spirinia Gerlach, 1963 is a relatively common genus and twelve valid species have been described to date (Silva *et al.* 2009). *Spirinia* is very similar to *Chromaspirina* Filipjev, 1918 in the shape and structure of the head region, amphids, pharynx, and male copulatory apparatus. The main distinction between the two genera is the shape of the buccal cavity: *Spirinia* is characterised by a narrow and lightly cuticularised buccal cavity with no or a small dorsal tooth and no or minute subventral teeth, whereas *Chromaspirina* is characterised by a more heavily cuticularised, medium to large buccal cavity with a large dorsal tooth and smaller subventral teeth. A third genus, *Perspiria* Wieser & Hopper, 1967, is also similar to *Spirinia* and differs from the latter in having a conico-cylindrical to filiform tail and by the amphids not completely surrounded by body annulations. Species of these three genera show considerable overlap in the above-mentioned morphological characteristics, however, and the boundaries between them have become somewhat blurred. Here, we describe *S. verecunda* sp. nov. from the Chatham Rise crest (350 m water depth) on the continental margin of New Zealand, and we propose several nomenclatural changes to species of *Spirinia* and *Chromaspirina* to allow a more consistent distinction between the genera based on buccal morphology. Several nomenclatural changes are also suggested for the genus *Perspiria* to reflect differences in tail shape and body annulations of the head region between the latter genus and *Spirinia*.

Stygodesmodora Blome, 1982 is a relatively uncommon genus comprising only two species: *S. bacillicauda* (Gerlach, 1963) and *S. epixantha* Blome, 1982. These species were described from shallow sandy sediments in the Indian Ocean and North Sea, respectively. Here a third species, *S. confusa* sp. nov., is described from the continental margin of New Zealand (350 and 686 m water depth); it is the first species of the genus to be described from the deep sea.

Material and Methods

Samples for species descriptions were obtained from the Chatham Rise and southern Hikurangi margin, Southwest Pacific. The Chatham Rise is a submarine ridge that extends eastwards from the South Island of New Zealand, over water depths ranging from *ca.* 250 to 3000 m. The highly productive Subtropical Front (STF) is geographically constrained near the southern flank of the rise at around 44° S (Murphy *et al.* 2001). The southern Hikurangi margin lies on the south-eastern extremity of the North Island of New Zealand and is situated north of the Chatham Rise. Samples were collected at two sites: one situated near the head of a canyon on the southern Hikurangi margin during National Institute of Water and Atmospheric Research (NIWA) cruise TAN1004 (station 92, 686 m water depth, 41.8921° S, 174.6347° E), and one at an open slope site located near the central Chatham Rise crest during NIWA cruise TAN1103 (station 69, 350 m water depth; 43.331° S, 178.288° E).

Sediment samples were collected using an Ocean Instruments MC-800A multicorer (MUC; core internal diameter = 9.52 cm). Each sample consisted of one subcore of internal diameter 26 mm taken to a depth of 5 cm. Samples were fixed in 10% formalin and stained with Rose Bengal. Samples were subsequently rinsed on a 1 mm sieve to remove large particles and on a 45 µm sieve to retain nematodes. Nematodes were extracted from the remaining sediments by Ludox flotation and transferred to pure glycerol (Somerfield & Warwick 1996). Species descriptions were made from glycerol mounts using differential interference contrast microscopy, and drawings were made with the aid of a camera lucida. Specimens for scanning electron microscopy (SEM) were transferred to a 2% glutaraldehyde solution with sodium cacodylate buffer overnight, then transferred to a 4% osmium tetroxide solution for 2 hours. They were then gradually transferred to pure ethanol using a graded ethanol series, critical point dried, and mounted onto stubs before being coated with gold using a sputter coater. Observations were made using a Hitachi TM3000 tabletop SEM at high vacuum mode.

All measurements are in µm, and all curved structures are measured along the arc. Type specimens are held in the NIWA Invertebrate Collection, Wellington, New Zealand. Abbreviations in the text are as follows:

a	=	body length/maximum body diameter
abd	=	anal body diameter
b	=	body length/oesophagus length
c	=	body length/tail length
cbd	=	corresponding body diameter
%V	=	vulva distance from anterior end of body × 100/total body length.

Results

Phylum Nematoda Cobb, 1932
Class Chromadorea Inglis, 1983
Subclass Chromadoria Adamson, 1987
Order Desmodorida de Coninck, 1965
Suborder Desmodorina de Coninck, 1965
Superfamily Desmodoroidea Filipjev, 1922

Family **Desmodoridae** Filipjev, 1922

Type genus

Desmodora de Man, 1889 (synonym: *Mastodex* Steiner, 1921).

Emended diagnosis (modified from Lorenzen 1981 and Decraemer & Smol 2006)

Member of Desmodoroidea. Cuticle with coarse annulations, at least in anterior body region, or striated. Specialised somatic setae, such as adhesion tubes (Draconematidae) or ambulatory setae (Epsilonematidae) not present. Amphideal fovea usually spiral, unispiral, cryptospiral, or cryptocircular, and sometimes loop- or pore-shaped. Buccal cavity with small to large dorsal tooth and smaller or indistinct subventral teeth. Pharynx usually with round muscular posterior bulb, with or without cuticularised lumen; sometimes bulb is pyriform, elongated, or indistinct. Gubernaculum without dorsal apophyses. Males usually monorchic, except in *Onepunema* Leduc & Verschelde, 2013 (diorchic). Vulva usually posterior to middle of body. Tail conical.

Remarks

Armenteros *et al.* (2014) recently argued that the genus *Onepunema* Leduc & Verschelde, 2013 is more closely related to the Microlaimidae than to the Desmodoridae because of the presence of two testes (a feature common in the former but not found in the latter). The Microlaimidae, however, are characterised by the presence of outstretched ovaries, while *Onepunema* possesses reflexed ovaries, a characteristic of the Desmodoridae. This trait, in combination with the presence of a cephalic capsule in *Onepunema* (an important characteristic for the classification of supraspecific taxa within the Desmodorida and never found within the Microlaimoidea) suggests closer affinities with the Desmodoridae (and Desmodorinae in particular) than with the Microlaimidae. Armenteros *et al.* (2014: 26) incorrectly stated that "… the presence of reflexed ovaries […] suggest a relationship with *Bolbolaimus* Cobb, 1920, *Calomicrolaimus* Lorenzen, 1976 and *Microlaimus* de Man, 1880". Reflexed ovaries are not found in any of these genera; perhaps *Bolbolaimus* was confused with *Bolbonema* Cobb, 1920 (see comments under *Chromaspirina longisetosa* Jensen, 1985 below), which belongs to the Desmodoridae due to the presence of reflexed ovaries and a globular cephalic capsule.

Subfamily **Spiriniinae** Chitwood, 1936

Type genus

Spirinia Gerlach, 1963 (synonym: *Spira* Bastian, 1865).

Diagnosis (from Decraemer & Smol 2006)

Member of Desmodoridae. Body cuticle with fine annulations. Head not demarcated as a cephalic capsule. Amphideal fovea usually a simple spiral. Buccal cavity small or large, with a distinct or a minute dorsal tooth; two small ventrosublateral teeth may be present or absent.

Differential diagnosis

This subfamily is similar to the Desmodorinae but can be differentiated from the latter by the absence of a head capsule.

Remarks

The tribe Spirinacea was originally established by Chitwood (1936), before being raised to subfamily, family and superfamily rank by De Coninck (1965). According to the Principle of Coordination (ICZN, 1999, Article 36), "a name established for a taxon at any rank in the family group is deemed to have been simultaneously established for nominal taxa at all other ranks in the family group; (…). The name has the same authorship and date at every rank." The appropriate authorship for Spiriniinae should therefore be attributed to Chitwood (1936), and not Gerlach & Murphy (1965), as is often done (e.g., Armenteros *et al.* 2014).

Valid genera (Cavalcanti *et al.* 2009; Armenteros *et al.* 2014)

Alaimonema Cobb, 1920
Chromaspirina Filipjev, 1918
Metachromadora Filipjev, 1918
Onyx Cobb, 1891
Papillonema Verschelde *et al.*, 1995
Paradesmodora Schuurmans Stekhoven, 1950
Parallelocoilas Boucher, 1975
Perspiria Wieser & Hopper, 1967
Polysigma Cobb, 1920
Pseudometachromadora Timm, 1952
Sigmophoranema Hope & Murphy, 1972
Spirinia Gerlach, 1963
Spirodesma Cavalcanti *et al.*, 2009
Stygodesmodora Blome, 1982

We provide emended diagnoses and updated lists of valid species for *Chromaspirina*, *Spirinia*, and *Perspiria* below to clarify the distinctions between them. These lists are based on the original and any subsequent descriptions; as many of the original descriptions (especially early ones) lack type material, we often had to rely on short descriptions and/or illustrations.

Genus ***Chromaspirina*** Filipjev, 1918

Type species

Chromaspirina pontica Filipjev, 1918.

Emended diagnosis (modified from Maria *et al.* 2009 and Muthumbi *et al.* 1995)

Robust body with rounded or rectangular head region and short conical tail; fine body annulations. Annulated or non-annulated head region; body annulations envelop the amphideal fovea partly, entirely, or only begin at posterior edge of amphids. Amphideal fovea uni-, crypto- or multispiral, or loop-shaped. Cephalic setae at anterior edge or further posteriorly alongside amphids. Distinct buccal cavity, conspicuously cuticularised; large dorsal tooth and smaller subventral teeth; ventral field of denticles may be present. Pharynx with round or oval posterior bulb, sometimes weakly developed, and without cuticularised lumen. Precloacal supplements may be present.

Valid species

C. chabaudi Boucher, 1975. Main diagnostic characters: "… a medium-sized dorsal tooth and two small subventral teeth. Anterior portion with twelve chelorhabdia." Drawings show a distinct cuticularised dorsal tooth and subventral teeth (Boucher 1975: fig. 2).

C. crinita Gerlach, 1952. Main diagnostic character: "The dorsal tooth is massive and extraordinarily strong."

C. cylindricollis (Cobb, 1920). Main diagnostic characters: Only an 'onchium' is mentioned, without further comment, but a distinct cuticularised dorsal tooth is visible on the drawing (Cobb 1920: 325).

C. dubia Inglis, 1968. Drawing shows a very large dorsal tooth (Inglis 1968: fig. 53).

C. gerlachi Blome, 1982. Drawing shows a distinctly cuticularised dorsal tooth and subventral teeth (Blome 1982: fig. 20).

C. indica Gerlach, 1963. Drawing shows a very large dorsal tooth and thickly cuticularised subventral teeth (Gerlach 1963: fig. 1).

C. inglisi Warwick, 1970. Main diagnostic character: "... a prominent heavily cuticularized dorsal tooth..."; also clear in drawing (Warwick 1970: fig. 12).

C. lunatica Gerlach, 1965. Main diagnostic character: "...a strong dorsal tooth …"; also clear in drawing (Gerlach 1965: fig. 13).

C. madagascarensis Gerlach, 1953. Main diagnostic characters: "...a rather large dorsal tooth …"; drawing shows a large and thick dorsal tooth and subventral teeth (Gerlach 1953: fig. 4).

C. modesta Bussau, 1993. Drawing shows a large, heavily cuticularised dorsal tooth and subventral teeth (Bussau 1993: figs 55–57).

C. multipapillata Jayasree & Warwick, 1977. Main diagnostic characters: "Buccal cavity bearing a dorsal tooth and two small subventral teeth"; drawing unclear. It is difficult to draw a firm conclusion from the description and hence this species is left in this genus.

C. parapontica Luc & De Coninck, 1959. Main diagnostic characters: "Large dorsal tooth..."; drawing shows a large dagger-like dorsal tooth (Luc & De Coninck 1959: fig. 35).

C. pellita Gerlach, 1954 (synonym: *C. renaudae* Boucher, 1975). Drawing shows a very large dorsal tooth and conspicuous subventral teeth (Gerlach 1954: fig. 4).

C. pontica Filipjev, 1918. Main diagnostic character: "Three teeth…: one large dorsal …, and two small subventrals."

C. thieryi De Coninck, 1943. Main diagnostic character: "…with strong dorsal tooth... "; also clear in drawing (De Coninck 1943: fig. 5).

C. vanreuselae Verschelde & Vincx, 1996. Main diagnostic characters: "...with huge dorsal tooth and smaller single ventral tooth"; also clear in drawing (Verschelde & Vincx 1996: fig. 1).

Species inquirendae

C. amabilis (De Man, 1922). Considered *species inquirenda* by Jensen (1978).

C. cobbi Chitwood, 1938. Considered *species inquirenda* by Jensen (1978).

C. robusta Wieser, 1954. Considered *species inquirenda* by Wieser & Hopper (1967).

Remarks

Chromaspirina cylindricollis (Cobb, 1920) (synonym: *Mesodorus cylindricollis* Cobb, 1920). This species was in our opinion rightly transferred to *Chromaspirina* as it was described with an annulated head region. Maria *et al.* (2009), however, mistakenly showed a drawing of the head capsule of *Pseudochromadora quadripapillata* Daday, 1899 (synonym: *Micromicron cephalatum* Cobb, 1920), in their key to *Chromaspirina* species, instead of the non-annulated head region of *C. cylindricollis* (Maria *et al.* 2009: fig. 26).

Chromaspirina denticulata (Gerlach, 1953). This species was originally described as *Bolbolaimus denticulatus*. Luc & De Coninck (1959) later synonymized *Bolbolaimus* with *Chromaspirina*. However, Jensen (1978) rightly re-established the genus and re-instated *Bolbolaimus denticulatus* Gerlach, 1953 as a valid combination. However, he gave the wrong species authority in his identification key (Cobb, 1920 instead of Gerlach, 1953). Maria *et al.* (2009) included *C. denticula* (Gerlach, 1953) (*erratum*?) in their list of invalid *Chromaspirina* species. In conclusion, we recognize the re-establishment of *Bolbolaimus denticulatus* Gerlach, 1953 as a valid combination (as suggested by Jensen 1978) with the synonyms *Chromaspirina denticulata* (Gerlach, 1953), *Bolbolaimus denticulatus* Cobb, 1920 syn. nov., and *Chromaspirina denticula* (Gerlach, 1953) syn. nov.

Chromaspirina dimorpha (Hopper, 1961) and *C. inflexa* (Wieser, 1954) were rightly transferred back to *Desmodora* by Wieser & Hopper (1967).

Chromaspirina longisetosa Jensen, 1985. This species was transferred to *Bolbolaimus* (*erratum*?) by Muthumbi *et al.* (1995), but later transferred to *Bolbonema* by Verschelde *et al.* (1998). Because this species clearly has a head capsule and cephalic setae located posterior to the amphids, we confirm that it belongs to *Bolbonema*. Thus, we propose the combination *Bolbonema longisetosum* (Jensen, 1985) with the synonyms *Chromaspirina longisetosa* Jensen, 1985 and *Bolbolaimus longisetosus* (Jensen, 1985).

Chromaspirina lunatica Gerlach, 1965 (synonym: *C. lunata* Gerlach, 1965 syn. nov.). Gerlach (1965) described *C. lunatica*, but the species is incorrectly labelled as *C. lunata* in a figure caption (p. 132).

Chromaspirina pellucida (Cobb, 1920) and *C. punctata* (Cobb, 1920) were rightly transferred back to *Bolbolaimus* by Jensen (1978).

Chromaspirina rabosa (Gerlach, 1956). This species was originally described as *Desmodora rabosa* and later transferred to *Chromaspirina* by the same author (Gerlach 1963). Muthumbi *et al.* (1995) argued that the species should be left in the genus *Desmodora* because it has "... a well-developed head capsule with the amphids situated outside the rings." We note, however, that the cephalic setae are situated posterior to the amphids, and this species is therefore transferred to the genus *Bolbonema*. Thus, we propose the combination *Bolbonema rabosum* (Gerlach, 1956) comb. nov. with the synonyms *Desmodora rabosa* Gerlach, 1956 and *Chromaspirina rabosa* (Gerlach, 1956).

Chromaspirina spinulosa (Wieser, 1959). This species was transferred to *Metadesmodora* by Gerlach (1963) and then to *Echinodesmodora* by Blome (1982). Thus, we propose the combination *Echinodesmodora spinulosa* (Wieser, 1959) with the synonyms *Chromaspirina spinulosa* (Wieser, 1959) and *Metadesmodora spinulosa* (Wieser, 1959).

Genus ***Perspiria*** Wieser & Hopper, 1967

Type species

Perspiria hamata Wieser & Hopper, 1967.

Emended diagnosis (modified from Vincx & Gourbault 1989)

As for *Spirinia* (see below), but amphids almost never completely surrounded by body annulations (seldom more than half of amphids surrounded by body annulations), amphids are entirely surrounded by annuli in some rare cases; tail conico-cylindrical to filiform. Precloacal supplements usually present.

Valid species

P. elongata (Castro *et al.*, 2006) comb. nov. (synonym: *Spirinia elongata* Castro *et al.*, 2006). Main diagnostic characters: "...one very small subventral tooth and two small dorsal teeth." The drawings show amphids completely surrounded by body annulations, but the authors describe a conico-cylindrical tail with annuli which are more evident in the tail region compared to those along the rest of the body; thus, we transfer this species to *Perspiria* (Castro *et al.* 2006: figs 3c, 4).

P. flagellata Vitiello, 1971. Main diagnostic characters: "Amphid situated anterior to cuticle striations or partially surrounded by them"; very long and slender, almost filiform tail (Vitiello 1971: fig. 29).

P. hamata Wieser & Hopper, 1967. Drawing shows amphids only partially located within the body annuli; long and slender, almost filiform tail (Wieser & Hopper 1967: fig. 18).

P. lara (Silva *et al.,* 2009) comb. nov. (synonym: *Spirinia lara* Silva *et al.*, 2009). Main diagnostic characters: "Buccal cavity with one dorsal tooth and two minute ventrosublateral teeth" and "Cuticle with transverse striae..., less obvious in head region than in tail." Drawing shows amphid with annulations to mid-level of amphid only, and tail is described as conico-cylindrical. This species is therefore transferred to *Perspiria* (Silva *et al.* 2009: figs 2–3).

P. megamphida Vincx & Gourbault, 1989. Drawing shows substantial portion of the amphids is located on the non-annulated lip region; conico-cylindrical tail with more prominent annuli (Vincx & Gourbault 1989: fig. 1).

P. mokii Coles, 1987. Main diagnostic character: "...cuticle appears smooth…", hence amphids are not surrounded by annuli; long and slender filiform tail (Coles 1987: fig. 5).

P. papillata Vincx & Gourbault, 1989. Drawing shows anterior half of amphids on non-annulated lip region and posterior half located within the annulated head region; long, conico-cylindrical tail (Vincx & Gourbault 1989: fig. 2).

P. septentrionalis (Cobb, 1914) comb. nov. (synonyms: *Spirinia septentrionalis* (Cobb, 1914) and *Spira septentrionalis* Cobb, 1914). Main diagnostic characters: "From the anus the tail tapers for two-thirds of its length, then becomes cylindroid to the swollen terminus…" Drawings by Wieser (1954: fig. 126) clearly show the amphids in the head region with only their posterior half within the body annulations, leaving the anterior half on the non-annulated head region. The tail is conico-cylindrical. Thus, this species is transferred to the genus *Perspiria.*

P. striaticaudata (Timm, 1962). Tail long (11–17 abd) and slender with distinct annuli; thus it is transferred to *Perspiria.* Synonym: *Spirina striaticaudata* Timm, 1962.

Species inquirenda

Perspiria sabulicola (Filipjev, 1918/1921) comb. nov. (synonym: *Spirina sabulicola* Filipjev, 1918/1921). This species was described with a 'long tail (length 4 × width)' and hence should be transferred to the genus *Perspiria.* However, as the species was described based on only one female, it is regarded as a *species inquirenda.*

Genus ***Spirinia*** Gerlach, 1963

Type species

Spira parasitifera Bastian, 1865.

Emended diagnosis

Robust body with rounded or conical head region and conical tail, fine body annulations. Annulated head region; body annulations normally completely surround the amphideal fovea, with few exceptions. Spiral amphideal fovea. Buccal cavity narrow, lightly cuticularised; no or small/inconspicuous dorsal tooth, no or minute subventral teeth. Pharynx with small pyriform, oval, or rounded posterior bulb without cuticularised lumen. Precloacal supplements usually absent. Short conical tail.

Valid species

Spirinia gerlachi (Luc & De Coninck, 1959). Main diagnostic characters: "Annulations beginning halfway down the lateral organ…long attenuated conical tail"; also clear in drawings (Luc & De Coninck 1959: fig. 25). Despite the presence of an amphid only partially surrounded by cuticle annulations, this species is left within *Spirinia* because of the shape of the tail (conical), which is the primary trait for differentiating between *Spirinia* and *Perspiria.*

Spirinia gnaigeri Ott, 1977. Amphids almost entirely located within the body annulations in one male, but only the posterior half of another male's amphid is surrounded by body annulations (compare figs 42 and 43 in Ott 1977); however, as the species has a conical tail, it belongs within *Spirinia.*

S. guanabarensis (Maria *et al.*, 2009) comb. nov. (synonym: *Chromaspirina guanabarensis* Maria *et al.*, 2009). Main diagnostic characters: "Buccal cavity large with an acute dorsal tooth and two smaller ventrosublateral teeth.*"* However, later in the text it is stated that: *"Chromaspirina guanabarensis* sp. nov. belongs to the group of species characterized by a poorly developed dorsal tooth." Drawings also show a small buccal cavity with a small dorsal tooth and minute subventral teeth (Maria *et al.* 2009: fig. 1). Tail is conical. This species is therefore transferred to *Spirinia.*

S. hopperi Coles, 1987. Main diagnostic characters: "Cuticle without striations... small dorsal tooth is present in the narrow mouth cavity." Drawings show a conical tail (Coles 1987: fig. 3).

S. inaurita (Wieser & Hopper, 1967) comb. nov. (synonym: *Chromaspirina inaurita* Wieser & Hopper, 1967). Main diagnostic characters: "Buccal cavity with three small teeth (these teeth are smaller than in most other species of *Chromaspirina* and stress the proximity of the genus to *Spirinia*)."; the drawings are inconclusive, but the description clearly describes small teeth. This species is therefore transferred to *Spirinia.*

S. laevioides Gerlach, 1963. The drawing shows a small buccal cavity, a small dorsal tooth, and amphids located within the annulated head region (Gerlach 1963: fig. 2).

S. laevis (Bastian, 1865) (synonym: *S. pilosa* Allgén, 1935). The drawing shows small teeth, amphids located entirely within the annulated head region, and a short conical tail (Bastian 1865: fig. 204).

S. okemwai (Muthumbi *et al.*, 1995) comb. nov. (synonym: *Chromaspirina okemwai* Muthumbi *et al.*, 1995). We studied the type material, which shows that this species has a cylindrical pharynx with only a slightly swollen terminal bulb, whose shape cannot be described. The teeth are difficult to distinguish and hence cannot be called 'large' or 'distinct'; therefore, we transfer this species to the genus *Spirinia*.

S. parasitifera (Bastian, 1865). Some descriptions of this species show the amphids being only partially surrounded by annulations (Gerlach 1963; Vincx & Gourbault 1989), and others describe specimens with amphids entirely surrounded by annulations (Coles 1987; Vincx & Gourbault 1989); furthermore, additional descriptions and drawings depict either a conical tail (Gerlach 1963: fig. 1; Coles 1987: fig. 3) or a prolonged conical to long conical tail (Wieser & Hopper 1967: fig. 17). Hence, until further examination and comparison of these different specimens can be done, we leave this species within *Spirinia*.

S. parma (Ott, 1972) comb. nov. (synonym: *Chromaspirina parma* Ott, 1972). Main diagnostic characters: "... stoma conical, with a small acute dorsal tooth and two minute ventral teeth..."; this species is therefore transferred to the genus *Spirinia.*

S. schneideri (Villot, 1875). Luc & Deconinck (1959) describe this species with a smooth cuticle, and Coles (1987) describes it "without cuticular striations". The SEM micrograph by Coles (1987: fig. 9a), however, shows striations surrounding the amphids. The tail is short and conical.

S. sophia Silva *et al.*, 2009. Main diagnostic characters: "Buccal cavity with one dorsal tooth and two minute, ventrosublateral teeth"; clear in drawings, which also show the amphid completely surrounded by cuticle annulations (Silva *et al.* 2009: fig. 6). Tail conical.

Species inquirendae

Spirinia granulata (Allgén, 1929) comb. nov. (synonym: *Spirina granulata* Allgén, 1929). This species was described based on two juveniles only, with very poor drawings and inadequate descriptions. Hence, we have to regard this species as *species inquirenda.*

S. paucispira (Schuurmans Stekhoven, 1950) comb. nov. (synonym: *Chromaspirina paucispira* Schuurmans Stekhoven, 1950). This species was considered *species inquirenda* by Wieser & Hopper (1967) and Muthumbi *et al.* (1995), because the original description was based on a single female specimen. It was stated that the buccal cavity of this species has an "indication of a dorsal tooth". Based on this character, which is sufficient to distinguish between *Spirinia* and *Chromaspirina*, this species is transferred to the genus *Spirinia* (but remains *species inquirenda* due to the incomplete nature of the original description).

S. similis (Cobb, 1898) (synonym: *Spira similis* Cobb, 1898). Considered invalid by Coles (1987) due to lack of illustrations.

S. tenuicauda (Allgén, 1959) (synonym: *Spirina tenuicauda* Allgén, 1959). This species has to be regarded as a *species inquirenda*, as the only description Allgén made was to write "differing from *Sp. parasitifera*," and his drawings are inadequate.

Spirinia verecunda sp. nov.

urn:lsid:zoobank.org:act:83F4CF38-29CF-4540-B05D-2245120A489B

Figs 1–3; Table 1

Diagnosis

Spirinia verecunda sp. nov. is characterised by a short, stout body, cuticle covered in minute hair-like structures, unispiral amphideal aperture and cryptospiral amphideal fovea, buccal cavity with a small dorsal tooth and minute subventral teeth, eight elongated glands surrounding the anterior portion of the pharynx, large sperm cells, spicules with weak capitulum, and absence of precloacal supplements.

Type material

Holotype

NEW ZEALAND: ♂, NIWA 88379, 20 Feb. 2011, NIWA cruise TAN1103, station 69, central Chatham Rise, 43.331° S, 178.288° E, 350 m.

Paratypes

NEW ZEALAND: 2 ♀♀, NIWA 88380, same data as holotype.

Etymology

The species name is derived from the Latin word *verecundus* (bashful, shy, coy), in reference to the anterior part of the head region deeply folded inwards.

Description

Male

Body stout, cylindrical, pale orange colour, tapering slightly towards both ends. Annulated cuticle, no lateral differentiation; annulations 0.6–0.8 µm apart; cuticle slightly thicker in pharyngeal region than elsewhere (2.2–2.5 *vs* 1.6–2.0 µm). Cuticle covered with dense layer of minute hair-like structures from posterior to amphid to near tail tip (barely visible under light microscope; Fig. 3A–B); larger, 1 × 5 µm bacteria-like structures with narrow base also attached in pharyngeal region (Fig. 1D). A single protist, 74 µm long and 18 µm wide attached subdorsally near cloaca (Fig. 3E). Short somatic setae arranged in eight longitudinal rows along entire body length.

Blunt, slightly rounded annulated head region with annulations completely surrounding amphid (Fig. 1A–B). Four cephalic setae near anterior edge of amphid; inner and outer labial sensillae not observed because labial region is folded inwards from level of cephalic setae. Cryptospiral amphideal fovea with slightly cuticularised outline and unispiral amphideal aperture (Fig. 3A–B).

Buccal cavity with slightly cuticularised dorsal tooth and minute subventral teeth. Eight elongated glands, 12–17 µm in length, distributed around anterior portion of pharynx just posterior to buccal cavity (Figs 1C–D, 3C). Pharynx muscular, oval-shaped posterior pharyngeal bulb, lumen not cuticularised. Nerve ring at 40–50% of pharynx length. Secretory-excretory system not observed. Cardia 10–12 µm long.

Reproductive system monorchic, with relatively large, outstretched testis situated to the left of intestine. Mature sperm large, globular to elongated in shape, up to 16 µm in length. Short, arcuate spicules with weak capitulum and tapering distally; velum present (Fig. 1G). Gubernaculum with cuneus broadest in middle portion and tapering proximally and distally; poorly developed lateral crurae (Fig. 1H). Precloacal supplements not observed. Tail conical, with two pairs of short subventral setae and one pair of subdorsal setae; three caudal glands restricted to tail region and spinneret present.

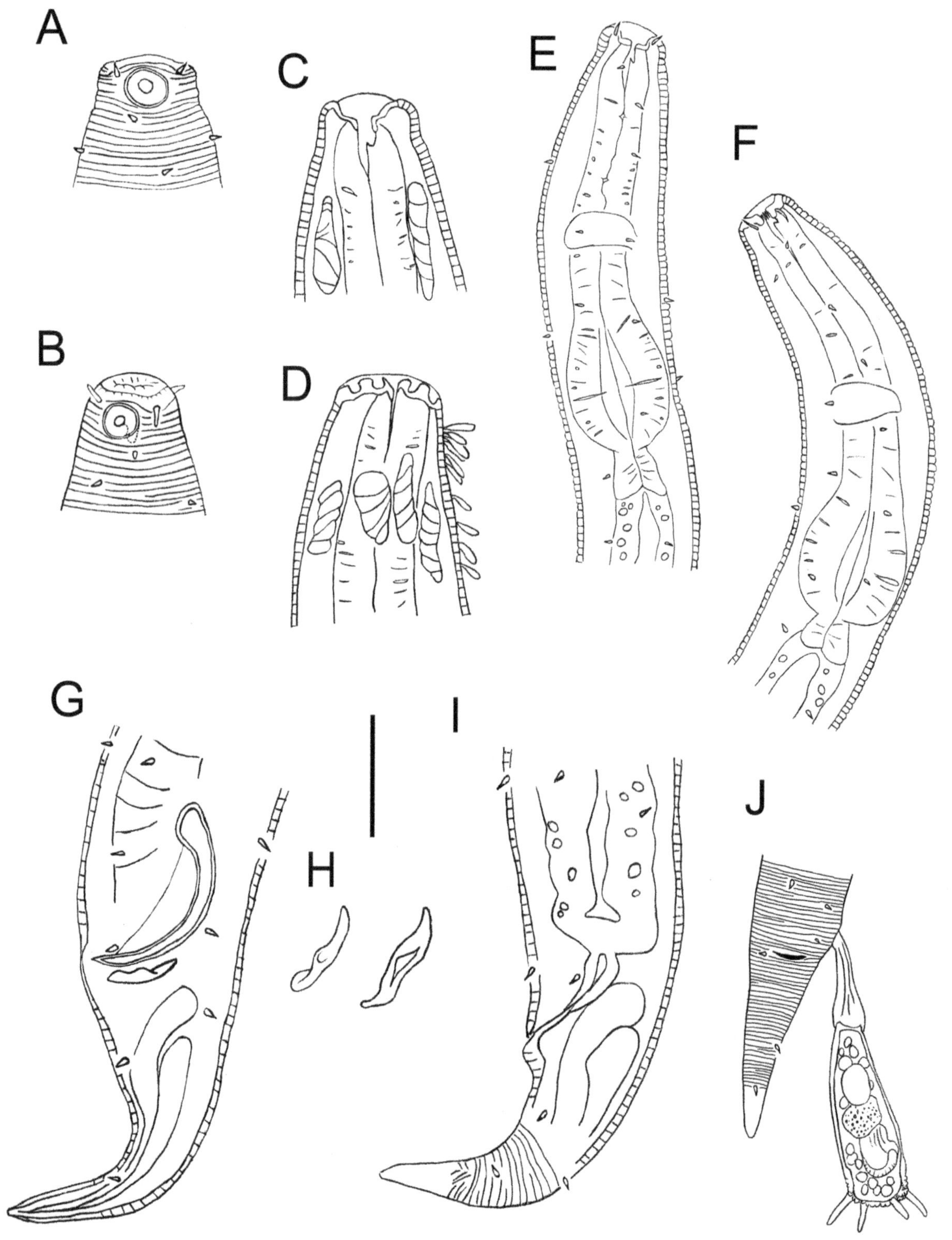

Fig. 1. *Spirinia verecunda* sp. nov. **A**. Surface view of male head. **B–C**. Male head, showing buccal cavity. **D**. Female head, showing bacteria attached to cuticle. **E**. Anterior body region of male. **F**. Anterior body region of female. **G**. Posterior body region of male. **H**. Gubernaculum, with details of crurae (left) and cuneus (right). **I**. Posterior body region of female. **J**. Posterior body region of female, showing attached protist. Scale bar: A–D, I = 20 μm, E–F = 30 μm, G = 16 μm, H = 12 μm, J = 35 μm.

Female

Similar to male, but with wider body. One specimen bears a single protist (suctorian) on tail. Reproductive system didelphic, amphidelphic with reflected ovaries situated either both on the right or both on the left of intestine. Vulva located slightly post median. Cuticular pars distalis vaginae and pars proximalis vaginae surrounded by constrictor muscle.

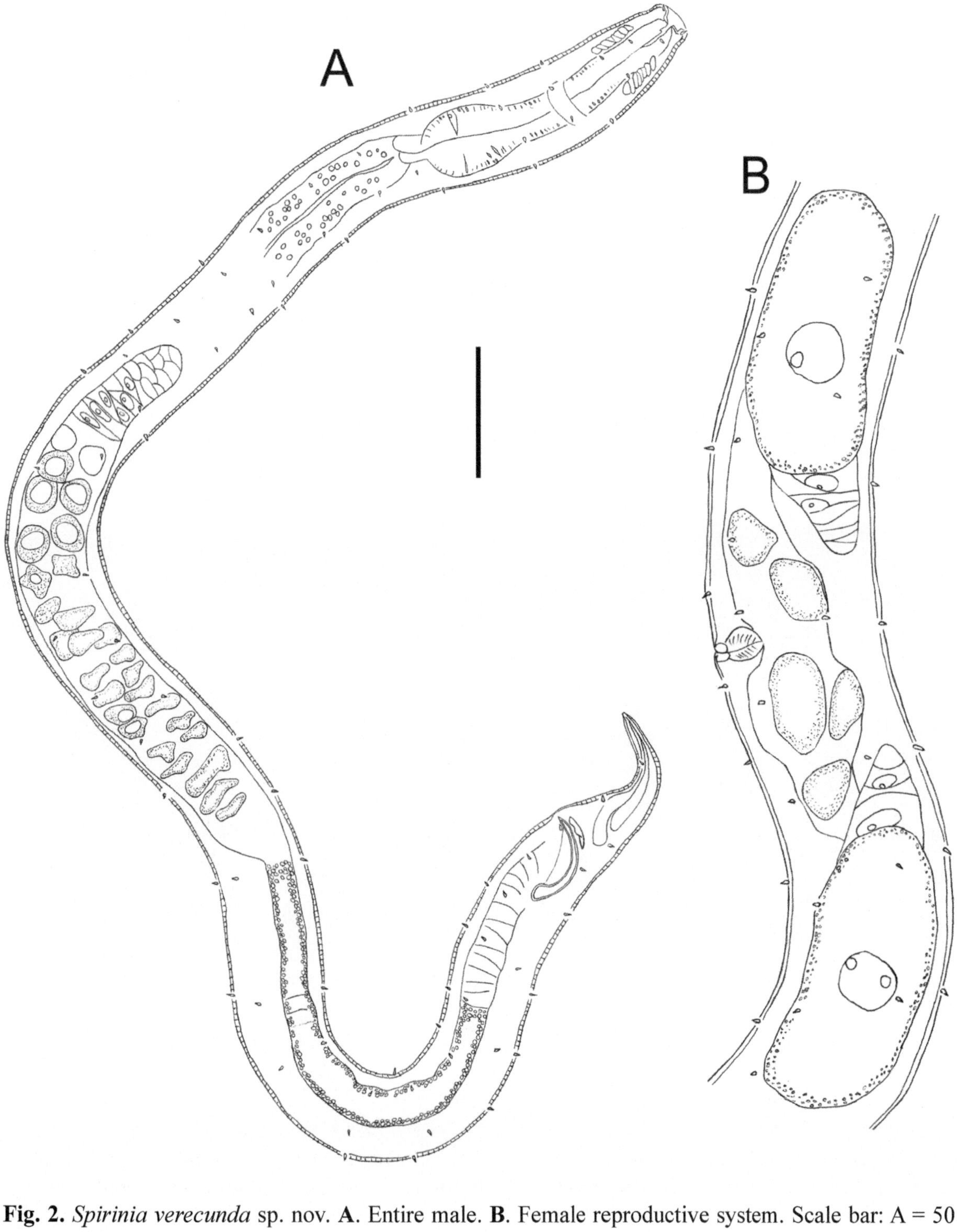

Fig. 2. *Spirinia verecunda* sp. nov. **A**. Entire male. **B**. Female reproductive system. Scale bar: A = 50 µm, B = 30 µm.

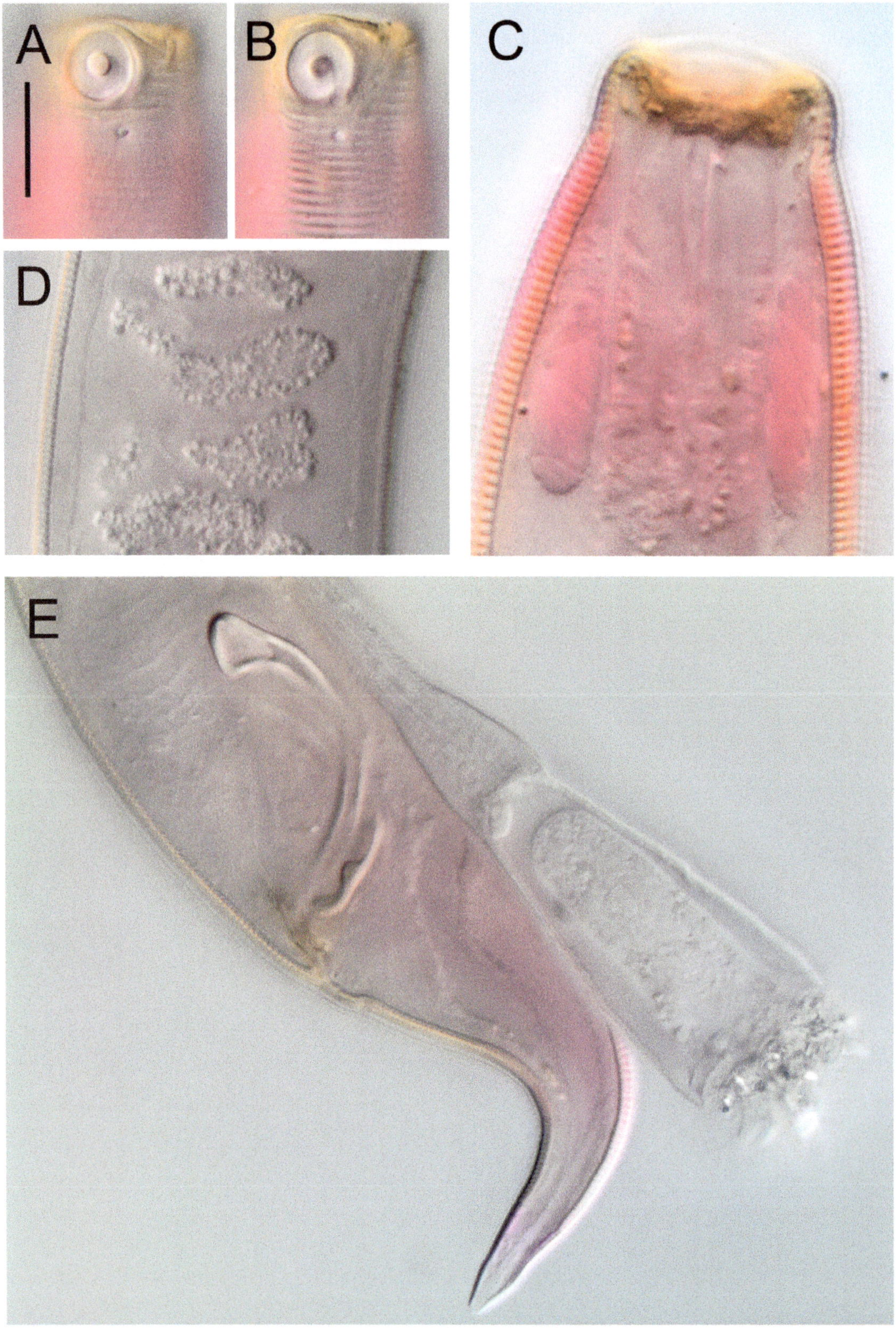

Fig. 3. *Spirinia verecunda* sp. nov., light micrographs. **A**. Male head, showing amphideal aperture and and cuticular hair-like structures. **B**. Male head, showing amphideal fovea. **C**. Male anterior body region showing buccal cavity and elongated glands. **D**. Mature sperm cells. **E**. Posterior body region of male, showing copulatory apparatus and attached protist. Scale bar: A–B, D–E = 10 μm, C = 8 μm.

Table 1. Morphometrics (µm) of *Spirinia verecunda* sp. nov. and *Stygodesmodora confusa* sp. nov. — a, body length/maximum body diameter; abd, anal body diameter; b, body length/pharynx length; c, body length/tail length; cbd, corresponding body diameter; V, vulva distance from anterior end of body; %V, V/total body length.

	Spirinia verecunda **sp. nov.**		***Stygodesmodora confusa*** **sp. nov.**		
	♂	♀♀	♂♂		♀♀
	Holotype	Paratypes	Holotype	Paratype	Paratypes
n	-	2	-	1	5
L	757	667, 675	1241	795	778–1029
a	22	17	35	28	22–34
b	8	6, 7	11	8	8–10
c	15	13	13	14	12–15
Head diameter	17	17, 18	18	14	14–16
Length of cephalic setae	2	2	7–8	6–9	5–9
Amphid height	6	6, 7	11	6	5–6
Amphid width	6	6, 7	11	6	6–7
Amphid width/cbd (%)	35	33, 41	61	43	40–47
Amphid from anterior end	2	3	4	5	3–7
Nerve ring from ant. end	49	45, 53	68	59	57–62
Nerve ring cbd	32	32, 33	29	25	24–25
Pharynx length	100	100, 107	109	95	92–105
Pharyngeal bulb diameter	26	27	22	20	20–22
Pharyngeal bulb cbd	31	33, 34	29	25	25–28
Max. body diameter	35	39, 40	35	42	30–37
Spicule length	38	-	39	42	-
Gubernaculum length	13	-	10	12	-
Anal body diameter	24	21, 22	28	21	17–20
Tail length	49	51, 52	93	57	62–85
Tail length/abd	2.0	2.4	3.3	2.7	3.5–4.5
V	-	347, 361	-	-	428–558
%V	-	51, 54	-	-	53–55
Vulval body diameter	-	39, 40	-	-	30–37

Remarks

Spirinia verecunda sp. nov. shows similarities with several *Spirinia* and *Chromaspirina* species, i.e., *Spirinia guanabarensis* comb. nov., *S. inaurita* comb. nov., *S. okemwai* comb. nov., *S. parma* comb. nov., *Chromaspirina chabaudi*, and *C. multipapillata.* Of these, *S. verecunda* sp. nov. resembles S. *okemwai* comb. nov. the most in cuticle ornamentation (annulated head region, cuticle with minute hair-like structures), size and shape of the amphids, presence of large sperm cells, and absence of precloacal supplements. *S. verecunda* sp. nov. differs from the latter in having shorter body length (667–757 *vs* 966–1425 µm), lower values of a (17–22 *vs* 22–27), shorter cephalic setae (2 *vs* 3–6 µm), shorter spicules (38 *vs* 41–50), and shorter gubernaculum (13 *vs* 19–25). *S. verecunda* sp. nov. also differs from S. *okemwai* comb. nov. in the presence of elongated glands in the anterior region of the pharynx (absent in S. *okemwai* comb. nov.) and the structure of the gubernaculum (hook absent *vs* dorsally directed proximal hook). *S. verecunda* sp. nov. can most readily be distinguished from *S. inaurita* comb. nov. and *S. parma* comb. nov. by the size of the amphids (in males: 35% cbd in *S. vercunda* sp. nov. *vs* >

60% cbd in *S. inaurita* comb. nov. and *S. parma* comb. nov.), from *Chromaspirina chabaudi* by the structure of the gubernaculum (not surrounding spicules in *S. verecunda* sp. nov. *vs* surrounding spicules distally in *C. chabaudi*), from *C. multipapillata* by the absence of precloacal supplements (present in *C. multipapillata*), and from *S. guanabarensis* comb. nov. by its small body size (< 800 µm in *S. verecunda* sp. nov. *vs* >1800 µm in *S. guanabarensis* comb. nov.) and low values of a (17–22 *vs* 30–48).

Genus ***Stygodesmodora*** Blome, 1982

Type species

Stygodesmodora epixantha Blome, 1982.

Emended diagnosis (modified from Blome 1982)

Cuticle with coarse annulations; numerous long somatic setae may be present; annulated head region (sometimes intraspecific variation, with the head of some specimens only partially annulated in dorsal and ventral regions). Amphid unispiral or spiral located on amphideal plate; small to medium sized buccal cavity with dorsal tooth and (usually) smaller subventral teeth. Pharynx with rounded or pyriform posterior bulb, without cuticularised lumen. Pre-cloacal supplements present or absent; spicules short, arcuate, and with well-developed capitulum; gubernaculum present.

Valid species

S. bacillicauda (Gerlach, 1963)
S. epixantha Blome, 1982

Species inquirenda

S. rotundicephala (Cobb, 1920)

Stygodesmodora confusa sp. nov.

urn:lsid:zoobank.org:act:44B8B268-21AE-4E30-8A7E-D707B1D54C8F
Figs 4–7; Table 1

Diagnosis

Stygodesmodora confusa sp. nov. is characterised by a relatively short body (778–1241 µm), spiral amphids with 1.0–1.25 turns, cephalic setae situated at or slightly posterior to mid-level of amphid, and males with four precloacal supplements consisting of short setae on wide bases.

Type specimens

Holotype

NEW ZEALAND: ♂, NIWA 88377, 24 Apr. 2010, NIWA cruise TAN1004, station 92, canyon axis on southern Hikurangi Margin, 41.8921° S, 174.6347° E, 686 m.

Paratypes

NEW ZEALAND: 1 ♂, 2 ♀♀, NIWA 88378, 20 Feb. 2011, NIWA cruise TAN1103, station 69, central Chatham Rise, 43.331° S, 178.288° E, 350 m; 3 ♀♀ (same data as other paratypes) were measured to provide morphometric data and then processed for scanning electron microscopy.

Etymology

The species name is derived from the Latin adjective *confusus,* meaning confused, perplexed or obscure (feminine form used, as *-dora* is feminine), and refers to the close affinities of this species with other desmodorid genera (see Remarks), which resulted in initial confusion in the placement of this species.

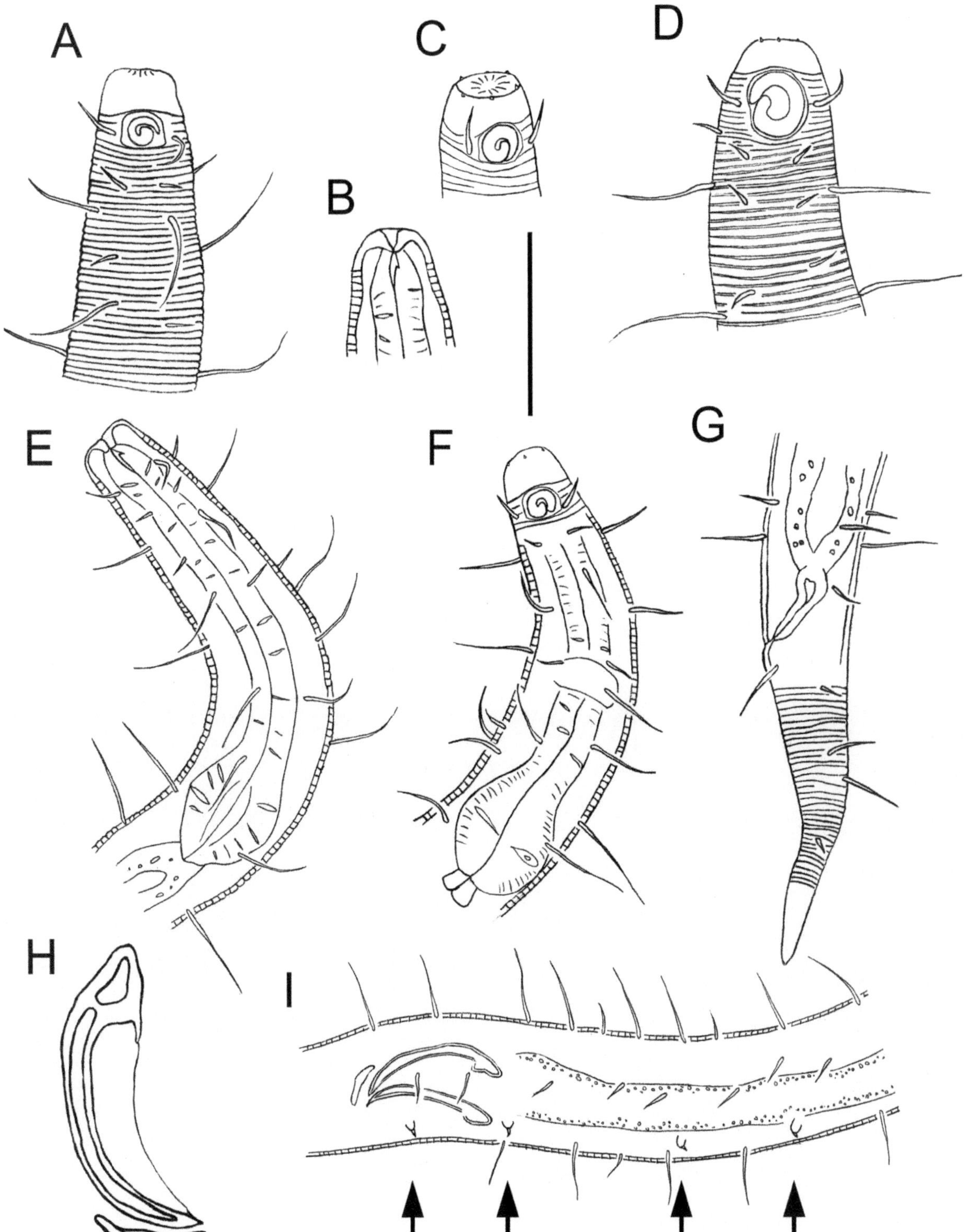

Fig. 4. *Stygodesmodora confusa* sp. nov. **A**. Anterior body region of female. **B**. Male head (holotype). **C**. Male head (paratype). **D**. Male head (holotype). **E–F**. Anterior body region of female. **G**. Female posterior body region. **H**. Right spicule and gubernaculum. **I**. Male copulatory apparatus. Arrows show position of pre-cloacal supplements. Scale bar: A–D = 30 µm, E–G = 36 µm, H = 20 µm, I = 40 µm.

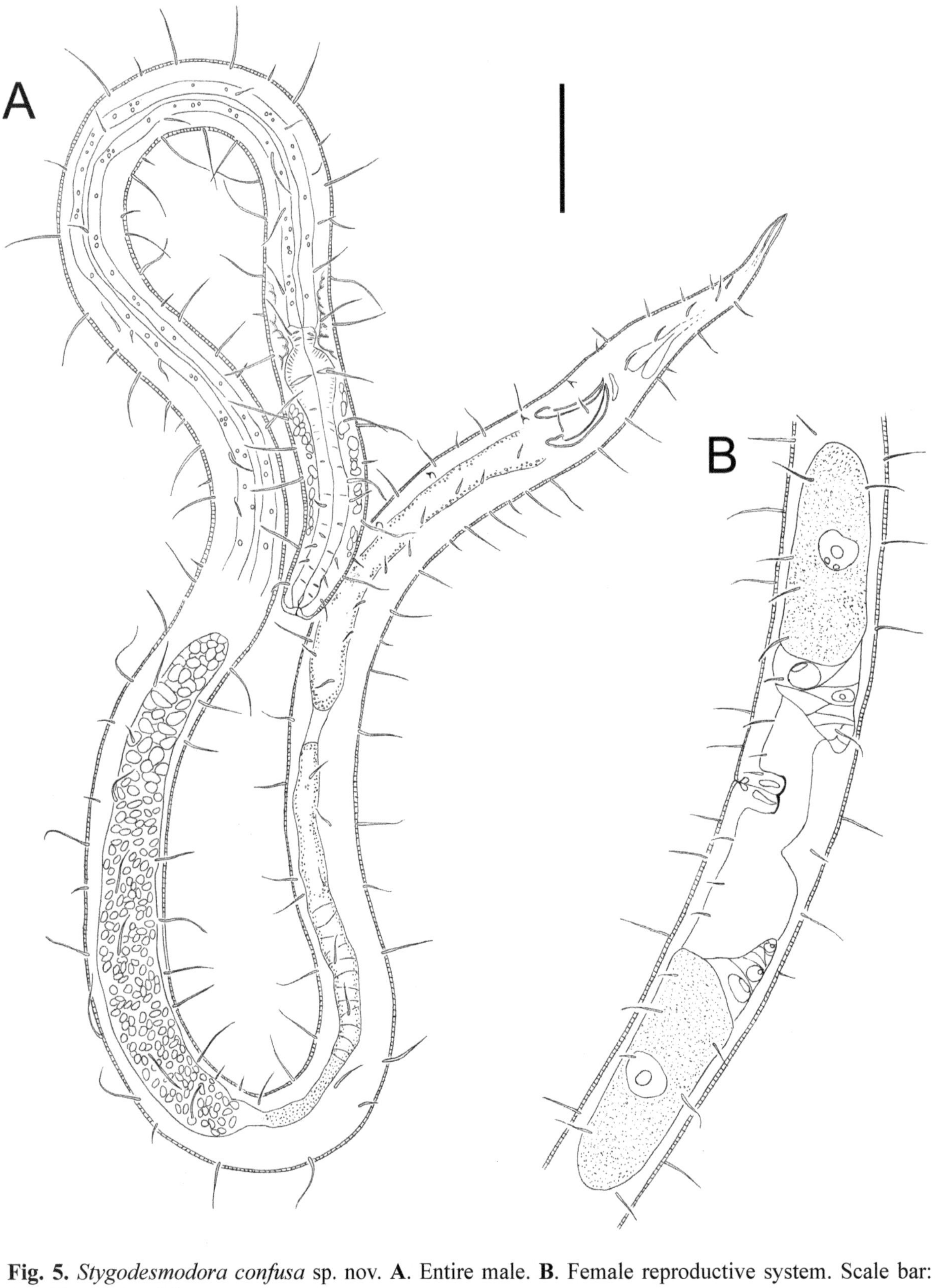

Fig. 5. *Stygodesmodora confusa* sp. nov. **A**. Entire male. **B**. Female reproductive system. Scale bar: A = 50 μm, B = 36 μm.

Description

Male

Relatively short cylindrical body, pale golden colour, tapering slightly towards both ends. Cuticle 0.9–1.4 µm thick, with coarse annuli, approximately 1 µm apart, no lateral differentiation. Ventral ala absent. Somatic setae of different sizes (4–19 µm long; Fig. 5A), set out in eight longitudinal rows in anterior one fifth of body length and in six longitudinal rows along rest of body except on tail, where they are irregularly arranged.

Annulated head region with annulations completely surrounding amphid (Figs 6A–C, 7C); in some specimens, there is a seemingly non-annulated head region, but some partial annulation of the dorsal

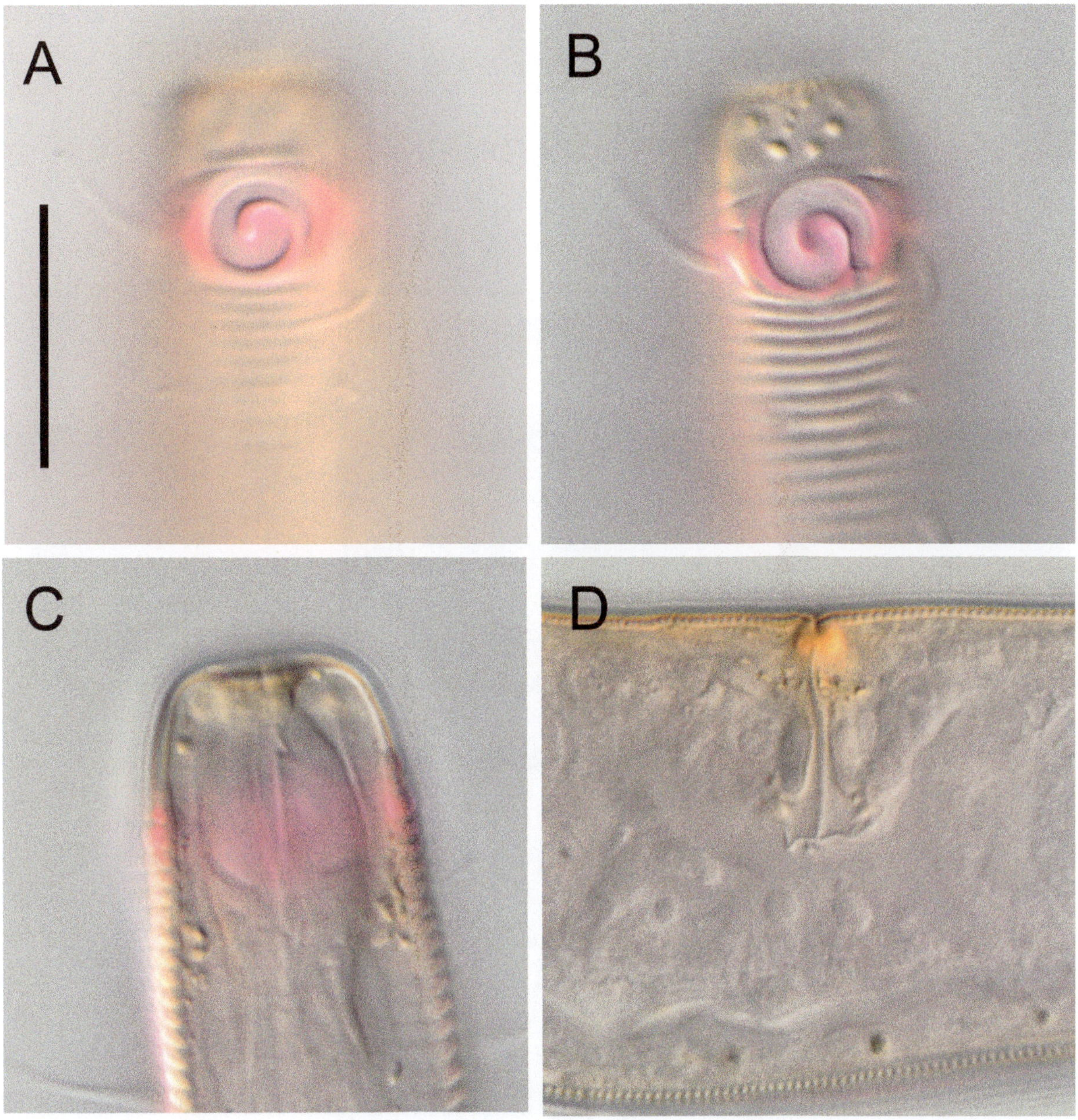

Fig. 6. *Stygodesmodora confusa* sp. nov., light micrographs of female. **A**. Head region showing amphideal aperture. **B**. Head region showing amphideal fovea. **C**. Head region showing buccal cavity. **D**. Vulva. Scale bar: A–C = 15 µm, D = 18 µm.

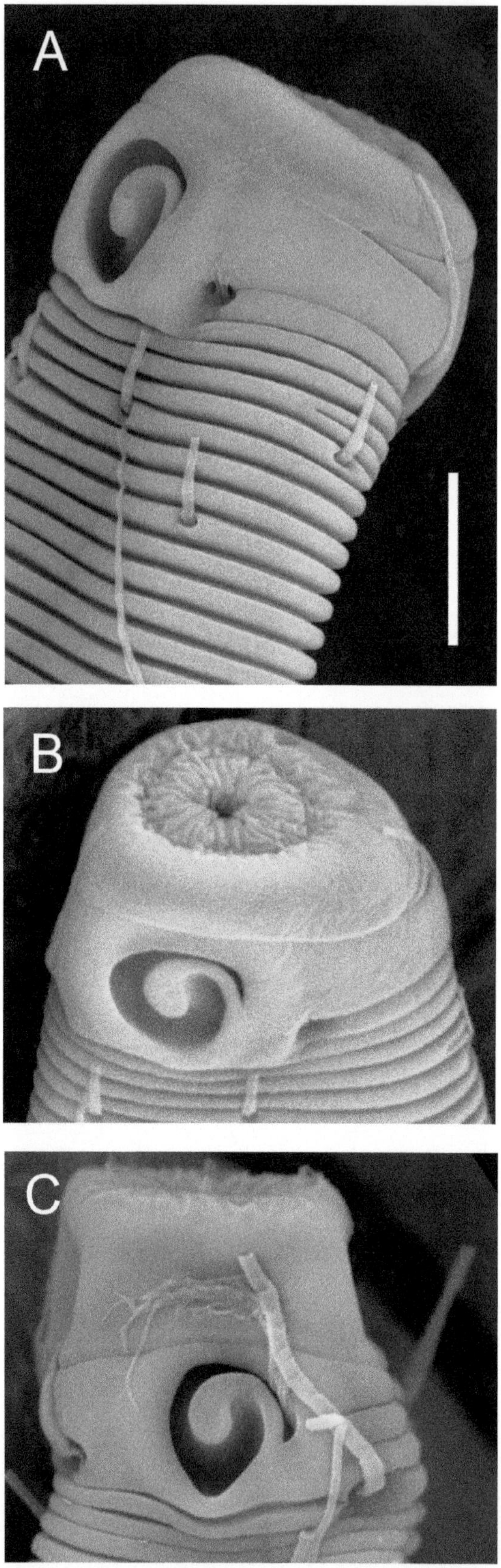

Fig. 7. *Stygodesmodora confusa* sp. nov. Scanning electron micrographs. Female head (three different specimens). Scale bar = 5 μm.

and ventral sides of the head can be observed (Fig. 7A–B); both forms are considered annulated head regions, even though the degree of annulation differs between the two morphotypes. Labial region folded inwards in all specimens. Six papilliform outer labial sensillae and four cephalic setae, 44–57% cbd long, situated at or slightly posterior to mid-level of amphid. Spiral amphideal fovea, 1.2–1.4 turns, located on a cuticularised amphideal plate (Figs 6B, 7B), larger in holotype than in paratype (~0.6 *vs* 0.4 cbd, respectively; Fig. 4C–D); amphideal aperture with similar shape to amphideal fovea but tapering proximally (compare Fig. 6A and 6B).

Buccal cavity small, with small dorsal tooth; subventral teeth not observed (Fig. 6C). Pharynx muscular, slightly swollen at anterior extremity, and with pyriform bulb at posterior extremity. Nerve ring at 60–62% of pharynx length. Secretory-excretory system not observed. Cardia short.

Reproductive system monorchic, with single anterior testis situated to the right of intestine, 53–63 μm long, outstretched and opposed. Mature sperm small, globular, 2–3 × 3–5 μm. Short, arcuate spicules tapering distally; well-developed capitulum and broad velum (Fig. 4H). Small, plate-shaped gubernaculum without crurae. Four pre-cloacal supplements consisting of short setae on broad bases (Fig. 4I), 23–45 μm apart, beginning 15 μm anterior to cloaca. Tail conical, with three small caudal glands and spinneret.

Female

Similar to males. Female reproductive system didelphic, amphidelphic with reflected ovaries situated either to the right or left of intestine, with anterior and posterior ovaries always situated on opposite sides. Vulva located slightly post median. Cuticular pars distalis vaginae and pars proximalis vaginae surrounded by constrictor muscle. Mature eggs 66–73 × 25 μm.

Remarks

Stygodesmodora confusa sp. nov. can most easily be differentiated from the other two species of the genus by the location of the cephalic setae at mid-level of amphids (*vs* anterior to amphids in *S. bacillicauda* and *S. epixantha*). *S. confusa* sp. nov. also differs from *S. bacillicauda* in the presence of pre-cloacal supplements (absent in *S. bacillicauda*) and the absence of lateral crurae on the gubernaculum (present in *S. bacillicauda*), and from *S. epixantha* by the presence of numerous long somatic setae (short somatic setae in *S. epixantha*).

Stygodesmodora confusa sp. nov. closely resembles species of other desmodorid genera, and distinguishing it from these other species requires careful examination of the head region. *S. confusa* sp. nov. resembles *Echinodesmodora moensi* Verschelde & Vincx, 1996 in the absence of a cephalic capsule, numerous long somatic setae, and cephalic setae at or slightly posterior to mid-level of amphids, but can be differentiated from the latter by the absence of pharyngeal lumen cuticularisation (present in *E. moensi*), amphid shape (≤ 1.25 turns *vs* 1.5–2.0 turns in *E. moensi*), presence of setose precloacal supplements (absent in *E. moensi*), and the absence of knotty protrusions on the tail (present in *E. moensi*). The most important distinction, however, is the presence of an amphideal plate in *S. confusa* sp. nov. This species is also superficially similar to species of *Bolbonema*, i.e., *B. brevicolle* (Cobb, 1920), *B. longisetosum* (Jensen, 1985) comb. nov., and *B. spiralis* Hourston & Warwick, 2010, in the presence of both short and long somatic setae, cephalic setae just posterior to mid-level of amphid, and in the structure of the copulatory apparatus. The amphideal plates in *S. confusa* sp. nov. give a swollen appearance to the head region laterally (see Fig. 7A), and in some specimens there is a gap in the body annulations in the head region (compare Fig. 7A and 7C), which may be wrongly interpreted as a non-annulated head region or even a head capsule. *S. confusa* sp. nov., however, is clearly defined by an annulated region, and not a head capsule, because (1) the swollen part of the head is mostly restricted to the area surrounding the amphids (i.e., the amphideal plates), (2) the cuticle is the same thickness in head and body regions (see Fig. 6C), and (3) annulations are present dorsally and ventrally at least to mid-level of amphids, with one continuous annulation anterior to the amphid (see Figs 6B, 7A–C). The genus *Bolbonema*, however, is characterized by a globular head capsule (i.e., conspicuously thicker cuticle and absence of annulations; Verschelde & Vincx 1996; Verschelde *et al.* 1998).

Discussion

Muthumbi *et al.* (1995) provided an emended diagnosis for *Chromaspirina* and, more recently, Silva *et al.* (2009) provided an emended diagnosis for *Spirinia.* Maria *et al.* (2009) briefly discussed the relationship between *Chromaspirina* and *Spirinia.* They identified the size of the buccal cavity and dorsal tooth, and the shape of the posterior pharyngeal bulb as the main distinguishing features between the two genera and noted that several *Chromaspirina* species have a small dorsal tooth which "emphasises the close relationship of this genus with *Spirinia*..." and that "further investigation is needed into the relationships [of] *Spirinia* with *Chromaspirina*" (Maria *et al.* 2009: 28). In his review of *Spirinia*, Coles (1987) identified buccal morphology as the main distinguishing feature between the two genera.

We argue that the shape of posterior pharyngeal bulb and the presence or absence of precloacal supplements show too much overlap to useful diagnostic features to distinguish *Spirinia* and *Chromaspirina*. For example, *S. okemwai* comb. nov. and *S. guanabarensis* comb. nov. lack precloacal supplements (a trait usually associated with *Spirinia*), but possess relatively weakly developed, oval-shaped posterior pharyngeal bulbs (a trait usually associated with *Chromaspirina*; see Maria *et al.* 2009). *S. sophia* Silva *et al.*, 2009 also appears to be intermediate between the two genera: it possesses precloacal supplements (absent in all other *Spirinia* species but present in some *Chromaspirina* species) and a round to oval-shaped posterior pharyngeal bulb (despite *Spirinia* being defined as having a round posterior pharyngeal bulb; see Silva *et al.* 2009). Differences in the morphology of the pharyngeal bulb and precloacal supplements are not reliable traits to distinguish between *Chromaspirina* and *Spirinia* because (1) differences in the shape of pharyngeal bulbs are often subtle (e.g., pyriform *vs* elongated or oval) and (2) precloacal supplements in these two genera are often difficult to distinguish and may have been omitted in some species descriptions. Buccal morphology provides the clearest and most consistent way to differentiate between *Spirinia* and *Chromaspirina* because it can easily be observed; morphological features of the head region are also the most important traits for differentiating between desmodorid genera.

Perspiria was originally described as a subgenus of *Spirinia* by Wieser & Hopper (1967). The only distinguishing feature of this subgenus given by the latter authors was "the more prominently striated and flagellate tail." Coles (1987) also gave the same distinguishing features between the two subgenera. Vincx & Gourbault (1989) later raised *Perspiria* to generic status and included two key characteristics in their diagnosis of the genus: amphids only partially surrounded by cuticle annulations and a "prominently striated" and "filiform" tail, which we interpret as a coarsely annulated, conico-cylindrical tail from the drawings (see figs 1 and 2 in Vincx & Gourbault 1989). Some *Spirinia* species, however, have coarsely annulated (e.g., *S. verecunda* sp. nov.) or conico-cylindrical tails (e.g., *S. septentrionalis* (Cobb, 1914)), and others have amphids not completely surrounded by the body annulations (e.g., *S. lara* Silva *et al.*, 2009). Here, we propose to differentiate between *Spirinia* and *Perspiria* primarily based on tail shape (conical *vs* conico-cylindrical to filiform), as well as position of amphids relative to cuticle annulations (entirely *vs* partially surrounded), and have suggested nomenclatural changes to reflect these distinctions.

The monospecific genus *Spirodesma* was recently described by Cavalcanti *et al.* (2009) from continental slope sediments in the Southwest Atlantic. The authors argued that the genus is most similar to *Chromaspirina* based on the rounded shape of the head, annulated cuticle, spiral amphideal fovea, and conical tail. We consider these arguments invalid because (1) an annlated cuticle is also a characteristic of some *Spirinia* (e.g., *Spirinia verecunda* sp. nov.) and *Perspiria* species (e.g., *P. papillata* Vincx & Gourbault, 1989), (2) the spiral amphideal fovea in *Spirodesma* is in fact very similar in shape to the cryptospiral amphideal fovea found in some *Spirinia* species (e.g., *S. verecunda* sp. nov.) and to the spiral amphideal fovea of *Perspiria papillata*, and (3) the tail of *Spirodesma* is conico-cylindrical (as stated in the original description), not conical. In addition, head shape is not a particularly informative trait for differentiating between genera. In our opinion, *Spirodesma* is most closely related to *Perspiria* because of the similar buccal cavity (narrow with small or minute teeth), the position of the amphids (only partially surrounded by cuticle annulations), and the shape of the tail (conico-cylindrical). The only difference between these genera appears to be the relative size of the teeth: *Perspiria* is characterised by a narrow, lightly cuticularised buccal cavity with no or a small/inconspicuous dorsal tooth and no or minute subventral teeth, while *Spirodesma* is characterised by a narrow, lightly cuticularised buccal cavity with three small, equal teeth. *Spirodesma* and *Perspiria* may therefore need to be synonymised in the future as more information becomes available on the former genus.

Stygodesmodora rotundicephala (Cobb, 1920) was originally described as *Amphispira rotundicephala* and was included in the World Register of Marine Species list of *Stygodesmodora* species at the time the present study was conducted (WoRMS Editorial Board 2014). The original description, which is based on a single juvenile specimen, clearly shows a head capsule, whereas no amphideal plate was described. Based on the drawing, the specimen/species could possibly belong in *Zalonema*, but, as the original species description of *Amphispira rotundicephala* was based only on one juvenile, we regard *A. rotundicephala* Cobb, 1920 to be a *species inquirenda*.

Acknowledgements

Funding was provided by NIWA's Coasts and Oceans Centre Research Programme 2 (2013/14 SCI) and the programme 'Impact of Resource Use on Vulnerable Deep-sea Communities' (CO1X0906) and "Consequences of Earth-Ocean Change" (C01X0702). We are grateful to Scott Nodder (NIWA) for facilitating sampling on Chatham Rise (TAN1103), and to Norliana Rosli for processing TAN1004 samples. We also acknowledge the other participants of NIWA voyages TAN1004 and TAN1103, and the officers and crew of RV *Tangaroa*.

References

Armenteros M., Ruiz-Abierno A. & Decraemer W. 2014. Revision of Desmodorinae and Spiriniinae (Nematoda: Desmodoridae) with redescription of eight known species. *European Journal of Taxonomy* 96: 1–32.

Bastian H.C. 1865. Monograph of the Anguillulidae, or free nematoids, marine, land, and freshwater; with descriptions of 100 new species. *The Transactions of the Linnean Society of London* 25: 73–184.

Blome D. 1982. Systematik der Nematoda eines Sandstrandes der Nordseeinsel Sylt. *Mikrofauna des Meeresbodens* 86: 1–194.

Boucher G. 1975. Nématodes des sables fins infralittoraux de la Pierre Noire (Manche Occidentale). I. Desmodorida. *Bulletin de Muséum National d'Histoire Naturelle* 285: 101–128.

Bussau C. 1993. *Taxonomische und Ökologische Untersuchungen an Nematoden des Peru-Beckens.* PhD Thesis, *Mathematisch-Naturwissenschaftliche Fakultät der Christian-Albrechts-Universität zu Kiel, Germany.*

Castro F.J.V., Bezerra T.N.C., Da Silva M.C. & Fonsêca-Genevois V. 2006. *Spirinia elongata*, sp. nov. (Nematoda, Desmodoridae) from Pina Basin, Pernambuco, Brazil. *Zootaxa* 1121: 53–68.

Cavalcanti M.F., Da Silva M.C., Da Fonsêca-Genevois V. 2009. *Spirodesma magdae* nov. gen. nov. sp. (Nematoda: Desmodoridae) from the Brazilian deep sea (Campos Basin, Rio de Janeiro, Brazil). *Zootaxa* 2096: 109–118.

Chitwood B.G. 1936. Some marine nematodes from North Carolina. *Proceedings of the Helminthological Society of Washington* 3: 1–16.

Cobb N.A. 1920. One hundred new nemas (type species of 100 new genera). *Contributions to a Science of Nematology 9: 217–*343.

Coles J.W. 1987. Observations on the marine nematode genus *Spirinia* Gerlach, 1963 (Desmodoridae: Spiriniinae) with descriptions of two new species. *Bulletin of the British Museum of Natural History* (*Zoology*) 53: 79–101.

De Coninck L.A. 1943. Sur quelques espèces nouvelles de nématodes libres des eaux et des terres saumâtres de l'Islande. *Biologisch Jaarboek* 10: 193–220.

De Coninck L.A. 1965. Classe des Nématodes: Systematique des nématodes et sous-classe des Adenophorea. *In*: Grassé P.P. (ed.) *Traité de Zoologie* 4: 586–681.

Decraemer W. & Smol N. 2006. Orders Chromadorida, Desmodorida and Desmoscolecida. *In*: Abebe E.-A., Transpurger W. & Andrassy I. (eds) *Freshwater Nematodes: Ecology and Taxonomy*: 497–573. UK CABI Publishing, Cambridge.

Gerlach S.A. 1953. Recherches sur la faune des eaux interstitielles de Madagascar. III. Sur quelques nématodes libres des eaux souterraines littorales de Madagascar. *Mémoires de l'Institut Scientifique de Madagascar* 8: 73–86.

Gerlach S.A. 1954. Nouveaux nématodes libres des eaux souterraines littorales françaises. *Vie et Milieu* 4: 95–110.

Gerlach S.A. 1956. Die Nematodenbesiedlung des tropischen Brandungsstrandes von Pernambuco. Brasilianische Meeres-Nematoden II. *Kieler Meeresforschungen* 12: 202–218.

Gerlach S.A. 1957. Marine Nematoden von der Kongo-Mündung. *Bulletin de l'Institut Royal des Sciences Naturelles de Belgique* 33: 1–28.

Gerlach S.A. 1963. Frielebende Meeresnematoden von den Malediven II. *Kieler Meeresforschungen* 19: 67–103.

Gerlach S.A. 1965. Freilebende Meeresnematoden aus der Gezeitenzone von Spitsbergen. *Veröffentlichungen des Instituts für Meeresforschung in Bremerhaven* 9: 109–172.

Gerlach S.A. & Murphy D.G. 1965. Die Familiengruppen-Namen der freilebenden Nematoden-Adenophorea (Aphasmidia). *Zoologischer Anzeiger* 175: 211–216.

Inglis W.G. 1968. Interstitial nematodes from St. Vincent's Bay, New Caledonia: Expédition française sur les récifs coralliens de la Nouvelle Calédonie. *Editions de la Fondation Singer-Polignac, Occasional Publications* 2: 29–74.

Jensen P. 1978. Revision of Microlaimidae, erection of Molgolaimidae fam. n., and remarks on the systematic position of *Paramicrolaimus* (Nematoda, Desmodorida). *Zoologica Scripta* 7: 159–173.

Leduc D. & Verschelde D. 2013. One new genus and two new free-living nematode species (Desmodorida, Desmodoridae) from the continental margin of New Zealand, Southwest Pacific Ocean. *Zootaxa* 3609: 274–290.

Lorenzen S. 1981. Entwurf eines phylogenetischen Systems der freilebenden Nematoden. *Veröffentlichungen des Instituts für Meeresforschung in Bremerhaven, Supplement* 7: 1–472.

Luc M. & De Coninck L.A.P. 1959. Nématodes libres marins de la region de Roscoff. *Archives de Zoologie Expérimentale et Générale* 98: 108–166.

Maria T.F., Esteves A.M., Smol N., Vanreusel A. & Decraemer W. 2009. *Chromaspirina guanabarensis* sp. n. (Nematoda: Desmodoridae) and a new illustrated dichotomous key to *Chromaspirina* species. *Zootaxa* 2092: 21–36.

Murphy R.J., Pinkerton M.H., Richardson K.M, Bradford-Grieve J.M. & Boyd P.W. 2001. Phytoplankton distributions around New Zealand derived from SeaWiFS remotely-sensed ocean colour data. *New Zealand Journal of Marine and Freshwater Research* 35: 343–362.

Muthumbi A., Verschelde D. & Vincx M. 1995. New Desmodoridae (Nematoda: Desmodoroidea): three new species from *Ceriops* mangrove sediments (Kenya) and one related new species from the North Sea. *Cahiers de Biologie Marine* 36: 181–195.

Ott J.A. 1977. New freeliving marine nematodes from the West Atlantic. I. Four new species from Bermuda with a discussion of the genera *Cytolaimium* and *Rhabdocoma* Cobb, 1920. *Zoologischer Anzeiger* 198: 120–138.

Silva M.C., Castro F.J.V., Cavalcanti M.F. & Fonsêca-Genevois V. 2009. *Spirinia lara* sp. n. and *Spirinia sophia* sp. n. (Nematoda, Desmodoridae) from the Brazilian continental margin (Campos Basin, Rio de Janeiro). *Zootaxa* 2081: 31-45.

Somerfield P. & Warwick R.M. 1996. *Meiofauna in Marine Pollution Monitoring Programmes: A Laboratory Manual*. Ministry of Ariculture, Fisheries, and Food, Lowestoft, U.K.

Verschelde D. & Vincx M. 1996. Four new species of the family Desmodoridae (Nematoda, Desmodoroidea) from Kenya. *Zoologica Scripta* 25: 1–20.

Verschelde D., Gourbault N. & Vincx M. 1998. Revision of *Desmodora* with descriptions of new desmodorids (Nematoda) from hydrothermal vents of the Pacific. *Journal of the Marine Biological Association of the United Kingdom* 78: 75–112.

Verschelde D., Nicholas W. & Vincx M. 2006. A review of the genera *Croconema* Cobb, 1920 and *Pseudochromadora* Daday, 1899 (Nematoda, Desmodoroidea): new species from the coasts of Kenya and Australia. *Hydrobiologia* 571: 17–40.

Vincx M. & Gourbault N. 1989. Desmodoridae from the Bay of Morlaix (Brittany) and the Southern Bight of the North Sea. *Cahiers de Biologie Marine* 30: 103–114.

Vitiello P. 1971. Nématodes libres marins des vases profondes du Golfe du Lion III. Monhysterida, Araeolaimida, Desmodorida. *Tethys* 2: 647–690.

Warwick R.M. 1970. Fourteen new species of free-living marine nematodes from the Exe estuary. *Bulletin of the British Museum of Natural History* (*Zoology*) 19: 137–177.

Wieser W. 1954. Free-living nematodes II. Chromadoroidea. *Lunds Universitets Arsskrift, N.F. Avd. 2* 50 (16): 1–149.

Wieser W. & Hopper B. 1967. Marine nematodes of the east coast of North America. I. Florida. *Bulletin of the Museum of Comparative Zoology* 135: 239–343.

WoRMS Editorial Board. 2014. World Register of Marine Species.

11

The millipede genus *Solaenodolichopus* Verhoeff, 1924 (Diplopoda, Polydesmida, Paradoxosomatidae). 1. New genus diagnosis and redescriptions of named species

Robert MESIBOV

Queen Victoria Museum and Art Gallery, Launceston, Tasmania, Australia 7250

Abstract. *Solaenodolichopus* Verhoeff, 1924 is redefined to include *S. pruvoti* (Brolemann, 1931), *S. rubriventris* Verhoeff, 1928, *S. sulcatus* (Verhoeff, 1928), *S. teres* (Verhoeff, 1924), *S. vittatus* (Verhoeff, 1924) and *S. walesius* Verhoeff, 1928, each of which is redescribed. Lectotypes are designated for *S. sulcatus*, *S. teres*, *S. vittatus* and *S. walesius*. *Parwalesoma* Verhoeff, 1937 is synonymised with *Solaenodolichopus* and *S. vittatus dorsalis* (Verhoeff, 1924) with *S. vittatus vittatus* (Verhoeff, 1924).

Keywords. Millipede, Paradoxosomatidae, Queensland, New South Wales, Australia.

Introduction

Solaenodolichopus Verhoeff, 1924 is an eastern Australian genus of Paradoxosomatidae with a distinctive gonopod structure and no obvious paranota on diplosegments. There are six named and many undescribed species native to Queensland (Mesibov 2008) and New South Wales (Car 2009). One of the named species, *S. pruvoti* (Brolemann, 1931), was first described from New Caledonia and has recently been reported from Western Australia (Car *et al.* 2013); more records are reported here from southeast Queensland, where *S. pruvoti* is likely to be native. Records of *S. pruvoti* from South Australia (Jeekel 1982, 2002) may be based on a misidentification (see below).

In this paper I review the history of the *Solaenodolichopus* genus concept, propose a new diagnosis, redescribe the six named species, designate several lectotypes, accept two previously proposed synonymies and propose two new synonymies. New species of *Solaenodolichopus* will be treated in future papers in this series.

Materials and Methods

'Male' and 'female' in the text refer to adult individuals. All specimens are stored in 70–80% ethanol in their respective repositories.

The photomicrographs in Fig. 11C and 11D were taken using an eyepiece video camera mounted on an Amscope binocular microscope. Other photomicrographs were taken with a Canon EOS 1000D digital SLR camera mounted on a Nikon SMZ800 binocular dissecting microscope equipped with a beam splitter. Measurements were made to the nearest 0.1 mm with the same microscope using an eyepiece grid and a reference scale. Most photomicrographs used in the figures are focus-stacked composites prepared with Zerene Stacker 1.04 software. Preliminary gonopod drawings were traced from printed copies of focus-stacked photomicrographs; details were added and corrected by reference to the actual gonopod. SEM images were acquired digitally with a FEI Quanta 600 ESEM at low vacuum using uncoated specimens which were air-dried, examined and returned to alcohol. Maps were generated using ArcView GIS 3.2. All figures were prepared using GIMP 2.8 image editing software; backgrounds in some photomicrographs have been edited to remove distracting highlights and artifacts.

Latitude/longitude figures are given with the WGS84 datum together with my estimate of the spatial uncertainty (Darwin Core coordinateUncertaintyInMeters). Data in square brackets are my additions; latitude/longitude figures not in square brackets are based on label data or the respective museum collection databases.

Details of specimen lots of the species redescribed in this paper are tabulated as an Excell file in Supplement 1.

Images of *Solaenodolichopus* types as alcohol specimens and slide mounts, including their labels, are compiled in a PDF file in Supplement 2.

Abbreviations

AM	=	Australian Museum, Sydney, Australia
NHRS	=	Naturhistoriska Riksmuseet, Stockholm, Sweden
NSW	=	New South Wales, Australia
QLD	=	Queensland, Australia
QM	=	Queensland Museum, Brisbane, Australia
SAM	=	South Australian Museum, Adelaide, Australia
ZMB	=	Museum für Naturkunde, Berlin, Germany
ZSM	=	Zoologische Staatssammlung München, Munich, Germany

Results

Order Polydesmida Pocock, 1887
Suborder Strongylosomatidea Brölemann, 1916
Family Paradoxosomatidae Daday, 1889
Subfamily Australiosomatinae Brölemann, 1916
Tribe Antichiropodini Brölemann, 1916

Solaenodolichopus Verhoeff, 1924

Solänodolichopus Verhoeff, 1924: 10 (in key as subgenus of *Antichiropus*), 11 (in key as subgenus of *Antichiropus*), 19 (first description).
Paraulacoporus Verhoeff, 1928: 88, 90 (in key), 91 (in key), 93. [Type species *Paraulacoporus sulcatus* Verhoeff, 1928, by monotypy.]
Parwalesoma Verhoeff, 1937: 139 (first description). [New synonymy. Type species *Parwalesoma castaneum* Verhoeff, 1937, by monotypy.]

Solaenodolichopus – Attems 1926: 143 (in key as subgenus of *Antichiropus*); 1937: 260 (synonymised with *Aulacoporus*). — Verhoeff 1932: 1574 (as subgenus of *Antichiropus*), 1595 (in key), 1597 (in key). — Jeekel 1965: 8; 1968: 29 (as synonym of *Aulacoporus*); 1971: 233 (type species designated); 2000: 37 (in key), 40–42; 2003: 42. — Hoffman 1980: 166 (as synonym of *Aulacoporus*). — Nguyen & Sierwald 2013: 1159.
Solänodolichopus – Verhoeff 1928: 88 (raised to genus), 90 (in key), 91 (in key), 94. — Brolemann 1931: 299. — Verhoeff 1937: 139. — Verhoeff 1941: 8 (in key).
Paraulacoporus – Attems 1929: 258. — Brolemann 1931: 295, 298. — Verhoeff 1932: 1595 (in key), 1598 (in key). — Attems 1937: 260 (synonymised with *Aulacoporus*). — Verhoeff 1941: 8 (in key). — Jeekel 1965: 8. — Jeekel 1968: 29 (as synonym of *Aulacoporus*). — Jeekel 1971: 230 (in genus catalogue). — Hoffman 1980: 166 (as synonym of *Aulacoporus*). — Jeekel 2000: 40, 42 (synonymised with *Solaenodolichopus*). — Nguyen & Sierwald 2013: 1159 (as synonym of *Solaenodolichopus*).
Parwalesoma – Attems 1940: 549 (Verhoeff's description reworded). — Verhoeff 1941: 8 (in key). — Jeekel 1965: 8. — Jeekel 1968: 29, 126. — Jeekel 1971: 231. — Hoffman 1980: 166. — Jeekel 2000: 37 (in key), 40-42. — Jeekel 2003: 42. — Nguyen & Sierwald 2013: 1157.

Type species

Antichiropus (Solaenodolichopus) teres Verhoeff, 1924, by subsequent designation.

Other assigned species

S. pruvoti (Brolemann, 1931), *S. rubriventris* Verhoeff, 1928, *S. sulcatus* (Verhoeff, 1928), *S. vittatus* (Verhoeff, 1924) and *S. walesius* Verhoeff, 1928.

History of *Solaenodolichopus*

The sometimes confusing development of the *Solaenodolichopus* concept over the past 90 years has involved five genus names and four diplopodologists.

Besides *Solaenodolichopus*, some of the species redescribed in this paper have also been assigned to *Antichiropus* Attems, 1911, *Aulacoporus* Verhoeff, 1924, *Paraulacoporus* Verhoeff, 1928 and *Parwalesoma* Verhoeff, 1937.

The four diplopodologists were the Austrian Carl Attems (1868–1952, in full, Carl Aug. Graf [Count] Attems-Petzenstein), the German Karl Wilhelm Verhoeff (1867-1944), the Frenchman Henry Wilfred Brölemann (1860-1933) and the Dutchman Casimir Albrecht Willem Jeekel (1922-2010). I use the spelling 'Brolemann' below rather than 'Brölemann', following that author's practice in the 1931 publication cited in this paper. Brolemann dropped the umlaut in his name from 1920 onwards (J.-J. Geoffroy, in litt., 30 Apr. 2009).

All four specialists viewed the polydesmidan gonopod as a modified walking leg on which it was possible to recognise podomere homologs. Their differing opinions on homology are mentioned only briefly below in the context of a chronology of *Solaenodolichopus*. My purpose here is to try to explain the reasoning behind particular generic assignments in the past, not to thoroughly review an enduring and still-unresolved controversy.

Solaenodolichopus was first proposed by Verhoeff (1924) as a subgenus of *Antichiropus* Attems, 1911. Attems (1911) had described seven new species from Western Australia with gonopod telopodites divided into three main parts: a small, densely setose, basal portion (*Femur*); a long, cylindrical, slender, central portion (*Tibia*) tipped with short prongs (*Zacken*); and a long, thin, coiled, apical portion (*Kanalast*) carrying the prostatic groove and its terminal opening. Attems assumed that the tarsus of the modified leg had fused with the tibia, and that the distal portion of the tibia was its tarsal section (*Tarsalabschnitt*).

In his view, the first joint of this tarsal section consisted of the *Kanalast* and a prong-like process 'a' at the end of the tibia and close to the prostatic groove. Attems suggested that another tibial process, 'c', was a remnant of a second tarsal joint. (For a discussion of the *Antichiropus* gonopod and its components, see Car *et al.* 2013.)

Verhoeff (1924) described the telopodites of the new central Queensland species *teres* and the southern Queensland *vittatus* (the latter divided into two subspecies) as likewise having a small, densely setose, basal portion (*Praefemur*); a long, cylindrical, slender, central portion (*Femur*) tipped with short accessory branches (*Nebenäste*); and a long, terminal process carrying the prostatic groove and its opening. Verhoeff regarded the latter as an evolutionarily novel structure (*Neubildung*) he called a *Solänomerit*, which he had recognised in many other Polydesmida and which had no homologue on a walking leg; in what follows I will use the modern term 'solenomere'. The two Queensland species were placed in *Antichiropus* (*Solaenodolichopus*) not because this arrangement was definitive (wrote Verhoeff), but because the two species seemed more closely related to those in *Antichiropus* than to any other species he had previously examined. The gonopod telopodites were fundamentally similar and the first male legs (bearing a short, rounded, ventral, femoral process) were almost identical.

According to Verhoeff (1924), the polydesmidan solenomere was primitively protected by a long tibiotarsus, but there were four secondary developments away from this pattern. In one development (Verhoeff's case 1), the tibiotarsus was reduced so much that the solenomere became the free end of the gonopod, as in *Antichiropus*. The tibiotarsus had either disappeared completely in *Antichiropus*, or had been reduced to one of the small accessory branches at the end of the femur (Attems' *Zacken*). In another development (case 4), the tibiotarsus and solenomere fused into a single process. The two developments could overlap (Verhoeff 1924: 6, "Die Fälle N. 1 und 4 gehen übrigens in einander über"). In the new Queensland species, Verhoeff believed he could detect a fusion in which the solenomere had dominated, so that the tibiotarsus was reduced to a small subterminal tooth on the *teres* solenomere and a larger but likewise distal process on the solenomere in *vittatus*.

In the same 1924 paper, Verhoeff proposed *Aulacoporus* as a new genus for three northern Queensland species. *Aulacoporus* was distinguished from *Antichiropus* (*Solaenodolichopus*) in two ways. First, the prostatic groove opened at the extreme tip of the solenomere in *Aulacoporus*, whereas in *Antichiropus* (*Solaenodolichopus*) the opening was carried on a near-apical process of the solenomere. The solenomere and tibiotarsus had fused in *Aulacoporus*, but the tibiotarsus had completely disappeared in one *Aulacoporus* species and in the other two was reduced to a distal bulge on the solenomere.

To explain Verhoeff's second diagnostic difference between *Aulacoporus* and *Antichiropus* (*Solaenodolichopus*), I need to digress briefly.

Polydesmidan species have been classified, in part, on the structure of the paranota, which are lateral extensions of the body wall. In many families the paranota are wide, dorsoventrally flattened and more or less horizontal. In some families there are species with no detectable paranota, at least on diplosegments, i.e. the body rings are simple circles in cross-section. Between these two extremes, paranota are often said to be 'reduced'. Verhoeff (1924) attempted to refine the description of paranota for classification purposes. Instead of *Kiel* (keel, carina) for paranota in general, Verhoeff proposed *Seitenwülste* (lateral bulges) for all lateral swellings and extensions. In species with wide, dorsoventrally flattened, wing-like *Seitenwülste*, the term *Seitenflügel* (lateral wings) would be appropriate. Each *Seitenwulst* was often accompanied medially by a narrow, dorsal, longitudinal furrow which Verhoeff called a *Seitenfurche* (lateral groove). Species with no *Seitenwülste* at all might or might not have *Seitenfurchen*.

Attems (1937) used these terms slightly differently. He observed that well-developed paranota, i.e., *Seitenflügel*, often had swollen lateral margins. In some species the paranota had narrowed so much that only these marginal thickenings were still present, in which case one could speak of *Seitenwülste*, and in species without *Seitenwülste* there might still remain the medial, demarcating grooves, the *Seitenfurchen*. Both Attems and Verhoeff understood these variations as an evolutionary series in which primitively wide paranota were progressively reduced to bulges and, before disappearing entirely, to short, fine, lateral grooves on the metazonites.

The *Aulacoporus* and *Antichiropus* (*Solaenodolichopus*) species described by Verhoeff had reduced but detectable paranota on the haplosegments, i.e. rings 2-4. Neither group had *Seitenwülste* on the diplosegments, but there were distinct *Seitenfurchen* in *Aulacoporus* and none at all in *Antichiropus* (*Solaenodolichopus*).

Attems (1926) accepted both *Aulacoporus* and *Solaenodolichopus* (as a subgenus of *Antichiropus*), but reiterated his view that beyond the coxa on the gonopod were only a femur, tibia and tarsus. Verhoeff's *Solänomerit* was simply a tibial extension (*Tibialfortsatz*). In a key to paradoxosomatid genera ('Strongylosomidae'), Attems (1926) separated *Aulacoporus* (and other genera) having a simple telopodite without side branches or a tarsal section, from *Antichiropus* (and other genera) having a multi-tipped telopodite. The key listed *Antichiropus* as having 1-3 short processes, probably tarsal remnants, near the base of the tibial extension (i.e., the solenomere).

Verhoeff (1928) raised *Solaenodolichopus* to a genus and added two new species from NSW, *rubriventris* and *walesius*. He also proposed *Paraulacoporus* for another new NSW species, *sulcatus*. In *Paraulacoporus* as in *Aulacoporus* there were short *Seitenfurchen* on most diplosegments, although the grooves were indistinct on rings 15-19. Unlike *Aulacoporus* but like *Solaenodolichopus*, *Paraulacoporus* was characterised by a prostatic groove opening proximal to the solenomere tip. Unlike the situation in *Solaenodolichopus*, the two accessory branches at the end of the femur in *Paraulacoporus* were of equal size, and the process lateral to the prostatic groove opening on the solenomere was forked. Verhoeff described this forked process as a 'tarsal branch'.

Brolemann (1931) assigned a new New Caledonian species, *pruvoti*, to *Paraulacoporus*, although he observed no lateral groove (*sillon*) past ring 5. Combining the ideas of Attems and Verhoeff, Brolemann described the *pruvoti* telopodite as consisting of a short, densely setose, basal femoroid (*fémoroïde*), a long, slender tibia and a terminal seminal branch (*rameau séminale*). He regarded the short, posteromedial process at the end of the tibia as a reduced tarsus separate from the seminal branch, and rejected Verhoeff's opinion that the seminal branch in *P. sulcatus* was a fusion of solenomere and tibiotarsus. Brolemann also objected to the placement of *Solaenodolichopus* within *Antichiropus* by Verhoeff and Attems. He suggested instead that *Paraulacoporus* and *Solaenodolichopus* could be grouped together because both had an erect, expanded, more or less divided seminal branch, not a long, coiled strap as in *Antichiropus*.

Verhoeff (1932) offered simple new keys to the known Australian paradoxosomatids ('Strongylosomidae') based either on trunk details or gonopod structure. In the latter key, *Aulacoporus*, *Paraulacoporus*, *Solaenodolichopus* and *Walesoma* Verhoeff, 1928 were grouped as genera in which a long solenomere led from the end of the femur, which also might have 1-2 accessory branches at its apex. *Aulacoporus* and *Walesoma* were paired as genera in which the opening of the prostatic groove was at the extreme end of the telopodite, with no accessory process either proximal or distal to the opening. *Paraulacoporus* and *Solaenodolichopus* were paired as genera in which the opening of the prostatic groove was proximal to the extreme end of the telopodite, with accessory processes both proximal and distal to the opening.

Attems (1937) revised his nomenclature of sections of the gonopod telopodite, renaming femur, tibia and tarsus as prefemur, femur and tibiotarsus. The process carrying the prostatic groove (*Samenrinne*) and its opening was the *Rinnenast* (groove-branch) and was a process of the femur. More importantly, in 1937 Attems synonymised both *Paraulacoporus* and *Solaenodolichopus* under *Aulacoporus*. He described the gonopod and paranota in the expanded genus as follows (Attems 1937: 260-261, my translation): "Gonopod femur long to very long and slender, at the end with one or no side-branch. *Rinnenast* very large and broad, with one, sometimes two side-teeth. The tibiotarsus is reduced to a small spine. *Seitenwülste* weakly developed, low, round, sometimes only present on anterior segments. Longitudinal grooves either present on most segments or lacking from segments 5 or 6 onwards". In his key to *Aulacoporus* species, Attems separated the former *Aulacoporus* species from the former *Paraulacoporus* and *Solaenodolichopus* species simply on the presence or absence of *Seitenfurchen* on most rings.

Verhoeff (1937) described the new species *castaneum* from NSW and assigned it to a new genus, *Parwalesoma*. He noted that *Parwalesoma* was close to both *Aulacoporus* and *Solaenodolichopus*, but differed from the former genus in lacking *Seitenfurchen* on most rings. *Parwalesoma* and *Solaenodolichopus* differed in a somewhat sharper demarcation between femur and solenomere, and in the solenomere tip in *P. castaneum* being spoon-shaped. *Parwalesoma castaneum* Verhoeff, 1937 was in fact *Solaenodolichopus walesius* Verhoeff, 1928 redescribed from new material (see below under *S. walesius*), but it was another 60+ years before Verhoeff's confusion was detected by Jeekel (2000; see below).

In his last contribution to the *Solaenodolichopus* story, Verhoeff (1941) unwittingly redescribed Brolemann's *S. pruvoti* (from New Caledonia) as *Solaenodolichopus annulatus* (from Brisbane). He also offered a key to genera in which he separated the taxa lumped by Attems (1937). The relevant portion of the key is as follows (Verhoeff 1941: 8, my translation with Verhoeff's terminology updated):

c) Solenomere divided at the end into two branches, the medial one extending as three protuberances, of which the middle contains the opening of the prostatic groove, the proximal one a bend in the prostatic groove. Trunk without *Seitenwülste* but with *Seitenfurchen* ......*Paraulacoporus* Verh.
d) Prostatic groove without bend before its end ....e, f

e) Solenomere before its end with a spoon-shaped concavity. Trunk with neither *Seitenwülste* nor *Seitenfurchen* ....*Parwalesoma* Verh.
f) Solenomere without this concavity ....g, h

g) End of the solenomere drawn out into two closely standing tips, of which the more proximal contains the opening of the prostatic groove. Trunk with neither *Seitenwülste* nor *Seitenfurchen* ....*Solaenodolichopus* Verh.
h) Opening of the prostatic groove in a simple, apical tip...i, j

i) Solenomere significantly shorter than femur, its base enlarged, its end hooked. Trunk without *Seitenwülste* but with *Seitenfurchen* ....*Aulacoporus* Verh.
j) Solenomere and femur about equally long, the solenomere base not swollen. Trunk with weak, posteriorly rounded *Seitenwülste* and *Seitenfurchen* ....*Walesoma* Verh.

Reviewing the Australian Antichiropodini, Jeekel (1968) accepted Attems' (1937) synonymies and Verhoeff's *Parwalesoma*. Thirty-two years later, he confessed that he regarded *Parwalesoma* as another *Aulacoporus* synonym, but had not published this synonymy because the *Parwalesoma* type species, *P. castaneum* Verhoeff, 1937, would have become a junior homonym of the *Aulacoporus* type species, *A. castaneus* Verhoeff, 1924. "As the broad concept of *Aulacoporus* was intended to be a temporary solution anyway, the introduction of a new name for the type-species of *Parwalesoma* seemed just too radical" (Jeekel 2000: 40).

Jeekel (2000) then reinstated *Solaenodolichopus*, leaving within it *S. teres*, *S. vittatus* and *S. vittatus dorsalis* and transferring to it the *Paraulacoporus* species *P. sulcatus* and *P. pruvoti*. He also recognised *S. annulatus* as a synonym of *pruvoti*. *Solaenodolichopus* now included species with and without *Seitenfurchen*, and with considerable variation in the size and position of the processes at the end of the telopodite femur. "The species of this genus are characterised by the solenomerite of the gonopods having a pre-apical latero-distal process paralleling the apex proper. This process distinguishes the genus from the others" (Jeekel 2000: 41).

One of those other genera was *Aulacoporus* and its three original species, which Jeekel (2000) regarded as distinct not because the solenomere was simple with the prostatic groove opening at its tip, but because the gonopod apex curved strongly cephalad and bore a large lobe laterally – a lobe which Jeekel presumed was the tibiotarsus.

The other genus Jeekel (2000) distinguished from *Solaenodolichopus* was *Parwalesoma*. He recognised the type species *P. castaneum* as a junior subjective synonym of *S. walesius*, thus establishing *P. walesium* (Verhoeff, 1928). He also transferred *S. rubriventris* to *Parwalesoma*, noting that "In this genus the gonopods are lacking a pre-apical process on the solenomerite, but otherwise are quite similar to those of *Solaenodolichopus*" (Jeekel 2000: 42).

Jeekel's distinction between *Parwalesoma* and *Solaenodolichopus* is questionable in the case of *S. walesius* (see the taxonomic notes following the *S. walesius* redescription, below), and for that reason I am making *Parwalesoma* a junior subjective synonym of *Solaenodolichopus*. This action broadens the circumscription of *Solaenodolichopus* and might allow it to be subsumed under *Aulacoporus*, as favoured by Attems (1937). However, after examining late 20th-century specimens of *Aulacoporus castaneus* Verhoeff, 1924, *A. yarrabahnus* Verhoeff, 1924 and three undescribed *Aulacoporus* spp., all from a small area in the Queensland Wet Tropics, I agree with Jeekel (2000) that this genus is best kept separate from *Solaenodolichopus*. All *Aulacoporus* forms have prominent lateral grooves and slight paranotal swellings on all diplosegments, and in four of the six *Aulacoporus* forms the solenomere curves anteriorly. None of the *Solaenodolichopus* I have examined (as diagnosed above) has either prominent lateral grooves on all diplosegments or an anteriorly curving solenomere.

Diagnosis

Australian Paradoxosomatidae without distinct paranota on posterior rings, and with a gonopod telopodite with a small, densely setose prefemur and a more or less straight, slender and cylindrical femorite ending in an unbranched medial process, an unbranched lateral process and a much larger solenomere with a flattened tip on which the prostatic groove usually opens medially.

Description

Male with vertex bare, frons and clypeus sparsely setose; vertigial sulcus distinct, ending anteriorly just above level of antennal sockets; post-antennal groove shallow; antennal sockets separated by 1.0-1.1X a socket diameter. Antenna filiform (Fig. 1A), reaching dorsally to rear of ring 2; antennomere with relative lengths typically (2,3,4) > (5,6), the differences small, and maximum distal widths subequal or with 6 very slightly wider. Head and collum about equal in width in dorsal view, about as wide as rings 2–4 or a little narrower; diplosegments a little wider, slightly increasing in width to midbody, width decreasing progressively on last rings. Collum D-shaped in dorsal view; anterior margin straight medially, gently curving laterally; posterior margin straight or very slightly emarginate medially; collum corner broadly rounded (Fig. 1A). Ring 2 paranotum (Fig. 1A) a slightly thickened, overhanging ridge, nearly straight and more or less longitudinal, corners rounded, set at *ca.* 1/4 ring height or a little less, below level of collum corner. Rings 3 and 4 paranota (Fig. 1A) at about level of collum corner, reduced to shallow furrows curving upwards anteriorly. Pleural keels present on rings 2–4 as slight thickenings with upwardly curving margins. Diplosegments without paranotal thickenings, rings circular in cross-section. Prozonites and metazonites smooth, bare; waist short, shallow, with suture at anterior edge, sculptured indistinctly as variably low, rounded, longitudinal ridges; limbus a short, thin, continuous sheet. Pore formula normal; ozopores small, round, flush with lateral surface of ring, set at mid-height

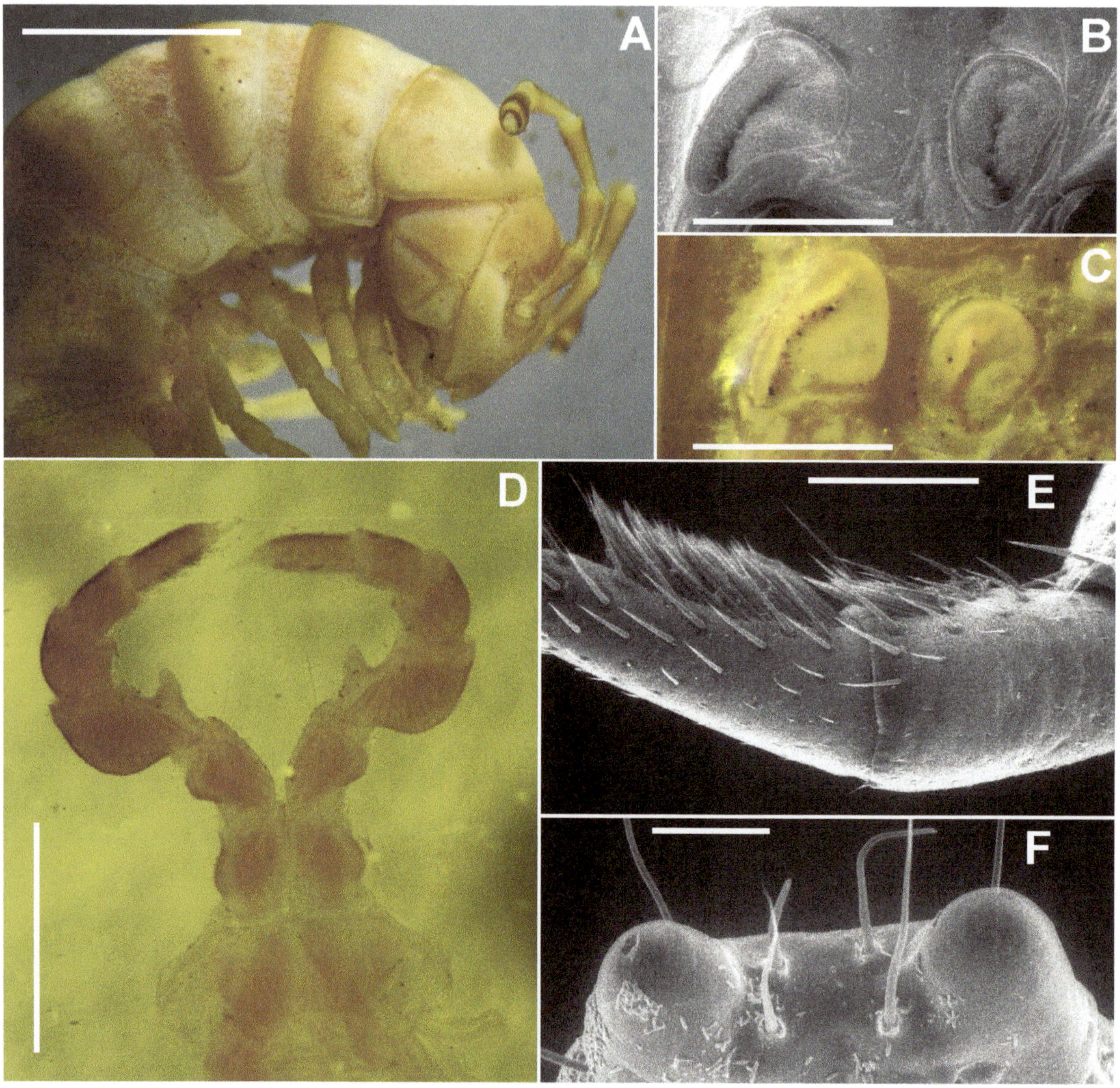

Fig. 1. — **A–B**. *Solaenodolichopus walesius* Verhoeff, 1928. **A**. Lectotype ♂, AM KS.76508, right anterior view. **B**. ♂ ex AM KS.16152, midbody spiracles, ventral view, anterior to left. — **C, E**. *S. vittatus* (Verhoeff, 1924). **C**. Paralectotype ♀, NHRS KAS1000000007, midbody spiracles, ventral view, anterior to left (image right-left reversed for ease of comparison with 1B). **E**. ♂ ex QM S74828, anterior leg tip with scopula, tarsus at left. — **D**. *S. teres* (Verhoeff, 1924), lectotype ♂ slide mount, NHRS KAS1000000005, legpair 1. — **F**. *S. pruvoti* (Brolemann, 1931), ♂ ex QM S74856, epiproct, posterior and slightly ventral view. Scale bars: A = 2.5 mm, B–C = 0.5 mm, D = 1 mm, E = 0.25 mm, F = 0.1 mm.

and mid-length on metazonite, in line with transverse furrow (when present). Diplosegment spiracles (Fig. 1B–C) just above and anterior to leg bases, more or less egg-shaped with smaller end ventral, anterior spiracle larger and extended anteroventrally; spiracles with thin, short rim; spiracular filter slightly emergent, forming rounded fold in inverted, tight U-shape in spiracular opening (Fig. 1B–C). Midbody sternites about as long as wide, sparsely setose; cross impressions wide, shallow, transverse impressions a little narrower and deeper than longitudinal. Leg 1 (Fig. 1D) with small, bluntly rounded, ventral, femoral process curving slightly distally, reaching to *ca.* 1/2 length of femur on ventral side. Midbody legs with relative podomere lengths femur > prefemur > (postfemur, tibia, tarsus) or tarsus very slightly longer than postfemur and tibia; femur 1.7–1.8X as long as tarsus; prefemora a little swollen dorsally. Dense brush (scopula) (Fig. 1E) of straight, finely pointed setae ventrally on tarsus and distal portion of tibia of most legs. Epiproct extending a little past anal valves, tapering in dorsal view with sides more or less straight, *ca.* 1/4-1/3 maximum width of preanal ring at apex; apex truncate, corners produced posteriorly as small, rounded bumps (Fig. 1F); spinnerets in rectangular array, wider than long. Hypoproct subtrapezoidal to broadly paraboloid. Gonopore small, round, opening ventrally on slight distomedial swelling of leg 2 coxa. Sternal lamella (Fig. 2) wide, *ca.* 80–90% of width between leg 4 bases, perpendicular to ring or slightly tilted anteriorly, with sparse long setae on posterior surface and dense, short setae on anterodistal surface. Aperture *ca.* 1/3 of ring 7 prozonite width, rounded-rectangular and tilted slightly posteriorly; prominent, rounded, transverse swelling just anterolateral to aperture on either side (Fig. 3A).

Female very similar to male but with legs proportionally shorter, prefemora not swollen; epigyne *ca.* 1/6-1/5 ring width, ends slightly raised and cradling leg 2 coxae, centre slightly raised in small, rounded triangle (Fig. 3C).

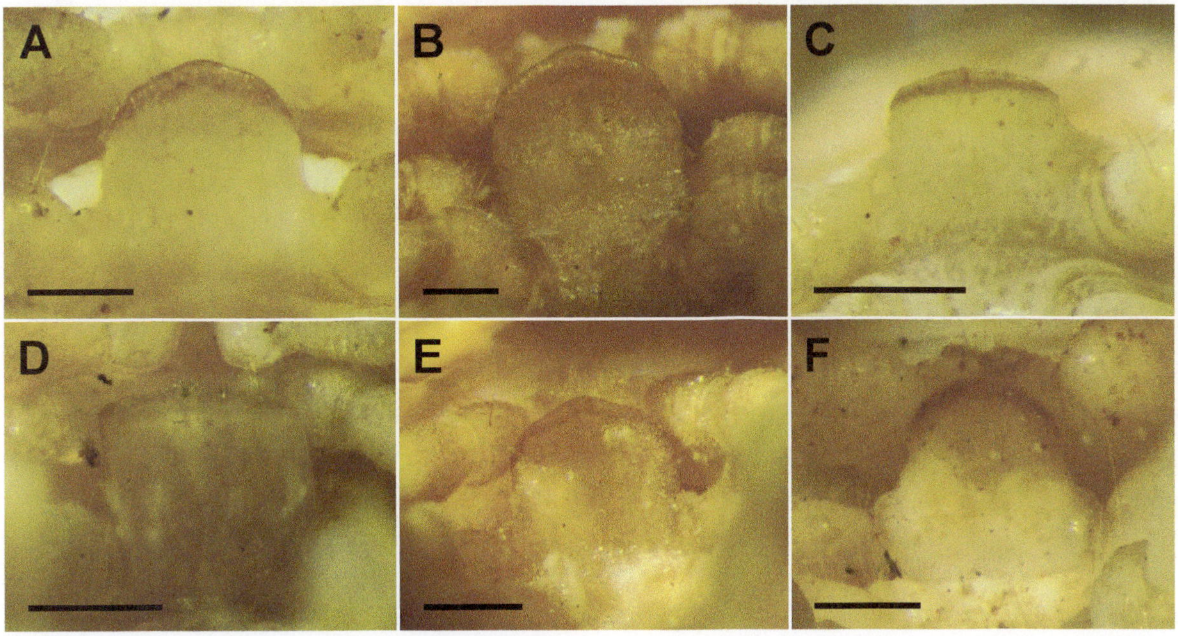

Fig. 2. Posterior views of sternal lamella in ♂♂. — **A.** *Solaenodolichopus pruvoti* (Brolemann, 1931), QM S74856. — **B.** *S. rubriventris* Verhoeff, 1928, QM S74690. — **C.** *S. sulcatus* (Verhoeff, 1928), lectotype, AM KS.76503. — **D.** *S. teres* (Verhoeff, 1924), lectotype in alcohol, ZSM A20052188. — **E.** *S. vittatus* (Verhoeff, 1924), QM S74829. — **F.** *S. walesius* Verhoeff, 1928, lectotype, AM KS.76508. Scale bars: 0.5 mm.

Species-diagnostic characters

Adults of the six named *Solaenodolichopus* species conform to the genus description above. Omitted from the genus description are diagnostic characters reported in the descriptions of individual species, below. Besides gonopod structure, these species-diagnostic characters are: colour, maximum midbody width, transverse furrow distinctness, longitudinal furrow details for diplosegments, distribution of scopulae on legs, sternal lamella details, leg spacing on rings 6 and 8, shape of the anterior margin of the gonopod aperture, and leg 2 coxa details in females.

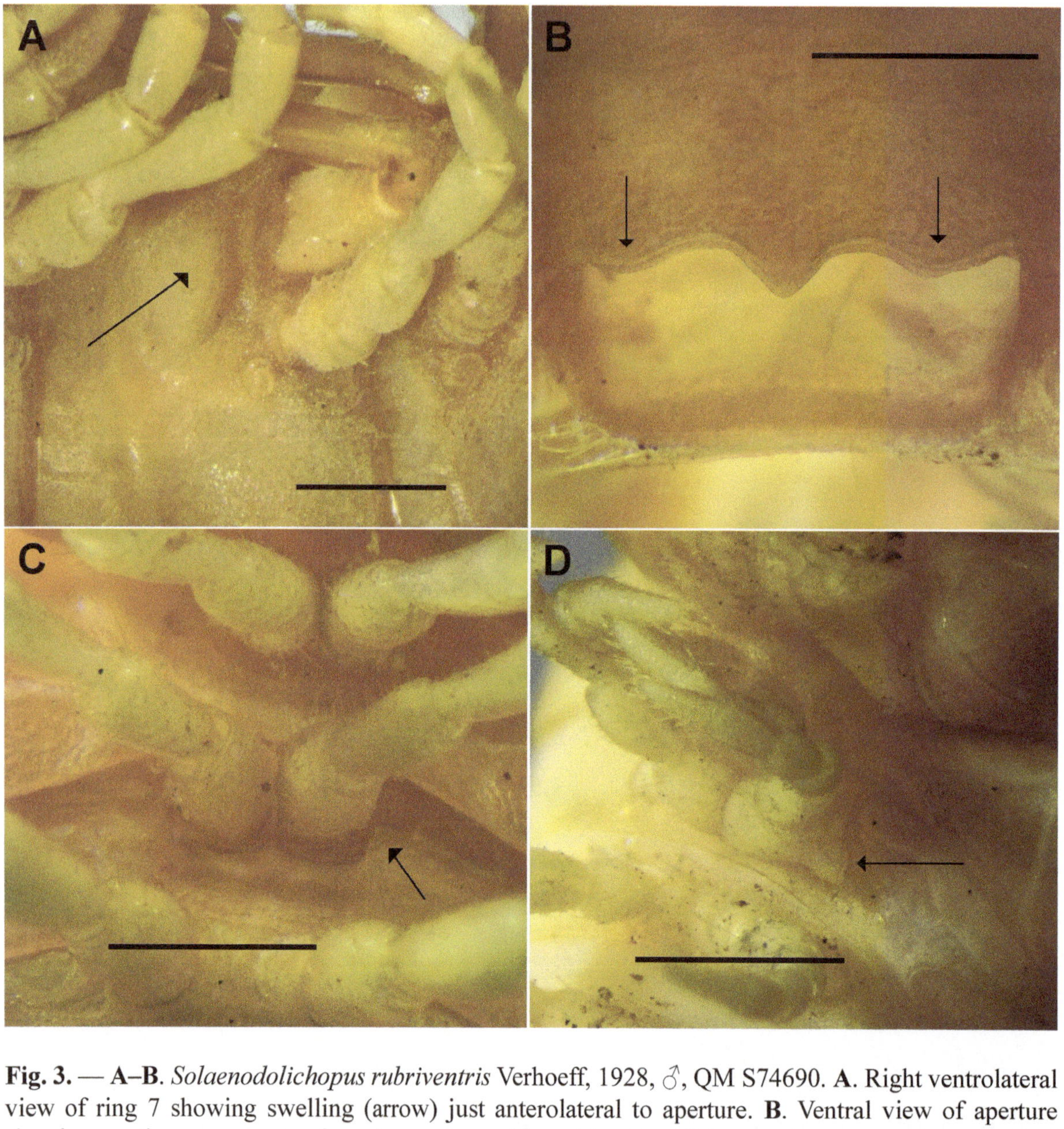

Fig. 3. — **A–B**. *Solaenodolichopus rubriventris* Verhoeff, 1928, ♂, QM S74690. **A**. Right ventrolateral view of ring 7 showing swelling (arrow) just anterolateral to aperture. **B**. Ventral view of aperture showing gently convex extensions (arrows) on either side of medial extension on anterior margin of aperture. — **C**. *S. walesius* Verhoeff, 1928, ♀, AM KS.16152, posteroventral view of epigyne showing raised end (arrow). — **D**. *S. teres* (Verhoeff, 1924), paralectotype ♀, NHRS KAS1000000007, left ventrolateral view of anterior rings showing distolateral process on leg 2 coxa (arrow) just medial to raised end of epigyne. Scale bars: 1 mm.

In the species descriptions I use the terms 'prefemur', 'femorite' and 'solenomere' in agreement with the usage of Car *et al.* (2013: 84) for *Antichiropus*; the terms are not intended to be implied statements of homology.

A brief dichotomous key to the six named *Solaenodolichopus* species follows. The key will need major revision as the many undescribed species are named and documented in future papers in this series.

1. Body pale yellow or light brown with narrow dark brown rings at the rear of each metazonite .... ***S. pruvoti*** (Brolemann, 1931)
– Body more or less uniformly dark in colour, with or without a lighter, medial dorsal, longitudinal band 2

2. Longitudinal furrows laterally at level of ozopore on diplosegments to about ring 15, medial and lateral femorite processes obvious, spine-like, about equal in length ...... ***S. sulcatus*** (Verhoeff, 1928)
– No longitudinal furrows laterally on any diplosegment, medial and lateral femorite processes greatly different in length or both inconspicuously small 3

3. Lateral femorite process much larger than medial .... ***S. walesius*** Verhoeff, 1928
– Medial femorite process much larger than lateral, or both femorite processes inconspicuously small 4

4. Distal 1/3 of femorite bent at *ca.* 45°, femorite processes inconspicuously small .... ***S. teres*** (Verhoeff, 1924)
– Femorite straight, medial femorite process large and obvious 5

5. Solenomere with several spine-like processes in basal half, no large indentation in distal half .... ***S. rubriventris*** Verhoeff, 1928
– Solenomere without spine-like process in basal half, with large, U-shaped indentation in distal half .... ***S. vittatus*** (Verhoeff, 1924)

Solaenodolichopus pruvoti (Brolemann, 1931)
Figs 1F, 2A, 4A–B, 5, 6A–B, 7, 10A, 11D

Paraulacoporus Pruvoti Brolemann, 1931: 295 (first description); figs 29–32 (p. 297), 33–35 (p. 299).
Solänodolichopus annulatus Verhoeff, 1941: 11 (first description); figs 1–4 (p. 23).

Solaenodolichopus pruvoti – Jeekel 2000: 40 (new combination), 42, 46. — Nguyen & Sierwald 2013: 1159.
Solaenodolichopus annulatus – Jeekel 2002: 60 (as synonym of *Solaenodolichopus pruvoti*).
Solänodolichopus annulatus – Nguyen & Sierwald 2013: 1159 (as synonym of *Solaenodolichopus pruvoti*).
Aulacoporus pruvoti – Attems 1937: 261 (new combination), 264 (Brolemann's 1931 description reworded); fig. 331 (p. 265; same as fig. 32 in Brolemann 1931). — Jeekel 1968: 18, 29; 1981: 47; 1985: 34.
Aulacoporus annulatus – Jeekel 1968: 20, 29; 1981: 49; 2000: 41 (synonymised with *Solaenodolichopus pruvoti*), 42.

non? *Solaenodolichopus pruvoti* – Jeekel 2002: 60 (partial redescription from new material), 77; fig. 1 (p. 62).
non? *Aulacoporus pruvoti* – Jeekel 1982: 124 (redescription from new material); figs 2, 3 (p. 126).

Material examined

Syntypes of *S. pruvoti*

At least 1 male and 1 female, Dumbéa, New Caledonia [22°09' S, 166°27' E, ±5 km], date unknown, A. Pruvot-Fol; specimens not located.

Syntypes of *S. annulatus*

At least 1 male, Brisbane, QLD [27°28' S, 153°01' E, ±10 km], Dec. 1936, J. Mauritzon; specimens not located.

Other material

QUEENSLAND: 1 ♂, Mt Glorious [27°20' S, 152°46' E, ±2 km], Oct. 1970, H. Williams, QM S5944; 1 ♂, Mt Nebo [27°23' S, 152°47' E, ±2 km], 19 Nov. 1978, A. Rozefelds, pitfall, wet sclerophyll, QM

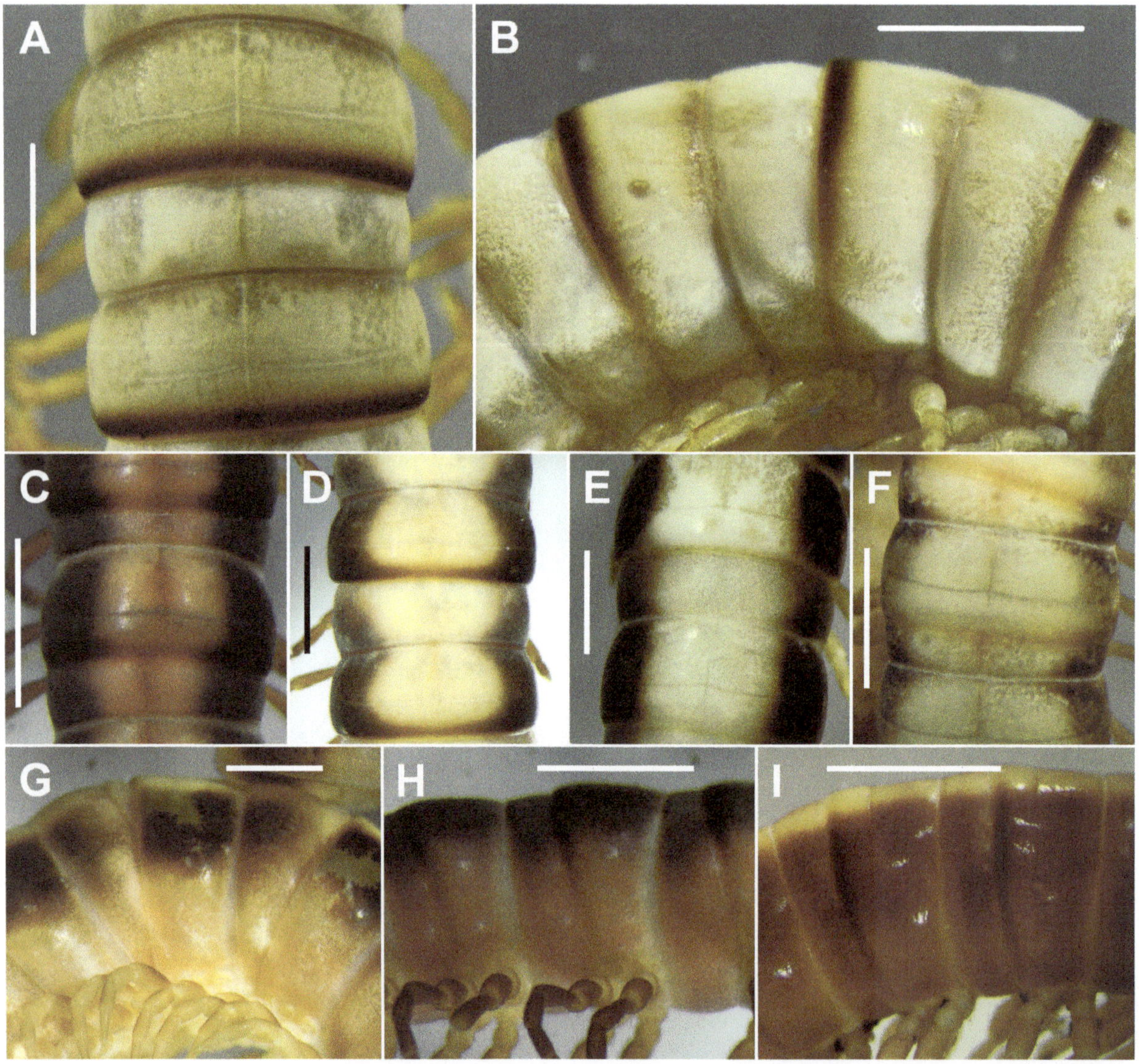

Fig. 4. — **A–B**. *Solaenodolichopus pruvoti* (Brolemann, 1931), QM S74856. Dorsal (A) and lateral (B) views of midbody rings of ♂. — **C–I**. *S. vittatus* (Verhoeff, 1924). Dorsal (C-F) and lateral (G-I) views of midbody rings of ♂; QM S74767 (C, H), QM S74855 (D), QM S74836 (E, G), QM S74876 (F), QM S5946 (I). Scale bars: 2 mm.

S5945; 2 ♂♂, 1 ♀, Camira, 27°37' S, 152°55' E [±2 km], 30 Jul.–Oct. 1990, R. Raven, open forest, pitfall, QM S19631; 1 ♂, Mt Glorious barracks, 27°19'54" S, 152°45'12" E [±500 m], 660 m a.s.l., 7 Dec. 1991–6 Mar. 1992, G. Monteith, rainforest, pitfall+intercept, QM S74833; 1 ♂, Mt Mee, 27°02'53" S, 152°40'49" E [±500 m], 300 m a.s.l., 3 Mar.–12 Apr. 1992, D. Cook, rainforest, intercept, QM S74834; 2 ♂♂, Stony Creek via Samford, 27°20'20" S, 152°47'52" E [±500 m], 200 m a.s.l., 22 Oct. 1994–2 Feb. 1995, H. Janetzki and G. Monteith, rainforest, pitfall, QM S74835; 1 ♂, 1 ♀, Camerons Scrub, knoll top, 27°30'27" S, 152°43'40" E [±500 m], 90 m a.s.l., 15 Sep.–11 Nov. 1998, G. Monteith, D. Cook and G. Thompson, vine scrub, pitfall, sample 7380, QM S74841; 3 ♂♂, 1 ♀, same details but 11 Nov. 1998–13 Jan. 1999, sample 7557, QM S74856; 1 ♂, same details but Camerons Scrub, top ridge,

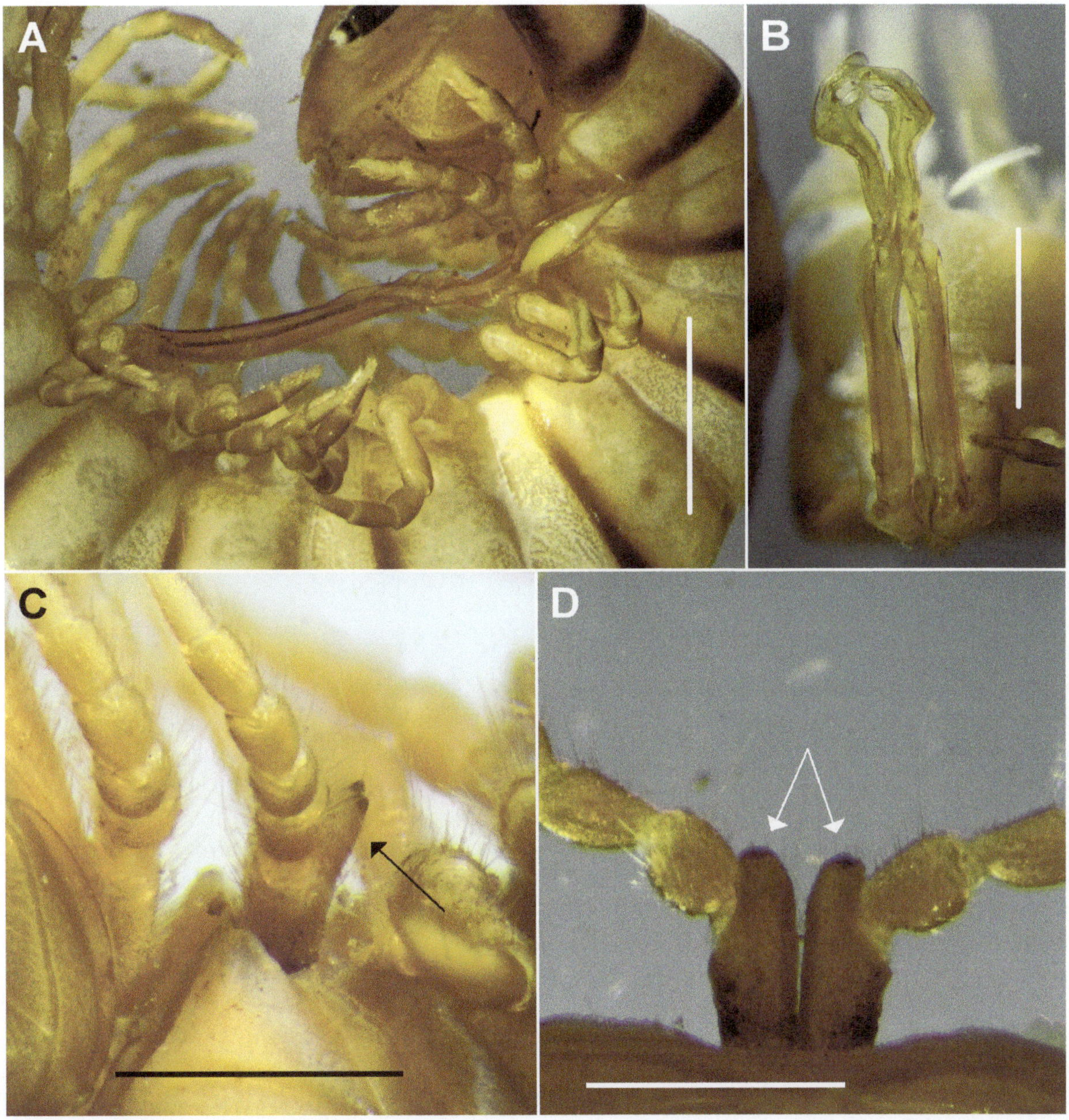

Fig. 5. *Solaenodolichopus pruvoti* (Brolemann, 1931). **A**. Anterior lateral view showing reach of gonopod telopodites, QM S74856. **B**. Posterior view of gonopods, WAM T52345. **C–D**. Lateral (C) and posterior (D) views of leg 2 coxal process in ♀, QM S74857. Scale bars: A–B = 2 mm, C–D = 1 mm.

27°30'29" S, 152°43'54" E [±500 m], 160 m a.s.l., 21 Sep.–11 Nov. 1998, sample 7381, QM S74842; 1 ♂, 1 ♀, same details but 13 Jan.–16 May 1999, sample 7666, QM S74864; 1 ♂, 1 ♀, same details but Camerons Scrub, below road, 27°30'45" S, 152°43'39" E [±500 m], 40 m a.s.l., 11 Nov. 1998–13 Jan. 1999, sample 7555, QM S74857; 1 ♂, The Knobby, 27°30'27" S, 152°35'18" E [±500 m], 240 m a.s.l., 11 Nov. 1998–13 Jan. 1999, G. Monteith, D. Cook and G. Thompson, vine scrub, pitfall, sample 7567, QM S74858; 1 ♂, Split Yard Creek, 27°22'41" S, 152°37'56" E [±500 m], 150 m a.s.l., 27 Dec. 1998–13 Jan. 1999, G. Monteith, D. Cook and G. Thompson, vine scrub, dung trap, sample 7568, QM S74859; 1 ♂, Boombana, Brisbane, 27°24'08" S, 152°47'22" E [±100 m], 440 m a.s.l., 7 Nov. 2003, R. Raven, B. Baehr and O. Seeman, day hand collection, ex S.C. 51695, QM S74874.

WESTERN AUSTRALIA: 1 ♂, 1 ♀, Curtin University, Environmental Biology Building, Bentley, [32°00'21" S, 115°53'40" E, ±200 m], 25 Apr. 2002, T. Cocket, WAM T47890; 1 ♂, same details but Physics Building, 32°00'23" S, 115°53'45" E [±50 m], 10 m a.s.l., 4 Nov. 2008, M.S. Harvey, in building, WAM T73493; 1 ♂, 1 ♀, Nursing Home, Hamersley Road, Subiaco [31°57'28" S, 115°49'25" E, ±100 m], 30 m a.s.l., 22 Jan. 2003, K. Debnam, inside nursing home, WAM T47891; 1 ♀, 7 Kingston Street, Nedlands, 31°57'56" S, 115°48'46" E [±100 m], 10 m a.s.l., 19 Apr. 2003, J.M. Waldock, found dead on front verge near driveway, WAM T52005; 1 ♂, same details but 21 Apr. 2003, J.M. Waldock, walking on footpath in front garden, WAM T52006; 1 ♂, same details but 19 May 2003, S. Slack-Smith, in garden, WAM T52345; 1 ♂, same details but 28 Sep. 2007, WAM T95077; 2 ♀♀, same details but 6–9 Nov. 2008, WAM T94065; 2 ♂♂, 1 juvenile, same details but 6 Nov. 2008, on path in garden after rain, WAM T94986; 2 ♀♀, 3 juveniles, same details but 8 Dec. 2008, on path in garden, WAM T94973; 1 ♂, 22 Windelya Road, Murdoch [32°04'14" S, 115°49'31" E, ±50 m], 30 m a.s.l., 3 Apr. 2005, O. Mueller, WAM T63211; 1 ♂, same details but 24 Apr. 2005, on verandah at night, WAM T65825; 1 ♀, Edward Street, Nedlands, 31°58' S, 115°48' E [±2km], 20 m a.s.l., 1 Mar. 2006, A. Baynes, WAM T67503; 2 ♂♂, Garfield Way, Greenwood, 31°49' S, 115°48' E [±2 km], 40 m a.s.l., 25 May 2006, T. Struthers, garden litter, WAM T78133; 1 ♂, 3 ♀♀, same details, WAM T112662; 1 ♂, 56 Edward Street, Nedlands [31°58'54" S, 115°48'24" E ±50 m], 20 m a.s.l., 3 May 2007, A. Baynes, WAM T82708; 1 ♂, University of Western Australia, Crawley, bus stop on Fairway near corner of Cooper Street, 31°58'41" S, 115°48'56" E [±100 m], 11 Jun. 2008, J. Wojcieszek, WAM T89288; 1 ♂, 7 Dunkeld Street, Floreat, 31°55'51" S, 115°47'24" E [±100 m], 13 Dec. 2012, J. Foss, WAM T129488.

Description

(Based mainly on Camerons Scrub, QLD specimens.) As for the genus. Maximum male/female midbody width *ca.* 3.8/4.4 mm. Colour in alcohol: "Colour brown-black, ringed tawny-pink on the anterior (nested) end of the prozonites and on the median portion of the metazonites. Anal valves black. Legs pale yellow" (Brolemann 1931: 295, my translation); "Bright gray-yellowish with brown posterior margin rings on diplosegments, head mainly dark brown" (Verhoeff 1941, p. 11, my translation); base colour in QLD and WA specimens varying from pale yellow to light brown, with narrow, dark brown ring at rear of haplosegments, diplosegments and apodous ring; darker specimens with very narrow, light brown rings bordering waist; narrow marginal ring of dark brown on collum, brown patches on vertex and clypeus of head and on anal valves, antennomeres brown (darker distally), legs yellowish (Fig. 4A–B).

Male with transverse furrows very shallow, narrow, sometimes indistinct and easily overlooked ("Pas de sillon tranverse sur les métazonites", Brolemann 1931: 296; "Querfurchen sehr fein", Verhoeff 1941: 10). No longitudinal furrows laterally on diplosegments (Fig. 4B). Sternal lamella (Fig. 2A) with sides more or less straight, corners broadly rounded, distal margin convex. Scopulae on legs 1 to 29, i.e. not present on last podous ring. Leg bases on rings 6 and 8 separated more widely than on other rings, sternites a little depressed. Anterior margin of aperture with rounded-triangular, medial extension and shorter, gently convex extension on either side (as shown in Fig. 3B for *S. rubriventris*).

Gonocoxa *ca.* 1/2 telopodite length, slightly flattened anteroposteriorly and slightly concave anteriorly, with sparse, long setae anterodistally. Cannula short, narrow, uniformly tapering towards prefemur. Gonopod telopodite (Figs 5A–B, 6A–B, 7, 10A) reaching leg 4 bases on uncoiled specimens. Prefemur small, rounded, densely setose posteromedially, demarcated from femorite laterally by small, narrow notch. Remainder of telopodite clearly divided into femorite and solenomere at *ca.* 1/2 telopodite length. Femorite slender, straight in posterior view, slightly convex anteriorly in lateral view, the posterior surface very slightly concave; tipped with knob-like, rounded shoulder just lateral to lateral femorite process, a short, wide, longitudinal groove running basally from shoulder. Lateral femorite process a short, rounded tab directed anterodistally, close to solenomere base, extending distally a little past femorite shoulder. Medial femorite process teardrop-shaped, the rounded base with thin, translucent cuticle, the point directed anterodistally, terminating close to solenomere at 1/4-1/3 solenomere length. Solenomere at base *ca.* 2/3 femorite width, expanded and anteroposteriorly flattened distally; a narrow, ridge-like flange (arrow in Fig. 7D) along whole of anterior surface of solenomere, diverting medially in short loop at *ca.* 1/2 solenomere length, curving sharply laterally at base to form small, rounded, anterior extension. Solenomere divided at *ca.* 2/3 length by large, U-shaped indentation into medially directed apex and distomedially directed, rounded process terminating just basal to solenomere apex; apex divided by narrow indentation into anteroposteriorly flattened, medially directed, truncate, subapical tab carrying opening of prostatic groove, and apical, basodistally flattened tab extending medially just past

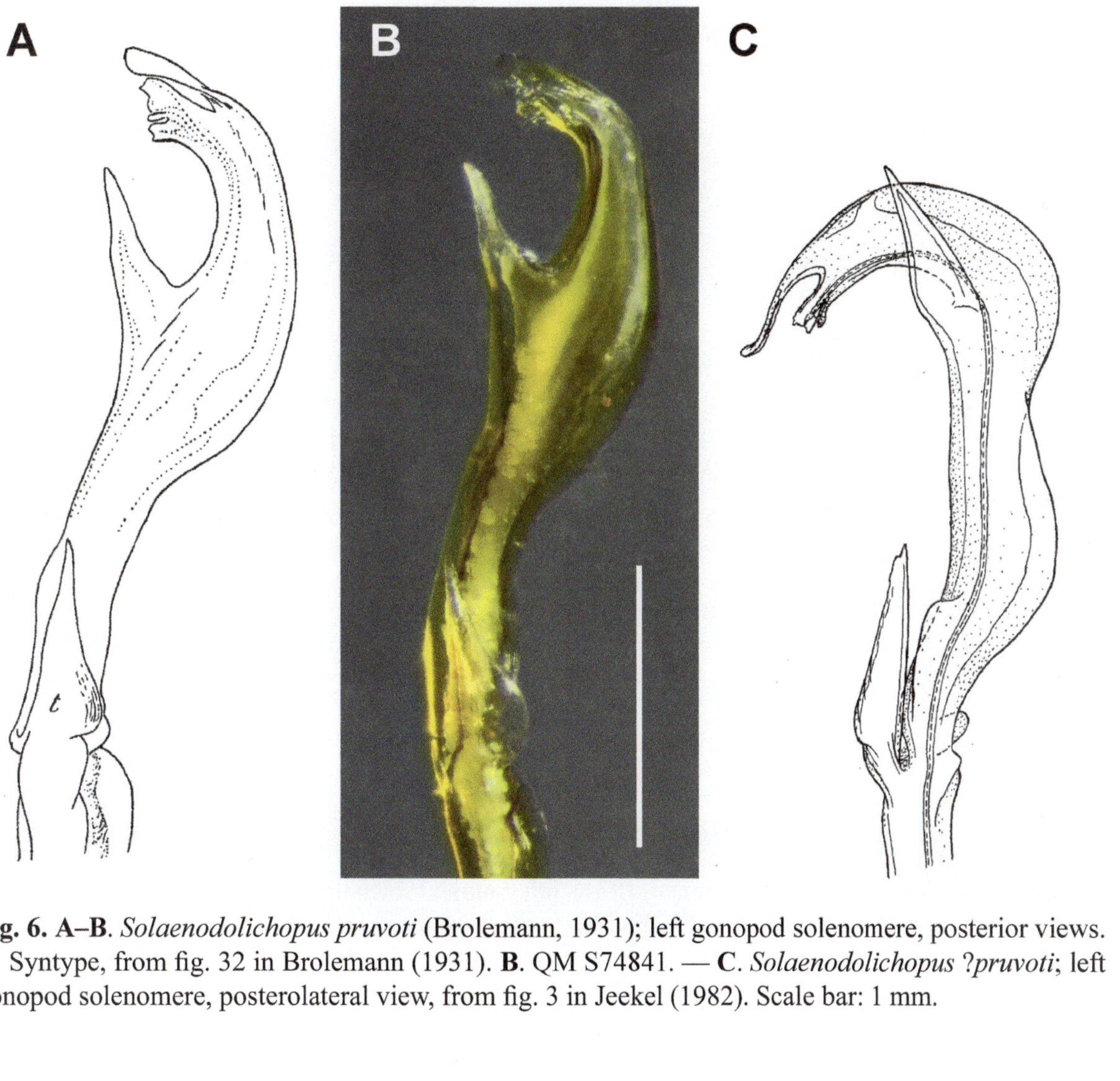

Fig. 6. A–B. *Solaenodolichopus pruvoti* (Brolemann, 1931); left gonopod solenomere, posterior views. **A**. Syntype, from fig. 32 in Brolemann (1931). **B**. QM S74841. — **C**. *Solaenodolichopus ?pruvoti*; left gonopod solenomere, posterolateral view, from fig. 3 in Jeekel (1982). Scale bar: 1 mm.

subapical tab. Prostatic groove (Fig. 7D) running on anterior surface of telopodite, following curve of solenomere medial to flange, terminating at midpoint on medial margin of subapical tab.

Female with leg 2 coxa extended posteroventrally as apically rounded process (Fig. 5C–D).

Distribution

Known from open forest, rainforest and vine scrub to at least 660 m a.s.l. in the D'Aguilar Range and adjoining areas west of Brisbane in southeast Queensland, from Mt Mee south to Ipswich, a north-south extent of *ca.* 65 km (map Fig. 12). This is almost certainly the native range of the species, as very similar, undescribed species are found elsewhere in southeast Queensland. Introduced and established in the Perth metropolitan area in Western Australia; so far reported from Kingsley south to Kardinya, a north-south extent of *ca.* 30 km. Introduced (and established?) in New Caledonia. Overlaps in range with *S. vittatus*.

Taxonomic notes

The types of *P. pruvoti* and *S. annulatus* have not yet been located, and I prefer not to designate lectotypes for these species based only on the published descriptions and illustrations of Brolemann (1931) and Verhoeff (1941). The Brisbane and Perth specimens listed above agree closely with those descriptions and illustrations (Figs 6A–B, 7A–B).

I suspect that the South Australian specimens described by Jeekel (1982, 2002) are an unnamed *Solaenodolichopus* species, not *S. pruvoti*. They differ from *S. pruvoti* in lacking pleural keels and in having a distinct transverse furrow on the metazonites (Jeekel 1982: 125) and a medial coxal process

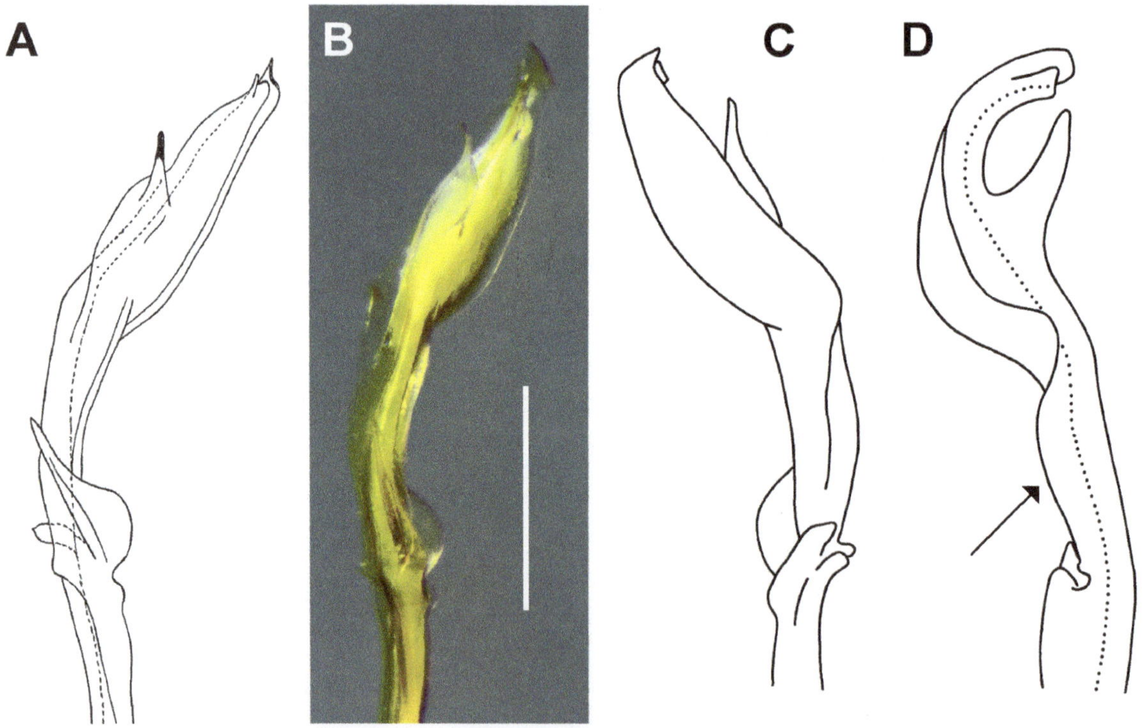

Fig. 7. *Solaenodolichopus pruvoti* (Brolemann, 1931), left gonopod solenomere. — **A–B**. Medial views, *S. annulatus* (= *S. pruvoti*) syntype from fig. 2 in Verhoeff (1941) (A) and QM S74856 (B). — **C–D**. Lateral (C) and anterior (D) views, QM S74857. In D, arrow points to flange on anterior surface and dotted line marks course of prostatic groove. Scale bar: 1 mm.

on male legs 6 and 7 (Jeekel 1982: 127). The specimens described in 2002 also differ from *S. pruvoti* in lacking dark transverse annulation (Jeekel 2002: 61). Both Brolemann (1931: 295, for *S. pruvoti*) and Verhoeff (1941: 11, for *S. annulatus*) note strongly contrasting annular rings of colour, and all non-type adults listed above are clearly ringed (Fig. 4A–B), apart from a decoloured male in QM S74859. Jeekel explained the difference in shape of the solenomere between *S. pruvoti* and the South Australian males (Fig. 6) by saying "It should be emphasised that the distal portion of the gonopod (i.e. the solenomerite) is quite flexible, and may be curving more or less widely in a caudal direction, and may be crooked at the base" (Jeekel 1982: 127). It is true that the *S. pruvoti* solenomere can be made to curve posteriorly by manipulation. However, in none of the preserved specimens I have so far examined of any *Solaenodolichopus* species is the solenomere 'naturally' curved as sharply as illustrated by Jeekel (Fig. 6C).

General notes

The collector of *S. annulatus*, Johan Mauritzon (1902-1950), was a lecturer in botany at Lund University when he travelled to Australia in 1936 in company with University of Uppsala zoologist Folke Linder (as reported in several contemporary Australian newspapers, e.g., *The Adelaide Chronicle*, 10 Sep. 1936: 52, http://trove.nla.gov.au/ndp/del/article/92463218, accessed 10 Oct. 2013). Mauritzon left Lund University soon after returning from his Australian trip (H. Wittzell, in litt. 10 Oct. 2013) and his Australian millipedes were later sent to Verhoeff from Lund by Torsten Gislén (Verhoeff 1941: 3). The types of *S. annulatus* and four other species described by Verhoeff (1941) from this material have not yet been located. The types may not have been returned to Lund, as none of the Mauritzon millipede specimens collected in Australia are currently in the Lund University Zoological Museum (J. Ekstrom, in litt. 13 Jan. 2014).

The reported type localities of two of Verhoeff's five 1941 species are incorrect, and it is unclear at which stage in the information chain from Mauritzon to Verhoeff the error appeared. *Hoplatessara (Walestessara) forceps* Verhoeff, 1941 (now *Cladethosoma forceps*) and *Hoplatria clavigera* Verhoeff, 1941 were both described from Mauritzon collections in Gippsland, a district in eastern Victoria. Both have since been collected in NSW west of Sydney, and neither has since been found in intensively collected parts of Gippsland. However, the 'Brisbane' type locality for *S. annulatus* is plausible, because Verhoeff (1941: 12) gives "XII.36" as the collecting date and there are "near Brisbane" locations for Mauritzon plant collections dated 22 Dec. 1936 in the NHRS Herbarium (http://www.nrm.se/english/researchandcollections/collections/databases/kryptos.8598_en.html, accessed 12 Oct. 2013) and in the Uppsala Museum of Evolution (http://130.238.83.220/botanik/home.php, accessed 12 Oct. 2013).

Brolemann's 1931 paper is devoted to species collected by 'Mme. Pruvot', i.e., the malacologist Alice Pruvot-Fol (1873-1972), but Brolemann explicitly named *Paraulacoporus pruvoti* for Pruvot-Fol's late husband, the zoologist Georges Pruvot (1852-1924) (Brolemann 1931: 298).

Solaenodolichopus rubriventris Verhoeff, 1928
Figs 2B, 3A–B, 8A–B, E–I

Solänodolichopus rubriventris Verhoeff, 1928: 94 (genus misspelled *Solanodolichopus*), 95 (first description, name printed in error as *Solandolichopus walesius*, corrected in journal Corrigenda); 114 (as *Solänodolichopus rubriventris* in list of species described in paper); figs 24–25 in pl. 10.

Aulacoporus rubriventris – Attems 1937: 61 (new combination), 263 (Verhoeff's 1928 description reworded); fig. 330 (p. 265; same as fig. 24 in Verhoeff 1928). —Jeekel 1968: 18, 29; 1981: 49.
Parwalesoma rubriventris – Jeekel 2000: 41 (new combination), 43. —Nguyen & Sierwald 2013: 1158.

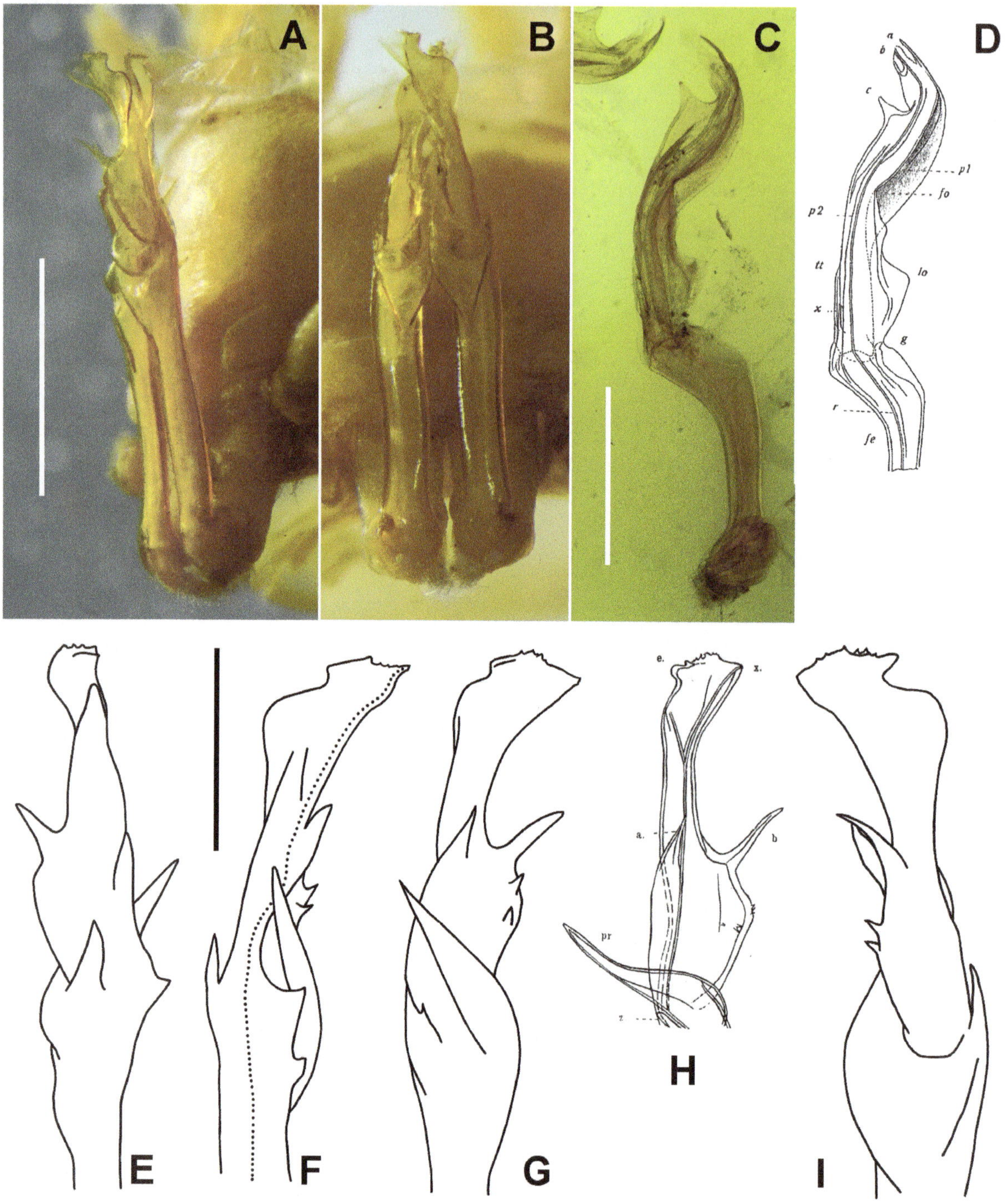

Fig. 8. — **A–B, E–I**. *Solaenodolichopus rubriventris* Verhoeff, 1928. Lateral (A) and posterior (B) views of gonopods *in situ*; left gonopod solenomere in lateral (E), anterior (F), medial (G), medial and slightly posterior (H) and posterior (I) views. A, B, E, F, G, I from QM S74690, H from fig. 24 in Verhoeff (1928). Dotted line in F marks course of prostatic groove. — **C–D**. *S. teres* (Verhoeff, 1924), anterior views of right gonopod telopodite of lectotype, from slide mount NHRS KAS1000000005 (C) and from fig. 7 in Verhoeff (1924) (D). Scale bars: A–B = 2 mm, C–I = 1 mm.

Material examined

Syntypes

1 male, Upper Richmond River, NSW [27°37 S, 153°00' E, ±10 km], Apr. 1890, R. Helms, specimen not located; 1 ♀ in alcohol with original Verhoeff label, broken between rings 11 and 12, same collection details, AM KS.76506 (formerly 47728); 1 female in alcohol with original Verhoeff label, broken between rings 7 and 8 and rings 11 and 12, same collection details, AM KS.76507 (formerly 47723); 1 female, same collection details, specimen not located.

Other material

QUEENSLAND: 4 ♂♂, Tamborine Mountain [27°58' S, 153°11' E, ±5 km], 22 Oct. 1912, H. Hacker, QM S74690; 1 ♂, 1 ♀ in fragments, [Lamington] National Park, Macpherson Range [28°15' S, 153°08' E, ±5 km], 15 Dec. 1926, P.A. Gilbert, AM KS.94916 (formerly K55468); 1 ♂, Eagle Heights [27°54' S, 153°12' E, ±5 km], M.B. Man, 23 Mar. 1955, under logs, QM S74691; 1 ♂, Mt Tamborine [Tamborine Mountain] [27°58' S, 153°11' E, ±5 km], pitfall 15 Jan.–4 Mar. 1979, J. Grimshaw, rainforest, QM S74692.

Description

(Based on Mt Tamborine males and syntype females.) As for the genus. Maximum male/female midbody width *ca.* 5.4/6.5 mm. Colour in alcohol: "Back brownish black, without rings, only the abdomen passing into a wine red colour" (Verhoeff 1928: 94); the least faded of the specimens examined are more or less uniformly red-brown, darker dorsally, with a pale waist and yellow legs.

Male with with distinct transverse furrow on metazonites, stopping a little above level of ozopore. No longitudinal furrows laterally on diplosegments. Sternal lamella (Fig. 2B) with straight sides, corners broadly rounded, distal margin a flat inverted V. Scopulae on legs 1 to 29, i.e., not present on last podous ring. Leg bases on ring 6 separated a little more than on ring 5; leg bases with normal separation on ring 8. Anterior margin of aperture with rounded-triangular, medial extension and shorter, gently convex extension on either side (Fig. 3B).

Gonocoxa *ca.* 2/3 telopodite length, slightly flattened anteroposteriorly and slightly concave anteriorly, with sparse, long setae anterodistally. Cannula short, narrow, uniformly tapering towards prefemur. Gonopod telopodite (Fig. 8A–B, E–I) reaching leg 6 bases when retracted. Prefemur small, rounded, densely setose posteromedially, demarcated from femorite laterally by small, narrow notch. Remainder of telopodite more or less straight, slender, clearly divided into femorite and solenomere at just over 1/2 telopodite length. Femorite straight, slightly flattened anteroposteriorly, parallel-sided in posterior view, apically thickened posteriorly as inverted triangle. Lateral femorite process small, pointed, tooth-like, directed distally, *ca.* 1/5 solenomere length. Medial femorite process teardrop-shaped, arising near lateral process, the base expanded as thin, translucent cuticle, curving anterodistally and closely pressed to triangular apical thickening of femorite and to solenomere base, tapering to point just anterior to solenomere at *ca.* 1/2 solenomere length. Solenomere arising from anterior side of femorite; slightly flattened anteroposteriorly with narrow, ridge-like thickening posteromedially; anteriorly with narrow flange of cuticle arising basomedially and produced basomedially as blunt point, and at *ca.* 1/3 solenomere length as short, distally acuminate process; thin, spine-like process arising at *ca.* 1/2 solenomere length from posteromedial ridge-like thickening, directed posterodistally and a little medially; ridge-like thickening extended slightly at *ca.* 1/4 solenomere length and bearing several fine, tooth-like projections. Solenomere expanded from *ca.* 3/4 length, anteroposteriorly flattened, the apex rounded laterally, medially extended as lamellar, posteromedially directed triangle, the apical margin of triangle expanded as narrow, basodistally flattened shelf tapering towards medial apex of triangle, and finely dentate medially. Prostatic groove (Fig. 8F) running just lateral to flange on anterior surface of solenomere, terminating at apex of triangular apical tab.

Female without process on leg 2 coxa.

Distribution

Known from the forested mountains behind the Gold Coast in southeast Queensland south across the Border Ranges to the Richmond River catchment in NSW, a north-south extent of at least 60 km (map Fig. 12). Overlaps in range with *S. vittatus*.

Taxonomic notes

Verhoeff (1928: 96) wrote that he examined one male and three females, but I have not located either the male or a third female. With the male missing, I am reluctant to designate one of the undescribed females as lectotype. My description of male *S. rubriventris* is based on Verhoeff's published notes and on the specimens listed above from Tamborine Mountain, *ca.* 80 km from the roughly approximated type locality (see below). The 1979 Tamborine Mountain male was identified as '*Atropisoma rubriventris*' by P.M. Johns during a 1987 visit to QM.

General notes

S. rubriventris, *S. sulcatus* and *S. walesius* were among millipede specimens supplied to Verhoeff by Charles Anderson (Verhoeff 1928: 79), Director of the Australian Museum in Sydney from 1921 to 1940. Verhoeff's German manuscript on the material included descriptions of 21 new millipede species and subspecies. It was translated into English by Anderson for the *Records of the Australian Museum* and the draft translation was sent to Verhoeff for approval (Verhoeff 1928: 79). Some typographical errors in nomenclature escaped proofreading (see synonymy above) and some morphological terms were mistranslated. For example, Anderson consistently translated *hinter* in the sense of 'behind, posterior', even when Verhoeff used *hinter* to mean 'beyond, distal' (as in English 'hinterland').

Although Richard Helms (1842–1914) had been working as a collector for the Australian Museum in Sydney in 1888 and 1889, he had gone to the Richmond River in 1890 "in the interest of a private syndicate" (Hedley 1915: 13). I have not been able to find a report by Helms on the Upper Richmond River trip. According to local historian Margaret Henderson (in litt., 9 Jan. 2006), "The 'Upper Richmond River' usually refers to the area from Casino to the source in the ranges". I place the collections very approximately at Kyogle, about halfway between Casino and the Border Ranges and *ca.* 30 km upstream from Casino.

Solaenodolichopus sulcatus (Verhoeff, 1928)
Figs 2C, 9

Paraulacoporus sulcatus Verhoeff, 1928: 93 (first description); fig. 11 in pl. 7.

Solaenodolichopus sulcatus – Jeekel 2000: 41 (new combination), 42, 43 (redescription from new material); figs 3-5 (p. 48) (see taxonomic notes, below). — Nguyen & Sierwald 2013: 1160.
Paraulacoporus sulcatus – Verhoeff 1932: fig. 969 (p. 1577).
Aulacoporus sulcatus – Attems 1937: 261 (new combination), 262 (Verhoeff's 1928 description reworded); fig. 326 (p. 262; same as fig. 11 in Verhoeff 1928). — Jeekel 1968: 18, 29; 1981: 49.

Material examined

Lectotype (here designated)
Male in alcohol with original Verhoeff label, broken between rings 8 and 9 and rings 11 and 12, North Dorrigo, NSW [30°16' S, 152°41' E, ±5 km], 4 Jan. 1923, A. Musgrave, AM KS.76503 (formerly K47706).

Paralectotype
1 male, details as for lectotype, not located (see taxonomic notes, below).

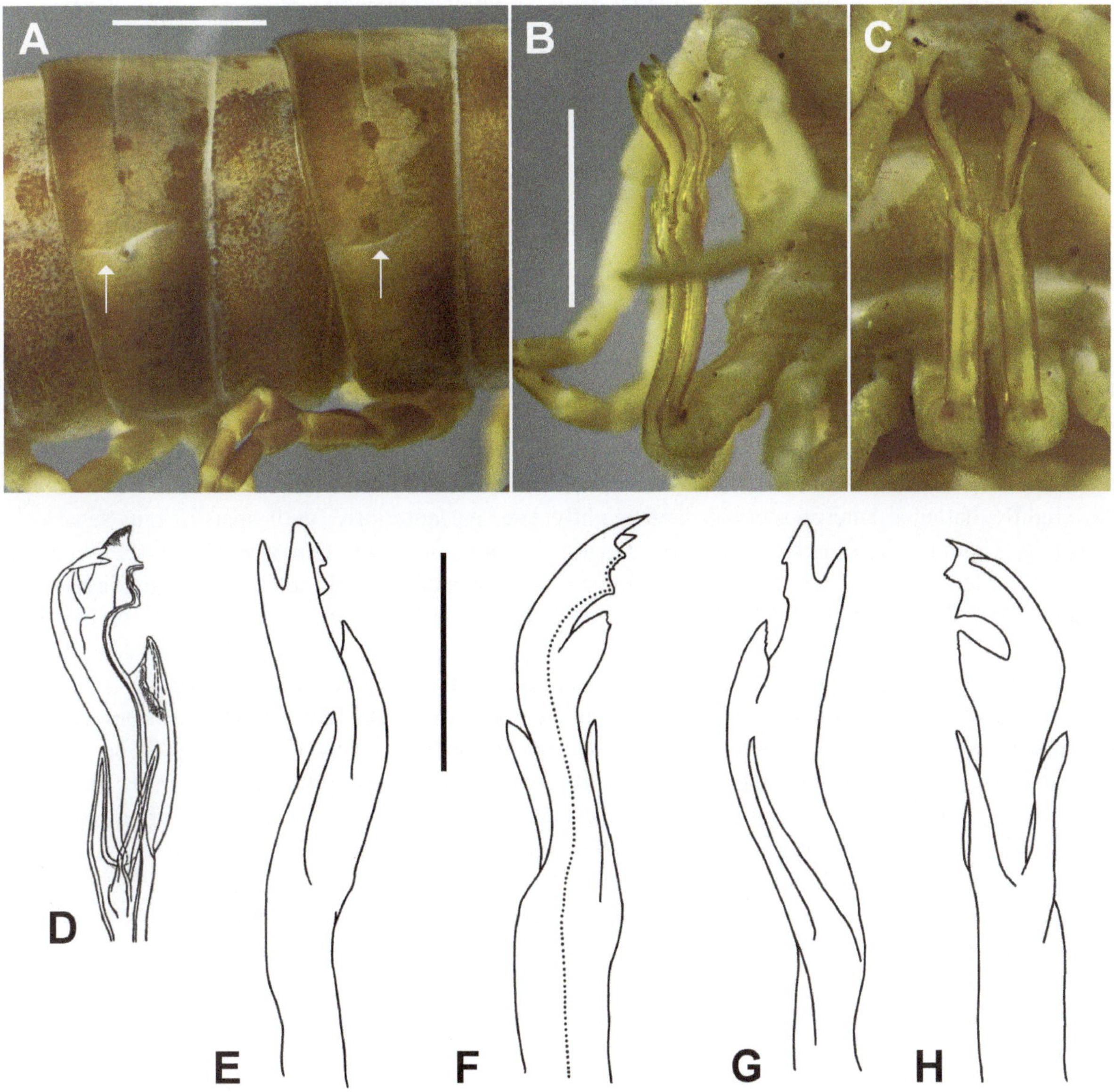

Fig. 9. *Solaenodolichopus sulcatus* (Verhoeff, 1928) ♂. **A**. Right lateral view of midbody rings, AM KS.77379. **B–C**. Lateral (B) and posterior (C) views of gonopods *in situ*, lectotype, AM KS.76503. **D**. Posterior and slightly medial view of right gonopod solenomere, possibly of missing paralectotype, from fig. 11 in Verhoeff (1928). **E–H**. Lateral (E), anterior (F), medial (G) and posterior (H) views of left gonopod solenomere, QM S75836. Arrows in A point to longitudinal grooves, dotted line in F marks course of prostatic groove. Scale bars: A–C = 2 mm, D–H = 1 mm.

Other material

NEW SOUTH WALES: 1 ♂, Dorrigo [30°19' S, 152°43' E, ±5 km], W. Heron, no date, SAM (specimen missing; described in Jeekel 2000); 1 ♂, Upper Styx State Forest - Petroi Forest [30°40'S 152°08'E ±10 km], 29 Jan. 1975, P. Johns and R. Killip, *Nothofagus moorei* forest, QM S75836; 1 ♂, Bruxner Forest Park, Coffs Harbour [30°'41" S, 153°06'03" E, ±2 km], 22 Mar.–13 Nov. 1980 [pitfall?], G. Monteith, sample GM 104 A/1, QM S75837; 1 ♂, Leagues Scrub Flora Reserve, Oakes State Forest, 30°36' S, 152°32' E [±2 km], 780 m, 20–21 Dec. 1998, D. Bickel, yellow pan, rainforest, AM KS.77379.

Description

As for the genus. Maximum midbody width *ca.* 5.0 mm. Colour in alcohol: "Body brownish black, greyish-yellow on the abdomen, back with rather broad yellowish median band, which continues over the pro- and meta-zonites [sic] to the telson. Legs black, brownish on the joints" (Verhoeff 1928: 93); lectotype colour now faded, prozonites light brown, metazonites and legs yellowish, pale mid-dorsal longitudinal band *ca.* 1/5 maximum ring width at midbody; more recent specimens also faded, patchy brown with pale mid-dorsal band as in lectotype.

Male lectotype with distinct transverse furrow on metazonites, stopping a little above level of ozopore (Fig. 9A). Distinct longitudinal furrows on diplosegments (Fig. 9A) to ring 15, rising slightly anteriorly, opening dorsally with very slight bulges below; furrow just above ozopore on pore-bearing rings. Sternal lamella (Fig. 2C) rounded-rectangular, distal margin very slightly convex. Scopulae on legs 1 to legs of ring 14 (lectotype) or 15 or 16 (other males examined). Leg bases on ring 6 separated a little more than on ring 5; leg bases with normal separation on ring 8. Anterior margin of aperture more or less straight with small, rounded-triangular, medial extension.

(Gonopod description based on lectotype and other males.) Gonocoxa a little more than 1/2 telopodite length, slightly flattened anteroposteriorly and slightly concave anteriorly, with sparse, long setae anterodistally. Cannula short, narrow, uniformly tapering towards prefemur. Gonopod telopodite (Fig. 9B-H) reaching leg 6 bases when retracted. Prefemur small, rounded, densely setose posteromedially, demarcated from femorite laterally by small, narrow notch. Remainder of telopodite more or less straight, slender, clearly divided into femorite and solenomere at just over 1/2 telopodite length. Femorite slightly flattened anteroposteriorly, slightly concave posteriorly in lateral view, parallel-sided in posterior view, apically a little thickened posteriorly. Femorite processes both about 1/3 solenomere length, tapering with rounded tips; lateral process curving a little anterolaterally, medial process curving a little anteriorly. Solenomere *ca.* 2/3 femorite width at base, curving slightly anterolaterally, then posteromedially, flattened to form trough-like hollow, concave posteromedially; basal half of anterior margin of hollow produced mediodistally as large, flat, extension with tapering, rounded tip at *ca.* 3/4 solenomere height; distomedially curving solenomere apex deeply notched to form rounded posterior point separate from anteroposteriorly flattened anterior margin of hollow, the latter produced apically as thin, subtrapezoidal tab, its medial margin roundly emarginate. Prostatic groove (Fig. 9F) running on anterior surface of telopodite, following curve of solenomere to basal corner of subapical, anterior margin tab, bending distally at right angle and opening at distal corner of tab.

Female not yet recognised.

Distribution

Rainforest (and eucalypt forest?) in northeast NSW from the Coffs Harbour area inland towards Armidale, an east-west extent of *ca.* 90 km, at elevations to at least 780 m a.s.l. (map Fig. 12). Overlaps in range with *S. walesius*.

Taxonomic notes

Verhoeff (1928: 94) reported that he examined two males, but I have not located a second specimen. The lectotype vial contains a label with 'LECTOPARATYPE / ♂ P.M. Johns / 19.xii.67', suggesting that Johns may have found another male or slide mount of male parts and labelled it 'lectotype'; this lectotypification has not been published.

In fig. 11 of Verhoeff (1928), reproduced here as Fig. 9D, the rounded posterior process at the solenomere apex is shown as apically notched. This appears to be an error, as the process is not notched at its tip in the lectotype or any of the other three available *S. sulcatus* males. Jeekel (2000) described a fourth male, from Dorrigo (NSW); his gonopod illustration (Jeekel's fig. 5: 48) also shows the posterior process without a notch. The Dorrigo male was borrowed from SAM but is currently missing from the Museum collection (K. Sparks, in litt., 2 Jan. 2014).

General notes

The collector of the types Anthony Musgrave (1895-1959) worked as an entomologist at the Australian Museum from 1920 until his death (http://adb.anu.edu.au/biography/musgrave-anthony-7716; accessed 30 Dec. 2013). He specialised in the taxonomy of arachnids, Diptera and Hemiptera, and is best known as the compiler of a monumental bibliography of Australian entomology (Musgrave 1932). Musgrave's 1923 North Dorrigo trip also yielded the types of *S. walesius* Verhoeff, 1928 (see below).

See also the general notes for *S. rubriventris* (above).

Solaenodolichopus teres (Verhoeff, 1924)
Figs 1D, 2D, 3D, 8C–D

Antichiropus (Solänodolichopus) teres Verhoeff, 1924: 12 (as *Antichiropus teres*), 20 (as *Solänodolichopus teres* in key), 21 (as *Antichiropus (Solänodolichopus) teres,* first description); figs 7–8 in pl. 1.

Aulacoporus teres Attems 1937: 261 (new combination), 263 (Verhoeff's 1924 description reworded); fig. 327 (p. 263; same as fig. 7 in Verhoeff 1924).

Solaenodolichopus teres – Verhoeff 1928: 94 (genus misspelled *Solanodolichopus*; new combination for *teres*, *vittatus* and *vittatus dorsalis*, referred to by Verhoeff as "the three forms previously described by me"). — Jeekel 2000: 40. — Nguyen & Sierwald 2013: 1159.
Aulacoporus teres – Attems 1937: 261 (new combination), 263 (Verhoeff's 1924 description reworded); fig. 327 (p. 263; same as fig. 7 in Verhoeff 1924). — Jeekel 1968: 18, 29; 1971: 233 (designated as type of *Solaenodolichopus*); 1981: 49.

Material examined

Lectotype (here designated)
Slide mount with original Verhoeff label containing male legpair 1, right gonopod telopodite, left gonopod telopodite missing prefemur, left gonopod prefemur attached to the distal portion of the gonocoxa, both legs 9 and a cuticle fragment which may be the anterior aperture margin from ring 7, NHRS-KAS1000000005; the rest of the male body missing legpair 1, ring 7 and parts of some legs, in alcohol but dried out in the past, somewhat distorted, ZSM A20052188; Colosseum, QLD [24°26' S, 151°35' E, ±5 km], E. Mjöberg, Nov.-Dec. 1912 (see taxonomic notes, below). The lectotype is the one male specimen examined by Verhoeff, i.e., the body parts on the slide mount plus the body parts in alcohol.

Paralectotype
Female, details as for lectotype, in alcohol, body broken between rings 8 and 9, NHRS-KAS1000000007.

Other material

None.

Description

As for the genus. Maximum midbody width *ca.* 3.0/3.5 mm (male lectotype/female paralectotype, but male distorted by drying out in past). Colour in alcohol: "Body gray-yellow, on each side with wide brown longitudinal band dorsally, becoming dark brown anteriorly. Between the brown longitudinal bands a washed-out ochre-yellow dorsal longitudinal band with a brown median stripe. Collum dark brown, medially more or less lighter" (Verhoeff 1924: 20, my translation); male lectotype largely colourless, female paralectotype dull yellow grading to light reddish-brown posteriorly on each ring with no trace of longitudinal banding.

Male with transverse furrow on metazonites, stopping well above level of ozopore. No longitudinal furrows laterally on diplosegments. Sternal lamella (Fig. 2D) rounded-rectangular, with near-vertical sides and more or less straight distal margin slightly raised as small, rounded projection in centre, and with wide, shallow concavity on posterior surface on either side. Aperture details uncertain due to Verhoeff's dissection, but if slide-mounted cuticle fragment is anterior margin of aperture, then with gently convex extensions to anterior aperture margin on either side of rounded-triangular, medial extension (as shown in Fig. 3B for *S. rubriventris*). Scopulae on legs 2-29, i.e., not present on first or last podous rings.

Gonopod telopodite reach on venter uncertain due to dissection, and not stated by Verhoeff (1924). Gonocoxa length relative to telopodite not known; gonocoxa fragment on lectotype slide with sparse, long setae anterodistally. Cannula short, narrow, uniformly tapering towards prefemur. Prefemur small, rounded, densely setose posteromedially, demarcated from femorite laterally by small, narrow notch. Femorite (Fig. 8C) *ca.* 40% of telopodite length, somewhat flattened anteroposteriorly, more or less uniformly wide to *ca.* 1/2 its length, then bending medially at *ca.* 45° and expanding, the end more or less truncate. Lateral femorite process represented by small, short, rounded ridge at femorite tip. Medial femorite process short, pointed, "thrust into a groove on the medial side of the base of the solenomerite and is therefore easily overlooked" (Verhoeff 1924: 21, my translation). Solenomere (Fig. 8C–D) at base *ca.* 1/2 femorite tip width, directed distally and curving laterally at midlength, distally flattened anteroposteriorly, the lateral profile gently convex; anterior surface with thin, ridge-like flange along almost entire length of solenomere, extended anterolaterally as large triangle near base and smaller triangle just below midlength. Solenomere distomedially with U-shaped indentation, the basal rim of the U extending as short, tooth-like projection. Solenomere distal from U curving medially, subdivided by narrow indentation into short, rod-like apical process and anteroposteriorly flattened, truncate, subapical process with opening of prostatic groove in middle of medial margin. Prostatic groove running on anterior surface of telopodite, following curve of solenomere on medial side of anterior surface flange.

Female paralectotype with distolateral process (Fig. 3D) on each leg 2 coxa extending *ca.* 1/2 coxa diameter laterally, with truncate lateral margin about coxa diameter in length; in situ the process resting on ring 3 ventral surface posterior to raised end of epigyne.

Distribution

So far known only from the type locality in central coastal Queensland (map Fig. 12).

Taxonomic notes

Verhoeff based his description on one male and one female. He also examined a female 20 mm long, but questioned whether it belonged to the same species (Verhoeff 1924: 22). The small female is a stadium

7 individual broken between rings 7 and 8 and is in the NHRS vial with the paralectotype. The lectotype slide was labelled 'lectotype' by P.M. Johns in 1967, but Johns did not publish this lectotypification.

Verhoeff (1924) did not give a collection date for the *S. teres* syntypes. Mjöberg's unpublished field diaries show that he arrived in Colosseum in the last week of November 1912 and left during the first week of December 1912 (Å. Ferrier, in litt., 16 Dec. 2013; see also Ferrier 2006).

The structure of the gonopod telopodite is a little unclear to me due to its flat mounting on a microscope slide, and I have not yet seen any other *Solaenodolichopus* specimens with gonopods perfectly matching those of *S. teres*. A male (QM S74768) pitfall-trapped in 1983 at Kroombit Tops, *ca.* 60 km west of Colosseum, has a similarly short and distally bent femorite and might be conspecific. However, the much-faded colouring of the Kroombit Tops male consists of brown transverse bands at the rear of the metazonites against a pale yellow ground, rather than the longitudinal banding seen by Verhoeff in *S. teres*. The medial femorite process in the Kroombit Tops male has a teardrop-shaped base as in *S. pruvoti* (for example); details of the base are unclear in the slide-mounted *S. teres* gonopods.

General notes

S. teres, *S. vittatus* Verhoeff, 1924 and *S. vittatus dorsalis* Verhoeff, 1924 were among specimens collected by the Swedish zoologist Erik Mjöberg (1882-1938) during two expeditions to Australia, 1910–1913 (Ferrier 2006). Verhoeff obtained the Mjöberg specimens from the Naturhistoriska Riksmuseet in Stockholm, described 43 new millipede species and subspecies from the material and published his results in the Swedish journal *Arkiv för Zoologi* (Verhoeff 1924). To Verhoeff's disappointment, the journal reproduced many of his drawings at a scale too small for clarity, among them those of *Aulacoporus* and *Solaenodolichopus* spp.: "Die Abbildungen sind leider etwas zu stark verkleinert worden, wodurch die Deutlichkeit mancher zarterer Teile gelitten hat" (Verhoeff 1924: 142).

Solaenodolichopus vittatus (Verhoeff, 1924)

Figs 1C–E, 2E, 10B–G

Antichiropus (Solänodolichopus) vittatus Verhoeff, 1924: 22 (first description); pl. 1, figs 5, 9, 10.
Antichiropus (Solänodolichopus) vittatus dorsalis Verhoeff, 1924: 21, 22 (first description); pl. 1, fig. 6 [New synonymy.]

Solaenodolichopus vittatus – Verhoeff 1928: 94 (genus misspelled *Solanodolichopus*; new combination for *teres*, *vittatus vittatus* and *vittatus dorsalis*, referred to by Verhoeff as 'the three forms previously described by me'). — Jeekel 2000: 40. — Nguyen & Sierwald, 2013: 1160.
Aulacoporus vittatus – Attems 1937: 261 (new combination), 263 (Verhoeff's description reworded); fig. 328 (p. 263; same as fig. 6 for *S. vittatus dorsalis* in Verhoeff 1924). — Jeekel 1968: 18, 29; 1981: 49.
Aulacoporus vittatus dorsalis – Verhoeff 1928: 94 (genus misspelled *Solanodolichopus*; new combination for *teres*, *vittatus vittatus* and *vittatus dorsalis*, referred to by Verhoeff as 'the three forms previously described by me'). — Attems 1937: 261 (new combination), 264 (Verhoeff's description reworded). — Jeekel 1968: 18, 29; 1981: 49.
Solaenodolichopus vittatus dorsalis – Jeekel 2000: 40. — Nguyen & Sierwald 2013: 1160.

Material examined

***S. vittatus* lectotype** (here designated)

Slide mount with original Verhoeff label, containing male legpair 1, right gonopod, left gonocoxa, left gonopod telopodite and 1 leg 9, NHRS KAS1000000004; and the rest of the male body in alcohol with original Verhoeff label, missing legpair 1, body broken between rings 3 and 4, rings 6 and 7, and rings

8 and 9, NHRS KAS1000000006; Mt Tambourine [Tamborine Mountain], QLD [27°58' S, 153°11 'E, ±5 km], E. Mjöberg, Oct. 1912 (see taxonomic notes, below). The lectotype is the one male specimen examined by Verhoeff, i.e. the body parts on the slide mount plus the body parts in alcohol.

***S. vittatus* paralectotype**

Female, details as for lectotype, body broken between rings 3 and 4 and rings 5 and 6, in alcohol, NHRS KAS1000000006.

***S. vittatus dorsalis* holotype**

Slide mount with original Verhoeff label, containing male legpair 1, both gonopod telopodites, 1 gonocoxa and both legs 9; and the rest of the male body in alcohol with original Verhoeff label, broken into head+collum and rings 3, 4+5, 6+7, 8, 9+10, 11-13 and 14-telson, missing legpair 1 and legs and gonopods from ring 7, aperture damaged; Glen Lamington [Lamington Glen], QLD [28°15' S, 153°01' E, ±5 km], E. Mjöberg, Nov. 1912 (see taxonomic notes, below). The holotype by monotypy is the one male specimen examined by Verhoeff, i.e. the body parts on the slide mount plus the body parts in alcohol, both in NHRS and registered as KAS1000000003.

Other material

QUEENSLAND: 2 ♂♂, Lamington National Park ('National Park') [28°14' S, 153°08' E, ±5 km], Dec. 1919, H. Hacker, QM S74826; 1 ♂, Tamborine Mountain ('Mt Tamborine') [27°58' S, 153°11' E, ±5 km], Oct. 1924, A. Musgrave and G. Geissmann, QM S74827; 1 ♂, same locality but 18 Mar. 1955, M.B. Wilson, in grass, QM S74764; 1 ♂, 1 ♀, same locality but 26 Mar. 1955, S.B. Gunn, QM S74829; 3 ♂♂, West Burleigh [28°07' S, 153°26' E, ±2 km], 12 Feb. 1953, QM personnel, QM S74828; 1 ♂, Warrie National Park [28°13' S, 153°16' E, ±2 km], 2 Jul. 1971, K.R. McDonald, rainforest, under logs, 'S30' on metal tag, QM S74830; 1 ♂, Bahrs Scrub, 27°44'59" S, 153°10'36" E [±500 m], 200 m a.s.l., pitfall, 29 Jan.–23 May 1981, G. Monteith, GM 1078/2, QM S74765; 1 ♂, Moreton Island, 27°11' S, 153°24' E [±10 km], 26 Apr. 1982, W. Houston, pitfall,, yellow patch, EIS, QM S5941; 1 ♂, Blue Lagoon, Moreton Island, 27°05'35" S, 153°26'27" E [±2 km], 21 Sep. 1982, W. Houston, EIS, pitfall, QM S5946; 1 ♂, Springbrook - N end, 28°09'47" S, 153°15'42" E [±500 m], 540 m a.s.l., pitfalls, 15 May–30 Aug. 1997, G. Monteith, rainforest, sample 5015, QM S74836; 1 ♂, Perrys Knob, 27°36'11" S, 152°36'17" E [±500 m], 200 m a.s.l., pitfall, 15 Sep–11 Nov. 1998, G. Monteith, D. Cook and G. Thompson, vine scrub, sample 7295, QM S74843; 1 ♂, same details but pitfall, 11 Nov. 1998–13 Jan. 1999, sample 7564, QM S74855; 1 ♂, Enoggera Reservoir site 4, 27°26'47" S, 152°55'23" E [±100 m], 110 m a.s.l., pitfall, 7 Aug.–16 Oct. 1999, G. Monteith and J. Holt, rainforest, sample 7854, QM S74871; 1 ♂, 1 ♀, 3 km SE of Kalbar, 27°57'46" S, 152°38'47" E [±100 m], 120 m a.s.l., pitfall, 2 Dec. 2000–7 May 2001, C. Burwell, sample 10160, brigalow scrub, QM S74766; 1 ♂, Karawatha Forest - site 6, 27°37'33" S, 153°5'24" E [±100 m], 60 m a.s.l., 30 Apr. 2003, S. Wright and E. Volschenk, eucalypt woodland, sample 51168, QM S74767; 1 ♂, Gold Creek Reservoir site 2, 27°28'05" S, 152°52'20" E [±100 m], 160 m a.s.l., 4 Nov. 2003, C. Burwell, dry rainforest, sifted litter berlesate, sample 51723, QM S74873; 2 ♂♂, Gold Creek Reservoir site 1, 27°27'53" S, 152°52'32" E [±100 m], 140 m a.s.l., pitfall, 2–30 Jan. 2004, Queensland Museum personnel, spotted gum open forest, sample 51819, QM S74875; 1 ♂, 2 ♀♀, same details but 30 Jan.–1 Mar. 2004, sample 51939, QM S74876; 2 ♂♂, Welsby Street, New Farm [27°27'58" S, 153°02'56" E, ±500 m], 1 Nov. 2010, M. Shaw, pitfall, QM S95213.

Description

As for the genus. Maximum male/female midbody width *ca.* 3.8 (mean of 24 males, range 2.9-4.7 mm)/4.8 mm (paralectotype female). Colour in alcohol: "Body reddish yellow with two broad, blackish, lateral longitudinal bands on the dorsum, separated by wide, reddish yellow median (band). Medial [longitudinal] stripe washed-out reddish, ventral flanks reddish-yellow" (*vittatus*, Verhoeff 1924: 20, my translation); "Trunk black, with brown legs and wide yellow reddish dorsal median band, significantly narrower than in *vittatus*. Ventral flanks black" (*vittatus dorsalis*, Verhoeff 1924: 21, my translation); colour in types faded and patchy, *dorsalis* medial longitudinal band *ca.* 1/3 ring width, *vittatus ca.* 2/3 ring width; colour and pattern variable in other specimens (Fig. 4C–I), but always with lighter colouring dorsally and darkest colouring on upper sides.

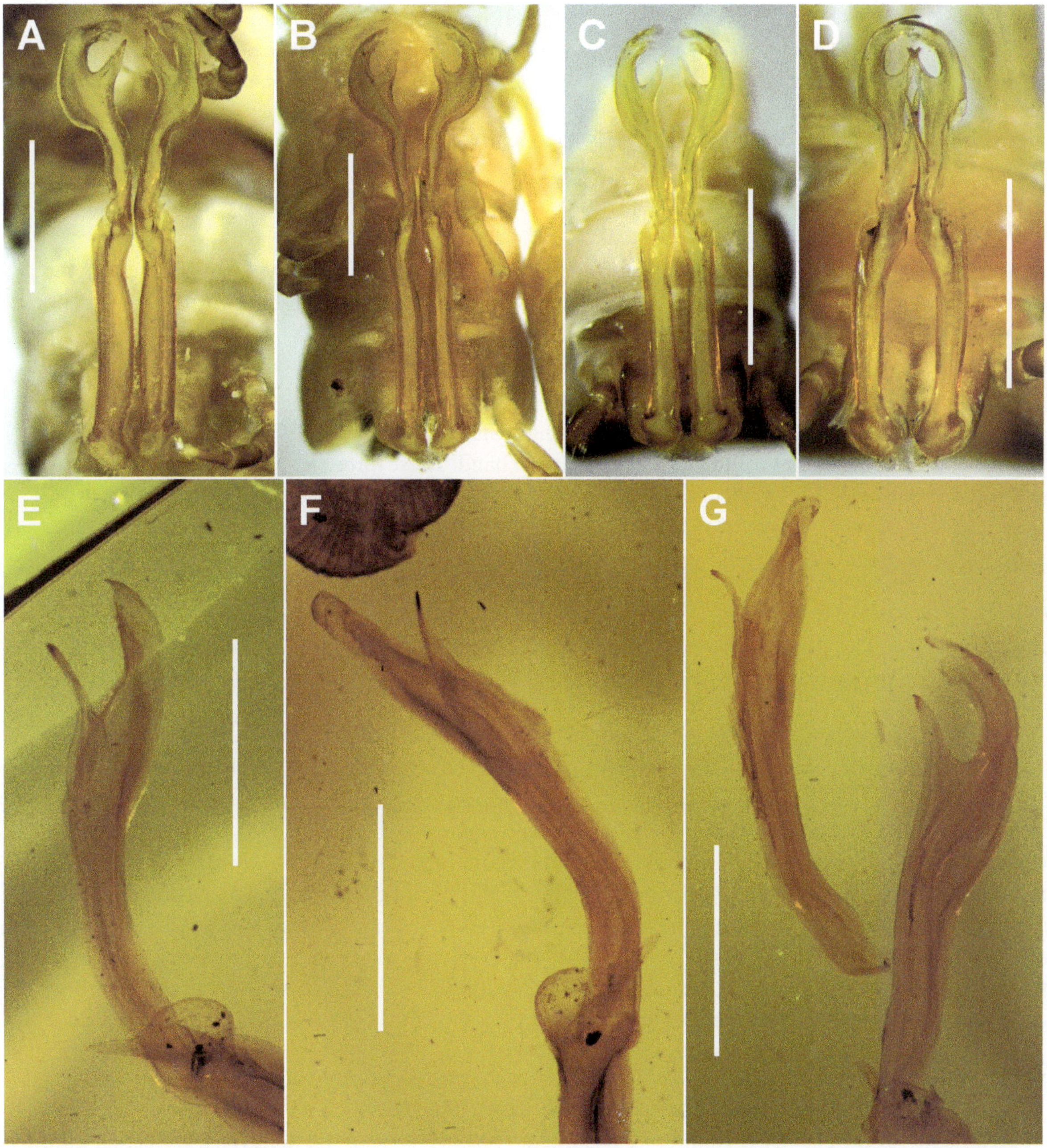

Fig. 10. A. — *Solaenodolichopus pruvoti* (Brolemann, 1931), posterior view of gonopods *in situ*, QM S74858. — **B–G.** *S. vittatus* (Verhoeff, 1924). **B–D**, posterior view of gonopods *in situ*; QM S74826 (B), QM S95213 (C), QM S74767 (D). **E–G**, views of slide-mounts; left (E) and right gonopod solenomere (F) of *S. vittatus* lectotype, NHRS KAS1000000004, and left (top left) and right (bottom right) gonopod solenomere (G) of *S. vittatus dorsalis* holotype, NHRS KAS1000000003. Scale bars: A–D = 2 mm, E–G = 1 mm.

Male with transverse furrows shallow, narrow, sometimes indistinct. No longitudinal furrows laterally on diplosegments. Sternal lamella (Fig. 2E) with sides more or less straight, corners broadly rounded, distal margin a flat inverted V. Scopulae on legs 1 to 29, i.e., not present on last podous ring. Leg bases on rings 6 and 8 separated more widely than on other rings, sternites a little depressed. Anterior margin of aperture with rounded-triangular, medial extension and shorter, gently convex extension on either side (as shown in Fig. 3B for *S. rubriventris*).

Gonopods usually as for *S. pruvoti* (Fig. 10A), but with femorite in some males thinner (Fig. 10B), and in other males shorter and bent medially near apex, and with medial femorite process larger, reaching to *ca.* 1/3-1/2 of solenomere length (Fig. 10D); intermediate forms occur (Fig. 10C).

Female with leg 2 coxa extended posteroventrally as apically rounded process (as shown for *S. pruvoti* in Fig. 5C–D).

Distribution

Forest and non-forest habitats in the Brisbane region, from Moreton Island *ca.* 130 km south to the Border Ranges, and inland *ca.* 80 km from the sea (map Fig. 12). Overlaps in range with *S. pruvoti* and *S. rubriventris*.

Taxonomic notes

The subspecies *S. vittatus dorsalis* was erected for a single male only differing from *S. vittatus vittatus* in size, colour of ventral flanks and width of the medial dorsal band. These three characters vary substantially in the 24 non-type males I have examined and do not appear to be correlated. I therefore regard the subspecies *dorsalis* as a synonym of the nominate subspecies *vittatus*. The gonopod telopodites of the types are virtually identical (Fig. 10E–G).

S. vittatus varies more across its range than other *Solaenodolichopus* (described and undescribed), in gonopod form as well as in size and colour pattern. It might better be called 'the *S. vittatus* group', and intensive sampling and genetic analysis across its range would be a worthwhile systematics project.

It is particularly interesting that the gonopods in *S. pruvoti* and some *S. vittatus* populations are indistinguishable, and that no colour pattern intermediates have so far been found. When redescribing *S. pruvoti* as *S. annulatus*, Verhoeff wrote: "Stands close to *vittatus* Verh. (contribution 3, figs 9 and 10) but is distinguished by different patterning and some special features of the gonopods: the rudiment of the tibiotarsus [rounded process on basal side of U-shaped indentation] is smaller in *vittatus* and not curved, the parsolenomere [apical tab] very similar but more transversely curved, the medial displacement of the solenomere is barely detectable, overall the solenomere is less curved and less clavate, the end in *vittatus* rounded" (Verhoeff 1941: 12, my translation). These small differences are largely contained within the variation seen in both *S. pruvoti* and *S. vittatus*. Until new diagnostic features are identified, *S. vittatus* and *S. pruvoti* are only separable on colour pattern.

Both Verhoeff and Mjöberg spelled the type locality of *S. vittatus* 'Mt Tambourine', which in the early 20th century referred to both a village and to the small basalt plateau on which the village was located; the modern name for the plateau is Tamborine Mountain. Verhoeff (1924) did not give a collection date for the *S. vittatus* types. Mjöberg's unpublished field diaries show that he collected at 'Mt Tambourine' in the last two weeks of October 1912 (Å. Ferrier, in litt., 16 Dec. 2013; see also Ferrier 2006). The lectotype slide of *S. vittatus* was labelled 'lectotype' by P.M. Johns in 1967, but Johns did not publish this lectotypification.

Verhoeff's one specimen of *S. vittatus dorsalis* was from 'Glen Lamington'. Lamington Glen was a locality and rainfall recording station on the Lamington Plateau in the first half of the 20th century, and Mjöberg's unpublished field diaries show that he collected on the 'Lamington Plateau' in mid-November 1912 (Å. Ferrier, in litt., 16 Dec. 2013; see also Ferrier 2006). The holotype slide was labelled 'holotype' by P.M. Johns in 1967.

General notes

See general notes for *S. teres* (above).

Solaenodolichopus walesius Verhoeff, 1928
Figs 1A–B, 2F, 3C, 11A–C, 11E–I

Solänodolichopus walesius Verhoeff, 1928: 94 (genus misspelled *Solanodolichopus*), 95 (first description, genus misspelled *Solandolichopus*), 114 (as *Solänodolichopus walesius* in list of species described in paper); pl. 10, figs 26–27.
Parwalesoma castaneum Verhoeff, 1937: 139 (first description); figs 6–7 (p. 138).

Parwalesoma castaneum – Attems 1940: 549 (Verhoeff's description reworded). — Jeekel 1968: 19, 29. — Jeekel 1971: 231 (noted as type of *Parwalesoma*). — Jeekel 1981: 49. — Jeekel 2000: 41 (synonymised with *Parwalesoma walesium*). — Nguyen & Sierwald 2013: 1157 (as synonym of *Parwalesoma walesium*).
Aulacoporus walesius – Attems 1937: 261 (new combination), 264 (Verhoeff's 1928 description reworded); fig. 329 (p. 264; same as fig. 26 in Verhoeff 1928). — Jeekel 1968: 18, 29; Jeekel 1981: 49.
Parwalesoma walesium – Jeekel 2000: 41 (new combination). — Nguyen & Sierwald 2013: 1157.

Material examined

***S. walesius* lectotype** (here designated)
Male in alcohol with original Verhoeff label, broken between rings 6 and 7 and rings 7 and 8, North Dorrigo, NSW [30°16'S 152°41'E ±5 km], 4 Jan. 1923, A. Musgrave, AM KS.76508 (formerly K47704).

***S. walesius* paralectotypes**
1 male in alcohol with original Verhoeff label, broken between rings 6 and 7, ZMB 7962; 1 female in alcohol without original Verhoeff label, broken between rings 5 and 6, AM KS.76509 (also formerly K47704); collection details for both as for *S. walesius* lectotype.

***P. castaneum* lectotype** (here designated)
Male in alcohol with original Verhoeff label, broken between rings 8 and 9, missing left antenna, ZMB 12652; Comboyne district, NSW [31°35' S, 152°28' E, ±10 km], date unknown, E.C. Chisholm (see general notes, below).

***P. castaneum* paralectotypes**
Slide mount with original Verhoeff label containing male legpairs 1 and 9 and both gonopods, ZSM A20033565; male rings 4–6 and an isolated diplosegment, and 1 female broken between rings 8 and 9, and 10 and 11, in alcohol with new labels (male and female in separate vials), ZSM 20052246; 1 female in alcohol with original Verhoeff label, broken between rings 6 and 7 and rings 12 and 13, AM KS76700; collection details for all three as for *P. castaneum* lectotype. I assume that the male rings in ZSM A20052246 and the slide-mounted parts in ZSM A20033565 are from the same individual, i.e., that together they represent a single male paralectotype specimen (see taxonomic notes, below).

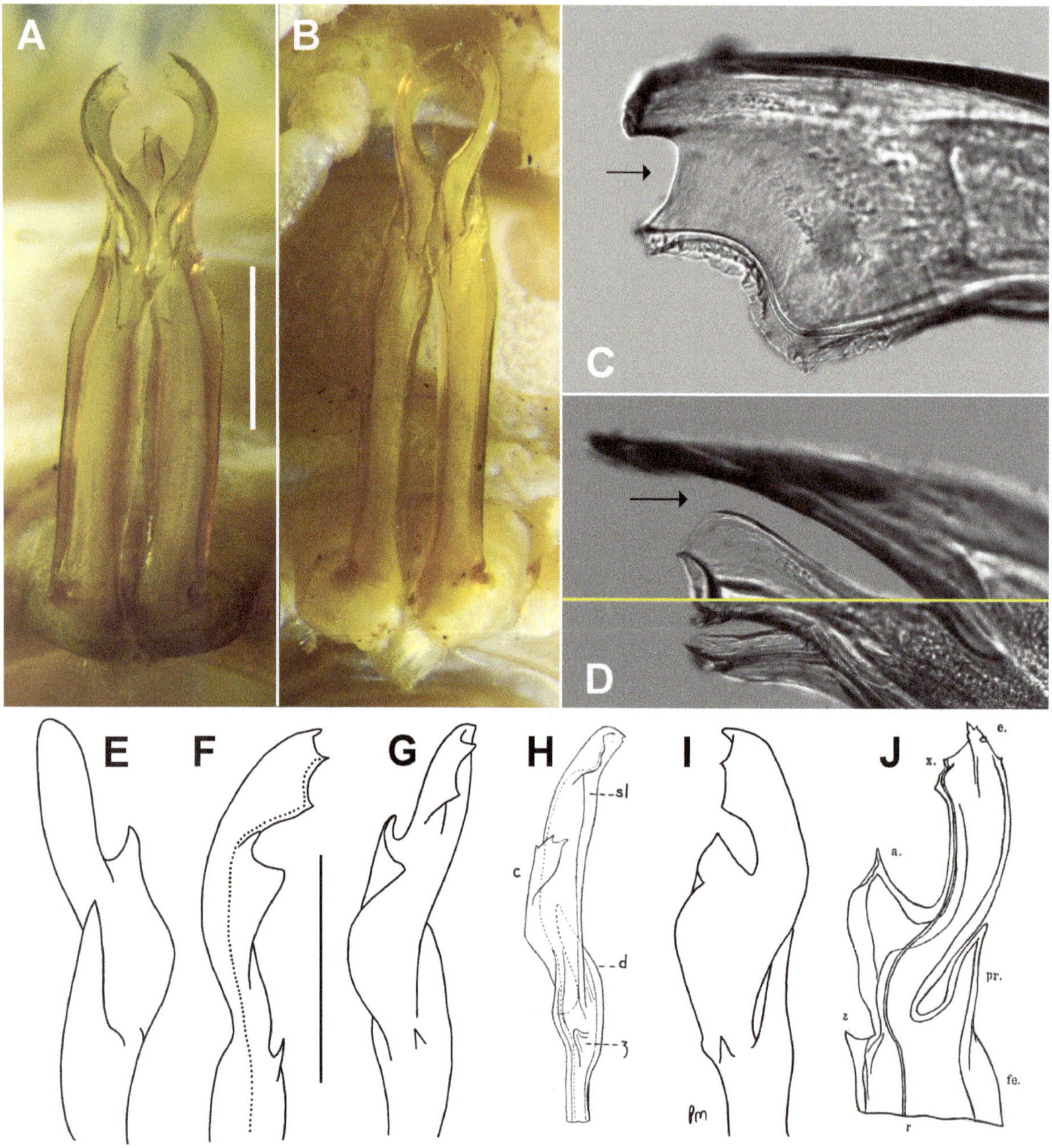

Fig. 11. — **A–C, E–J**. *Solaenodolichopus walesius* Verhoeff, 1928. **A–B**. Posterior views of gonopods *in situ* of lectotype, AM KS.76508 (A) and *Parwalesoma castaneum* Verhoeff, 1937 (= *S. walesius*) lectotype, ZMB 12652 (B). **C**. Posterior view of left gonopod solenomere tip, AM KS.16152. **E–G, I**. Lateral (E), anterior (F), medial (G) and posteromedial (I) views of left gonopod solenomere, AM KS.93755. **H**. Medial view of left gonopod solenomere of *P. castaneum* (= *S. walesius*), from fig. 6 in Verhoeff (1937). **J**. Posterior and slightly oblique view of left gonopod solenomere of *S. walesius*, from fig. 26 in Verhoeff (1928). — **D**. *S. pruvoti* (Brolemann, 1931), posterior view of left gonopod solenomere tip, QM S74856 (composite image at two focus levels). Dotted line in F marks course of prostatic groove. Scale bars: 1 mm.

Other material

NEW SOUTH WALES: 1 ♂, Comboyne [31°35' S, 152°28' E, ±10 km], Aug. 1926, E.C. Chisholm, AM KS.94102 (see taxonomic notes, below); 2 ♂♂, Comboyne Cave KSS-C4, 31°34'13" S, 152°23'38" E [±100 m], 21 Mar. 1971, C. Carter, AM KS.96065; 2 ♂♂, 2 ♀♀, Kerewong State Forest near Lorne, site 108, 31°36' S, 152°34' E [±2 km], pitfall, 7 Nov.–10 Dec. 1978, D. Milledge, AM KS.16152; 1 ♂, same details, KS.18483; 1 ♂, same details but site 81, 31°35' S, 152°38' E [±2 km], pitfall, 26 Sep.–12 Dec. 1978, AM KS.105104 (ex KS.96075); 1 ♂, 2 ♀♀, Boundary Creek Road, Boundary Creek State Forest, 29°56' S, 152°33' E [±500 m], pitfall, 4 Feb.–9 Apr. 1993, M. Gray and G. Cassis, AM KS.93755.

Description

(Based, unless otherwise indicated, on *S. walesius* lectotype and paralectotypes.) As for the genus. Maximum male/female midbody width *ca.* 4.5/4.5 mm. Colour in alcohol: "Body ringed, the metazonites for the most part dark brown and the prozonites wine red. / The wine-red colour of the prozonites extends on to the anterior bands of the metazonites, which are otherwise dark brown. The lower sides and abdomen are also wine-red, the legs yellowish" (Verhoeff 1928: 94/95); colouring of types now faded and patchy, but rings light brown, darker dorsally, lightening to yellow at the waist, darkening towards rear of metatergite; colouring of more recent specimens as noted by Verhoeff.

Male with distinct transverse furrow on metazonites, stopping a little above level of ozopore. No longitudinal furrows laterally on diplosegments. Sternal lamella (Fig. 2F) broadly paraboloid in outline. Scopulae on legs 1 to 29, i.e., not present on last podous ring. Leg bases on ring 6 separated a little more than on ring 5; leg bases with normal separation on ring 8. Anterior margin of aperture more or less straight with small, rounded-triangular, medial extension.

Gonocoxa (of male in AM KS.93755) about 1/2 telopodite length, slightly flattened anteroposteriorly, with sparse, long setae anterodistally. Cannula short, narrow, uniformly tapering towards prefemur. Gonopod telopodite (Fig. 11A–C, E–I) just reaching leg 6 bases when retracted. Prefemur small, rounded, densely setose posteromedially, demarcated from femorite laterally by small, narrow notch. Remainder of telopodite more or less straight, slender, clearly divided into femorite and solenomere at about 2/3 telopodite length. Femorite narrowing very slightly between ends, slightly flattened anteroposteriorly. Medial femorite process a small, triangular tab; lateral femorite process spine-like, between 1/4 and 1/3 solenomere length, arising more distally than medial process, curving slightly anteriorly towards solenomere. Solenomere arising anterior to femorite processes, *ca.* 1/2-2/3 femorite width at base, curving laterally then medially and flattening to form trough-like hollow, concave medially; basal half of anterior margin of hollow extending distomedially and a little anteriorly as large, flat, triangular tab reaching *ca.* 1/2 solenomere length, curving posteriorly, distal margin of tab with a few small teeth; hollow at apex of solenomere facing posteromedially, the anterior margin produced as small subapical tab with rounded notch medially on distal margin, the solenomere apex distal to the notch at right angles to the subapical tab. Prostatic groove running on anterior surface of telopodite and lateral to large, basal solenomere tab, following subapical tab margin and terminating at distal corner of subapical tab.

Female without process on leg 2 coxa.

Distribution

Known from eucalypt forest (and rainforest?) in northeast NSW from the ranges northwest of Coffs Harbour to the Comboyne district, a north-south extent of *ca.* 180 km (map Fig. 12). Overlaps in range with *S. sulcatus*.

Taxonomic notes

The lectotype and paralectotype *S. walesius* males from North Dorrigo are almost identical in size and morphological details to the lectotype *P. castaneum* male from the Comboyne district, 150 km to the south; they differ slightly in the slenderness of the femorite (Fig. 11A–B). It is a little puzzling that Verhoeff did not recognise them as the same species. His drawings of the *S. walesius* telopodite tip (figs 26 and 27 in Verhoeff 1928; Fig. 11J) appear to have been done from a posterior and slightly oblique view of the left gonopod of the lectotype, while the drawings of the *P. castaneum* telopodite tip (figs 6 and 7 in Verhoeff 1937; Fig. 11H) are from a medial view of the left gonopod telopodite on the lectotype slide mount. Verhoeff may have returned one of his *S. walesius* males to AM before he received the *P. castaneum* specimens, but he may still have held the male that found its way to ZMB. If not, it is possible that Verhoeff had no gonopod drawings of *S. walesius* to compare other than his published ones. Ironically, it was a comparison of those drawings that led Jeekel (2000) to propose that *P. castaneum* was a junior synonym of *S. walesius*.

In both *S. walesius* (Fig. 11C) and *S. pruvoti* (Fig. 11D) the solenomere tip ends in an L-shaped channel. One side of the 'L' is the thin, anteroposteriorly flattened tab at whose medial margin the prostatic groove opens. The other, thicker side of the 'L' is distal to this tab and lies in a plane at right angles to it. In *S. pruvoti*, a narrow notch (arrow in Fig. 11D) separates the two sides of the 'L', the distal side thus becoming a separate process, i.e., the 'pre-apical latero-distal process' of Jeekel (2000). In *S. walesius*, the two sides of the 'L' are joined at the tip by a shallow indentation (arrow in Fig. 11C). This difference does not seem to me to be important enough to separate two genera, which is why I have synonymised *Parwalesoma* with *Solaenodolichopus* (see above). In future papers I will document variations at the tip of the solenomere in other *Solaenodolichopus* species.

Verhoeff (1928, 1937) did not state how many specimens he examined, but gave male and female lengths for both *S. walesius* and *P. castaneum*. My lectotypifications imply that Verhoeff looked at two males and one female of *S. walesius* and two males (one dissected, most parts now missing) and two females of *P. castaneum*. However, the AM whole female in alcohol (KS.76509) and the ZSM male parts and whole female in alcohol (20052246) are missing any original Verhoeff labels. I include these among the types because I assume that the ZSM male parts are the remains of the male dissected for the ZSM slide mount, which does have a Verhoeff label, and because the collection details label for AM KS.76509 appears to be original. I have not made the male *S. walesius* in AM KS.94102 a paralectotype because it has only a recent, printed label; it may have been deposited directly in AM by collector E.C. Chisholm (see below). The ZSM slide of *P. castaneum* was labelled 'lectotype' by P.M. Johns in 1967, but this lectotypification was not published.

General notes

According to contemporary newspaper items, collector Dr E.C. Chisholm was a naturalist and Government Medical Officer in the Comboyne district from 1923 to 1935 (*The Port Macquarie News and Hastings River Advocate*, 4 Jun. 1927, p. 5, http://trove.nla.gov.au/ndp/del/article/ 112527391; *The Sydney Morning Herald*, 17 Jan. 1933, p. 6, http://trove.nla.gov.au/ndp/del/article/16945944; *The Maitland Daily Mercury*, 5 Jul. 1935, p. 3, http://trove.nla.gov.au/ndp/del/article/127096313; all accessed 24 Dec. 2013).

Chisholm presumably collected his Comboyne millipedes and sent them directly to Verhoeff (Verhoeff 1937: 133) during that 12-year period. As he had done with descriptions of NSW species in his 1928 paper, Verhoeff used the State museum journal, *Records of the Australian Museum*, as an outlet for his descriptions of *P. castaneum* and four other Comboyne millipede species. Verhoeff (1937) published in German, however, perhaps to avoid any misunderstandings arising from inadequate translation.

See also the general notes for *S. rubriventris* (above).

Discussion

As circumscribed above, *Solaenodolichopus* is one of Australia's most speciose millipede genera (Car 2009; Mesibov 2008), and combined genetic and morphological studies will be needed to resolve its internal classification. Additional collecting and genetic studies will also be needed to satisfactorily delimit the suspiciously variable *S. vittatus*, and to clarify its relationship with *S. pruvoti*, from which it can currently only be distinguished by colour pattern.

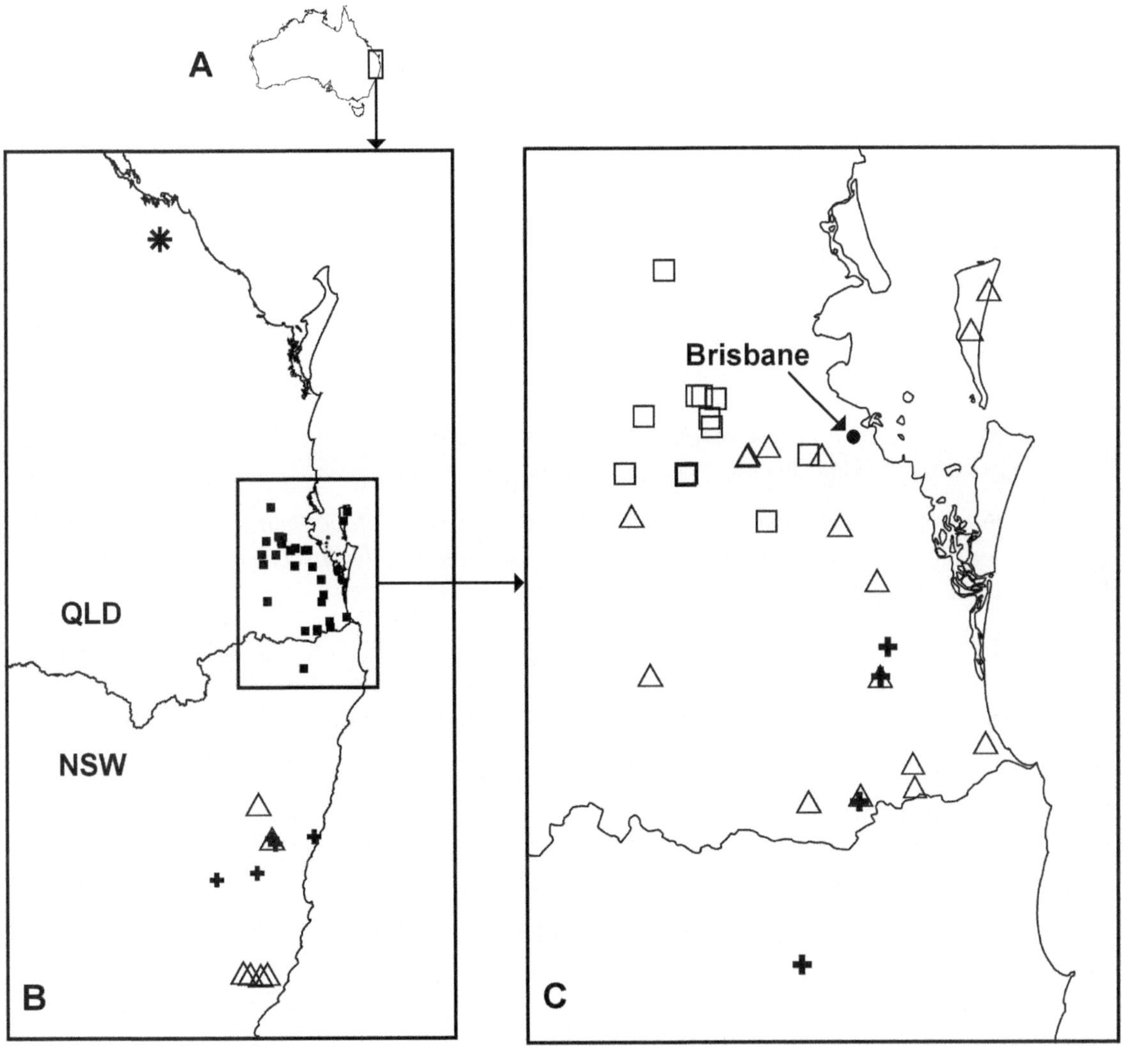

Fig. 12. Localities for named *Solaenodolichopus* species as of February 2014. — **A**. Australia showing location of map B (rectangle). — **B**. *S. teres* (Verhoeff, 1924) (star), *S. sulcatus* (Verhoeff, 1928) (crosses) and *S. walesius* Verhoeff, 1928 (open triangles), other species (filled squares); rectangle shows location of map C. — **C**. *S. pruvoti* (Brolemann, 1931) (open squares), *S. rubriventris* Verhoeff, 1928 (crosses) and *S. vittatus* (Verhoeff, 1924) (open triangles). Maps A–C are geographic projections; approximate north-south map extents are 900 km for B and 220 km for C.

Phylogenetic work would also be desirable in the larger group of eastern Australian genera, evidently closely related, to which *Aulacoporus*, *Solaenodolichopus* and *Walesoma* belong. The known forms of *Aulacoporus* and *Walesoma* (described and undescribed) are separated by *ca.* 1300 km, and this disjunction is unlikely to shrink with further collecting. The range of *Solaenodolichopus* (map Fig. 12) will expand considerably as new species are described and documented.

Acknowledgements

For loans of, or assistance with Verhoeff's types I am very grateful to Janet Beccaloni (Natural History Museum, London), Cathy Car (Western Australian Museum, Perth), Jason Dunlop (Museum für Naturkunde, Berlin), Jonas Ekström (Lund University, Lund), Karin Sindemark Kronestedt (Naturhistoriska Riksmuseet, Stockholm), Graham Milledge (Australian Museum, Sydney), Kate Sparks (South Australian Museum, Adelaide), Jörg Spelda (Zoologische Staatssammlung, München) and Verena Stagl (Naturhistorisches Museum, Vienna). I am especially grateful to Dr Åsa Ferrier (La Trobe University, Melbourne) for advice on Erik Mjöberg's Queensland expedition; to Margaret Henderson (Research Coordinator for the Richmond River Historical Society, Lismore, NSW) for advice on the Richmond River district; to Jean-Jacques Geoffroy (Muséum National d'Histoire Naturelle, Paris) for advice on H.W. Brolemann; and to Håkan Wittzell (Lund University Botanical Museum) and Jonas Ekstrom (Lund University Zoological Museum) for advice on Johan Mauritzon and his collections in Australia. I thank Sandrine Feig (University of Tasmania, Hobart) for assistance with SEM work. This study was funded by the author.

References

Attems C. 1911. Myriopoda exkl. Scolopendridae. *In*: Michaelsen W. & Hartmeyer R. (eds) *Die Fauna Südwest-Australiens. Ergebnisse der Hamburger südwest-australischen Forschungsreise 1905*. Vol 3 (6): 1–393. Gustav Fischer Verlag, Jena.

Attems C. 1926. Myriopoda. *In*: Kükenthal W. & Krumbach T. (eds) *Handbuch der Zoologie. Eine Naturgeschichte der Stämme des Tierreiches. 4(1). Progoneata, Chilopoda, Insecta I*: 1–402. Walter de Gruyter & Co., Berlin and Leipzig.

Attems C. 1929. Diplopoden des Belgischen Congo. I. Polydesmoidea. *Revue de zoologie et de botanique africaines* 17 (3): 253–378.

Attems C. 1937. *Das Tierreich. 68. Myriapoda. 3. Polydesmoidea. I. Fam. Strongylosomidae.* Walter de Gruyter and Co., Berlin.

Attems C. 1940. *Das Tierreich. 70. Myriapoda. 3. Polydesmoidea. III. Polydesmidae, Vanhoeffeniidae, Cryptodesmidae, Oniscodesmidae, Sphaerotrichopidae, Peridontodesmidae, Rhachidesmidae, Macellolophidae, Pandirodesmidae*. Walter de Gruyter and Co., Berlin.

Brolemann H.W. 1931. Myriapodes recueillis par Madame Pruvot en Nouvelle-Calédonie et aux Loyalty. *Archives de Zoologie Expérimentale et Générale* 72: 275–316.

Car C.A. 2009. Keeled millipedes (Diplopoda: Polydesmida: Paradoxosomatidae) in New South Wales. *General and Applied Entomology* 38: 1–7.

Car C.A., Wojcieszek J.M. & Harvey M.S. 2013. The millipede genus *Antichiropus* (Diplopoda: Polydesmida: Paradoxosomatidae), part 1: redefinition of the genus and redescriptions of existing species. *Records of the Western Australian Museum* 28: 83–118.

Ferrier Å. 2006. Dr Eric Mjöberg's 1913 scientific exploration of North Queensland's rainforest region. *Memoirs of the Queensland Museum Cultural Heritage Series* 4 (1): 1–27. [Available online from the Queensland Museum website, in 2013 as

Hedley C. 1915. Presidential address. *Journal and Proceedings of the Royal Society of New South Wales* 49:1–77.

Jeekel C.A.W. 1965. A new genus and a new species of the family Paradoxosomatidae from Australia (Diplopoda, Polydesmida). *Entomologische Berichten (Amsterdam)* 25: 7–14.

Jeekel C.A.W. 1968. *On the Classification and Geographical Distribution of the Family Paradoxosomatidae (Diplopoda, Polydesmida)*. C.A.W. Jeekel, Rotterdam.

Jeekel C.A.W. 1971. Nomenclator generum et familiarum Diplopodorum: A List of the genus and family-group names in the Class Diplopoda from the 10th edition of Linnaeus, 1758, to the end of 1957. *Monografieën van de Nederlandse Entomologische Vereniging* 5: i–xii, 1–412.

Jeekel C.A.W. 1981. Australia Expedition 1980; legit C.A.W. Jeekel and A.M. Jeekel-Rijvers. List of collecting stations, together with general notes on the distribution of millipedes in eastern Australia and Tasmania. *Verslagen en Technische Gegevens, Instituut voor Taxonomische Zoölogie (Zoölogisch Museum), Universiteit van Amsterdam* 30: 1–59.

Jeekel C.A.W. 1982. Millipedes from Australia, 1: Antichiropodini from South Australia (Diplopoda, Polydesmida, Paradoxosomatidae). *Bulletin Zoölogisch Museum, Universiteit van Amsterdam* 8 (14): 121–132.

Jeekel C.A.W. 2000. Millipedes from Australia, 12: Notes on the Paradoxosomatidae of New South Wales, with a key to the genera, and descriptions of a new genus and a new species (Diplopoda, Polydesmida). *Myriapod Memoranda* 2: 31–47.

Jeekel C.A.W. 2002. Millipedes from Australia, 14: A third contribution to the South Australian Paradoxosomatidae (Diplopoda, Polydesmida). *Myriapod Memoranda* 5: 60–77.

Mesibov R. 2008. Diversity of Queensland paradoxosomatid millipedes (Diplopoda: Polydesmida: Paradoxosomatidae). *Australian Entomologist* 35 (1): 37–46.

Musgrave A. 1932. *Bibliography of Australian Entomology 1775-1930: with Biographical Notes on Authors and Collectors*. Royal Zoological Society of New South Wales, Sydney.

Nguyen A.D. & Sierwald P. 2013. A worldwide catalog of the family Paradoxosomatidae Daday, 1889 (Diplopoda: Polydesmida). *Check List* 9 (6): 1132-1353. [Available online through the journal website,

Verhoeff K.W. 1924. Results of Dr. E. Mjöberg's scientific expeditions to Australia 1910-1913. 34. Myriapoda: Diplopoda. *Arkiv för Zoologi* 16 (5): 1–142.

Verhoeff K.W. 1928. On Diplopoda in the Australian Museum, Sydney. *Records of the Australian Museum* 16 (2): 79-115, pls 6-12. [Available online through the Australian Museum website; as of Dec. 2013 at

Verhoeff K.W. 1932. Klasse Diplopoda. *In*: *Dr H. G. Bronn's Klassen und Ordnungen des Tier-Reichs wissentschaftlich dargestellt in Wort und Bild. 5(II) 2(7-13). Gliederfüssler: Arthropoda*: 1073–2084. Akademische Verlagsgesellschaft m.b.H., Leipzig.

Verhoeff K.W. 1937. Ueber einige neue Diplopoden aus Australien. *Records of the Australian Museum* 20 (2): 133–149. [Available online through the Australian Museum website; as of Dec. 2013 at

Verhoeff K.W. 1941. Zur Kenntnis australischer Strongylosomiden und einiger anderen Diplopoden. *Lunds Universitets Årsskrift* (new series 2) 36 (17): 1–25.

12

On the Afrotropical genus *Holmelgonia* (Araneae, Linyphiidae), with the description of three new species from the Albertine Rift

Benoît NZIGIDAHERA & Rudy JOCQUÉ

Institut National pour l'Environnement et la Conservation de la Nature (INECN), B.P. 2757 Bujumbura, Burundi.
Corresponding author: Royal Museum for Central Africa (MRAC), B-3080 Tervuren, Belgium.

Abstract. Three new species of *Holmelgonia* were found in the mountain forest of Kibira National Park in Burundi: *H. afromontana* sp. nov. (♂♀), *H. bosnasutus* sp. nov. (♂♀) and *H. disconveniens* sp. nov. (♂). A key to the males in the genus, now containing 17 species, is provided.

Key words. Afromontane forest, Burundi, Kibira National Park, identification key, species swarm.

Introduction

Linyphidae have been popular among students of Afrotropical spiders and figure among the better known families of the continent, as far as the species living in the leaf litter layer is concerned. The interest in the family doubtlessly originated in the expertise several taxonomists had acquired in Europe, where the family Linyphiidae is by far the most speciose. Mainly Bosmans, Denis, Holm, Jocqué, Miller, Russell-Smith and Scharff all contributed to a great extent to the knowledge of the family thanks to the experience they had obtained by studying the European linyphiids. By the end of the last century, their extensive studies had yielded enough information about the family to result in an interactive key to the genera (Seyfulina & Jocqué 2007) and to allow generalisations about the distribution of the family, such as for instance the abundance of montane species (Scharff 1990, 1992, 1993) as compared to the paucity of the family in African lowland habitats and the small size of Afrotropical litter inhabiting linyphiids (Jocqué *et al.* 2013).

More recently, the fauna of the canopy yielded a high number of unknown linyphiid taxa (see Seyfulina & Jocqué 2009). Although the interest in Afrotropical linyphiids has apparently dwindled, the recent study by Frick & Scharff (2013) proves that the interest in Afrotropical linyphiids still slumbers.

The present study aims to stimulate renewed interest in African linyphiids and is the first of a series of papers on linyphiids from the southern part of the Albertine Rift, a mountain range with a particularly diverse fauna (Nzigidahera & Jocqué 2009; Jocqué *et al.* 2013). It provides an overview of the endemic Afrotropical genus *Holmelgonia* Jocqué & Scharff, 2007. The genus, originally described by Holm (1962) as *Elgonella*, received the replacement name *Elgonia* Platnick (1986), but eventually became *Holmelgonia* since *Elgonia* also turned out to be a homonym (Jocqué & Scharff 2007).

The genus *Holmelgonia* contains 14 species (Platnick 2013) and has a wide distribution (see Fig. 1) ranging from Côte d'Ivoire in the west to Kenya and Mozambique in the east. Although most species live at high altitude and have small endemic ranges (Jocqué & Scharff 1986; Scharff 1993), some of them (e.g., *H. holmi* Miller, 1970; *H. brachystegiae* Jocqué, 1981) have a wide distribution at mid-altitude.

Here we describe three new species from Burundi and provide a key to all the species for which males have been described.

The collections from which the type material of the new species originates are the result of an extensive sampling campaign carried out by the first author in protected areas of Burundi during three periods in 2003, 2005 and 2008. They covered altitudinal gradients between 1970 m and 2650 m above sea level (asl). The discovery emphasizes the biodiversity of the Albertine Rift spider fauna with high numbers of species in many spider groups (Nzigidahera & Jocqué 2009).

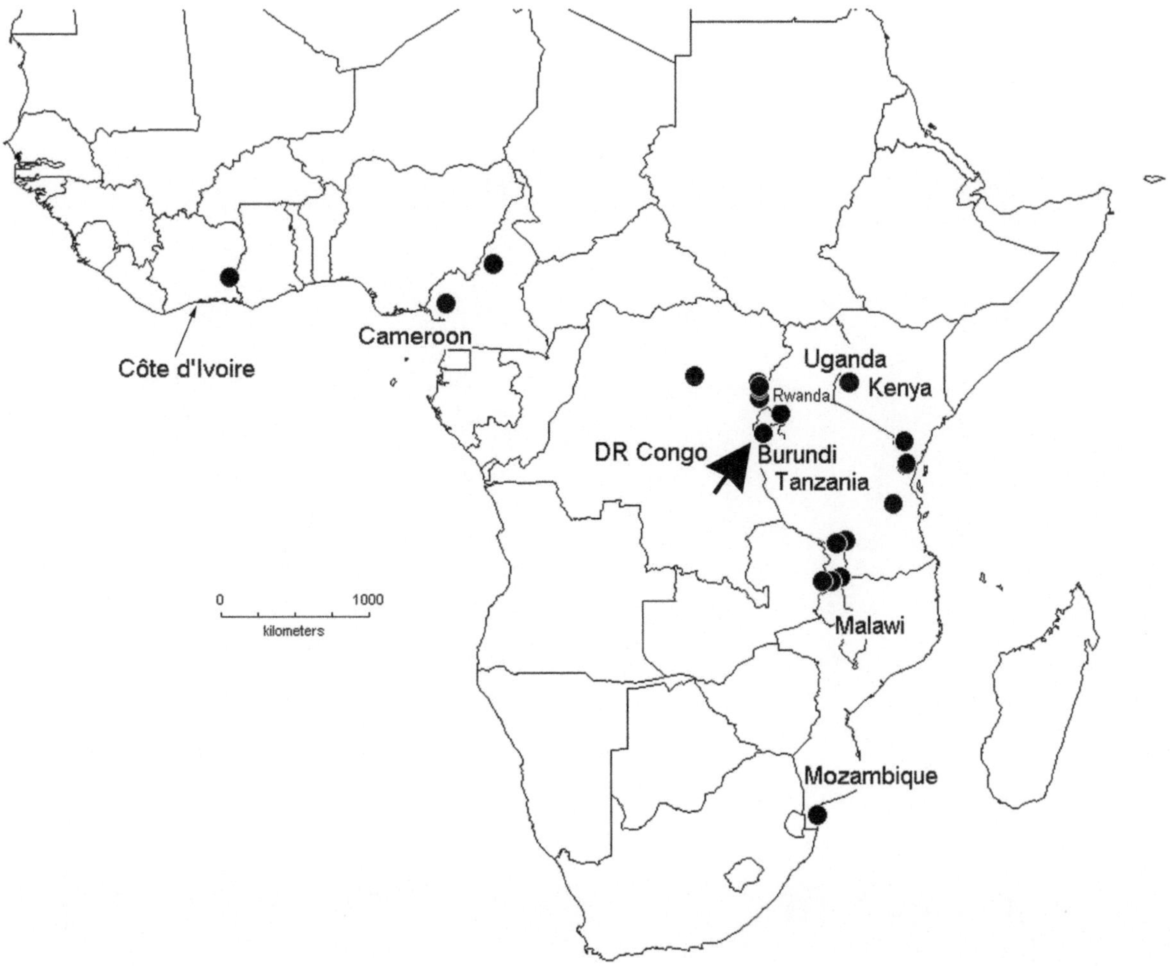

Fig. 1. Presently known distribution of *Holmelgonia* Jocqué & Scharff, 2007. Large arrow indicates Kibira National Park, type locality of the species described in this paper.

Material and methods

The specimens were collected by the first author in the afromontane forest of Kibira. Primary types are deposited in the Musée Royal de l'Afrique Centrale, Tervuren, Belgium (MRAC), whereas paratypes are divided among MRAC and the Institut National pour l'Environnement et la Conservation de la Nature, Bujumbura, Burundi (INECN). Specimens were observed, drawn and measured with a WILD M 10 stereo microscope. Details of the epigyne and male palps were observed with a Zeiss Stemi 2000 strereo microscope.

Epigynes were detached from the abdomen, cleared with methyl salicylate and temporarily mounted in a mixture of methyl salicylate and cedukol for examination. The same method was used to observe details of the male palp. These structures were photographed with a Leica MZ16 stereo microscope and subject to automontage with the Syncroscopy software. All measurements are in mm. As with the illustrations, they were taken from the holotype and paratypes unless stated otherwise. Coordinates are in decimal format.

Abbreviations

ALE	=	anterior lateral eyes
AME	=	anterior median eyes
AW	=	anterior width
d	=	dorsal
dl	=	dorsolateral
DT	=	diameter of tibia
ED	=	embolic division
Fe	=	femur
L	=	length
MA	=	median apophysis

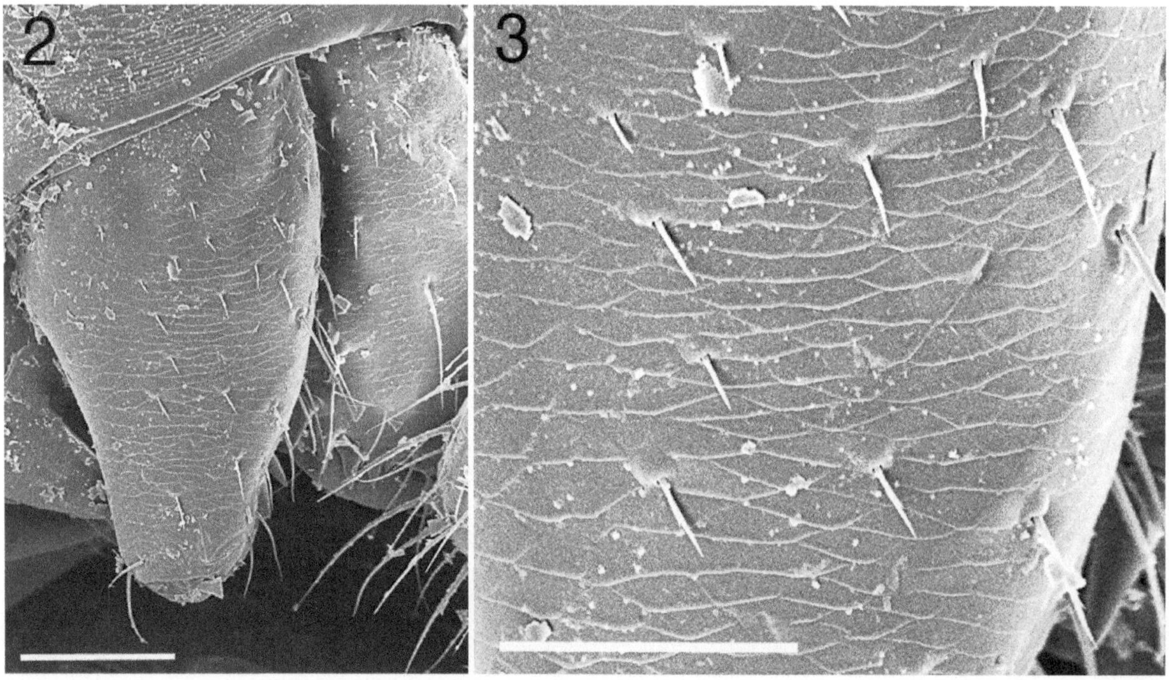

Figs 2-3. *Holmelgonia disconveniens* sp. nov. (♂, MRAC 227149). **2**. SEM micrographs of chelicerae, lateral view, showing absence of stridulating ridges. **3**. Detail of previous. Scale bars = 100 μm.

Mt	=	metatarsus
P	=	patella
Pl	=	prolateral
PLE	=	posterior lateral eyes
PME	=	posterior median eyes
Rl	=	retrolateral
t	=	tarsus
T	=	tibia
TM	=	position of trichobothrium on metatarsus expressed as ratio of distance to base and length of segment
V	=	ventral

Taxonomy

Class Arachnida Cuvier, 1812
Order Araneae Clerck, 1757
Family Linyphiidae Blackwall, 1859
Subfamily Erigoninae Emerton, 1882

Holmelgonia Jocqué & Scharff, 2007

Type species

Elgonella nemoralis Holm, 1962

Diagnosis

The generic definition is wide, as the original diagnosis by Holm (1962) had to be enlarged by Jocqué & Scharff (1986) to accommodate the species from Tanzanian mountain areas. The genus is characterized by the absence of cheliceral stridulating ridges (Figs 2-3); the presence of a double ventral row of setae on the femora; the long tibial spines, two to three times as long as the diameter of the segment; tibia I-IV with 2, 2, 1 and 1 spine, respectively; TMI between 0.32 and 0.7; males lack a cephalic lobe; the palpal tibia dorsally produced into a dorsal process (Figs 4-5, 7).

Key to the species of *Holmelgonia* (males only)

1. Dorsal tibial apophysis apically without teeth, smoothly rounded and with many long hairs (A) ....*H. afromontana* sp. nov. (Burundi)
– Dorsal tibial apophysis apically with teeth, sometimes with many hairs 2

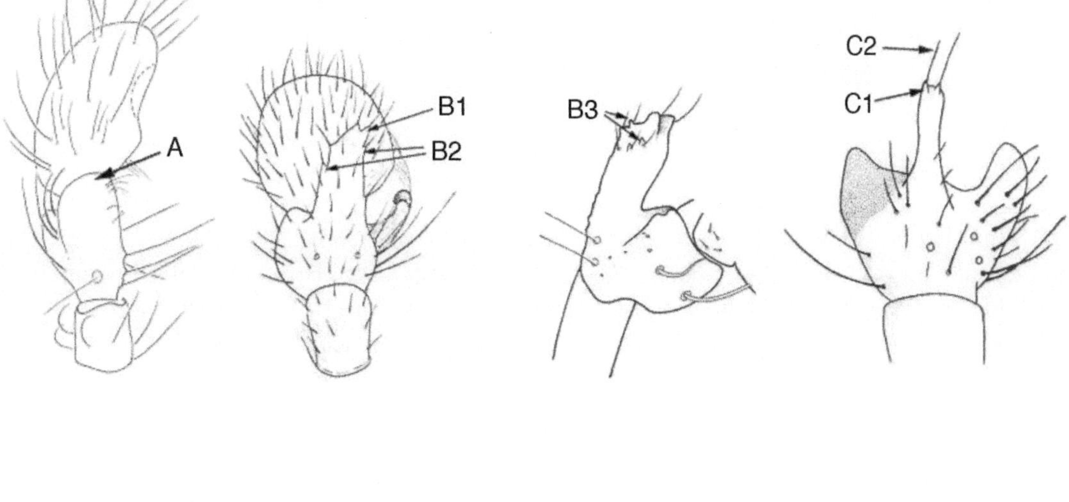

2. Dorsal tibial apophysis apically with two teeth (C1), each provided with hair (C2) but without teeth or undulation on the margins 3
– Dorsal tibial apophysis with teeth or undulation on the retrolateral margin (B2, dorsal view) or on distal part (B3, retrolateral view), apart from apical teeth (B1, dorsal view) 7

3. Dorsal tibial apophysis originating at the proximal base of tibia, straight, pointing upward (D1); retrolateral tibial apophysis absent (D2) ....
....*H. brachystegiae* (Jocqué, 1981) (Burundi, Malawi, Tanzania)
– Dorsal tibial apophysis obliquely oriented (E1); retrolateral tibial apophysis conical (E2), toothed (F) or rounded at the tip (G) 4

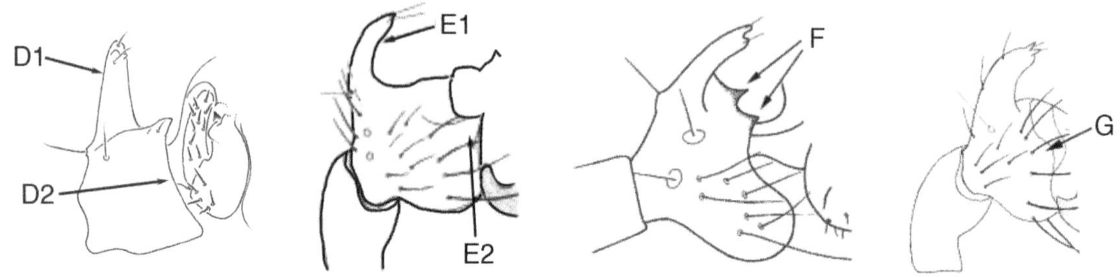

4. Retrolateral tibial apophysis conical (E2) or rounded at the tip (G) 5
– Retrolateral tibial apophysis with teeth (F) 6

5. Distal arm of paracymbium with indented posterior process (H); embolus semicircular or sickle-shaped (I) ....*H. falciformis* (Scharff, 1990) (Uzungwa Mts, Tanzania)
– Distal arm of paracymbium with simple posterior process (J); embolus with sharp bend (K) ...
....*H. annemetteae* (Scharff, 1990) (Uzungwa Mts, Tanzania)

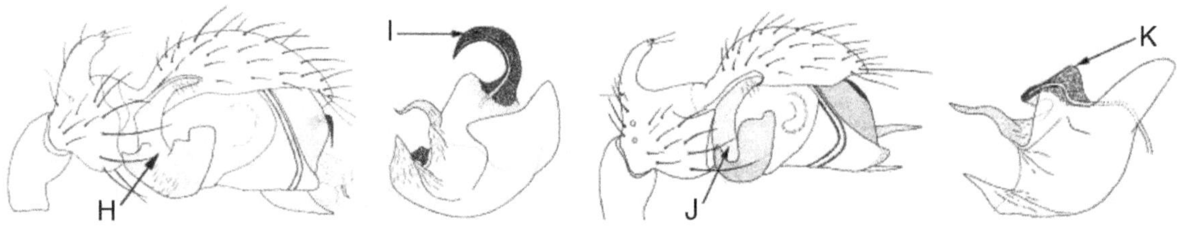

6. Retrolateral tibial apophysis with blackened teeth (F) on the transerverse ridge under the dorsal tibial apophysis ....*H. annulata* (Jocqué & Scharff, 1986) (Uluguru Mts, Tanzania)
– Retrolateral tibial apophysis with a single tooth on the lateral ridge (L) ....
....*H. basalis* (Jocqué & Scharff, 1986) (Usambara Mts, Tanzania)

7. Dorsal tibial apophysis with one tooth at the tip (M2); retrolateral tibial apophysis pointed at the tip (M1) ....*H. disconveniens* sp. nov. (Burundi)
– Dorsal tibial apophysis with two teeth at the tip (B1, B2, C1) 8

8. Dorsal tibial apophysis with a row of five teeth on the margin (N) ....
....*H. producta* (Bosmans, 1988) (Cameroon)
– Dorsal tibial apophysis with undulation on the prolateral margin (B2) or many small teeth near apical part (B3) 9

9. Dorsal tibial apophysis with undulation on the prolateral margin (B2) 10
– Dorsal tibial apophysis with many small teeth on the apical part (B3) 12

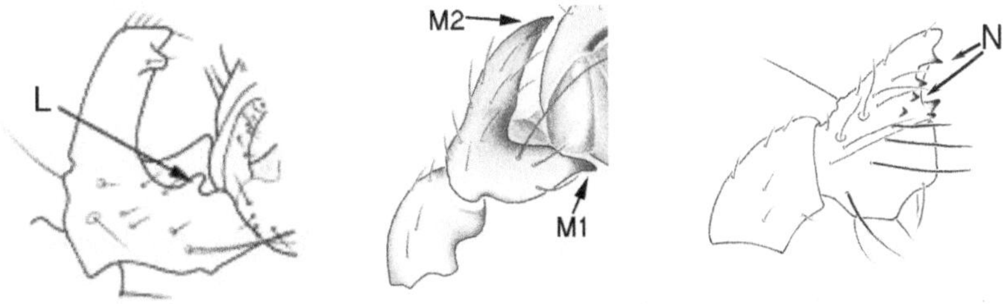

10. Dorsal tibial apophysis with a large prolateral lobe (O) ....
....*H. holmi* (Miller, 1970) (Congo, Cameroon)
– Dorsal tibial apophysis with small prolateral lobes or teeth on the margin (B2) 11

11. Distal arm of paracymbium rounded at tip (P1), proximal arm broad at base (P2) ....
....*H. bosnasutus* sp. nov. (Burundi)
– Distal arm of paracymbium hooked at the tip (Q1), proximal arm narrow at base (Q2) ....
....*H. nemoralis* (Holm,1962) (Kenya)

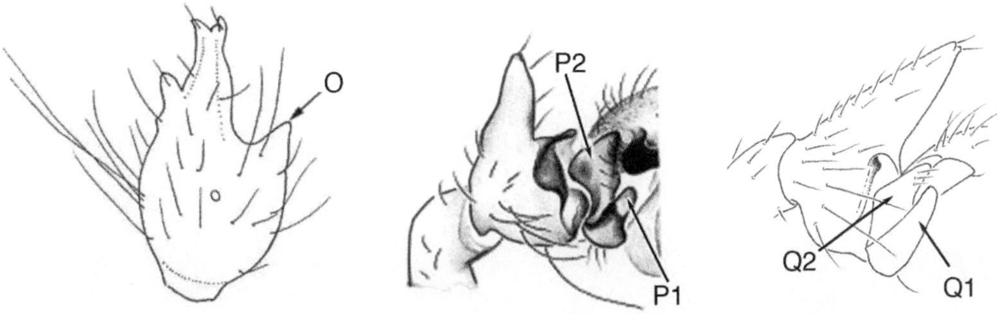

12. Embolic division ventrally with numerous small teeth (R1); frontal embolar apophysis very long (R2), central embolar apophysis very short (R3) and adjacent to embolus (R4) ....
....*H. perturbatrix* (Jocqué & Scharff, 1986) (Usambara Mts, Tanzania)
– Embolic division ventrally without teeth; frontal embolar apophysis very long and twisted (S1), central embolar apophysis on lateral side of division (S2) delimiting a concavity provided with numerous small warts (S3) ....*H. projecta* (Jocqué & Scharff, 1986) (Uluguru Mts, Tanzania)

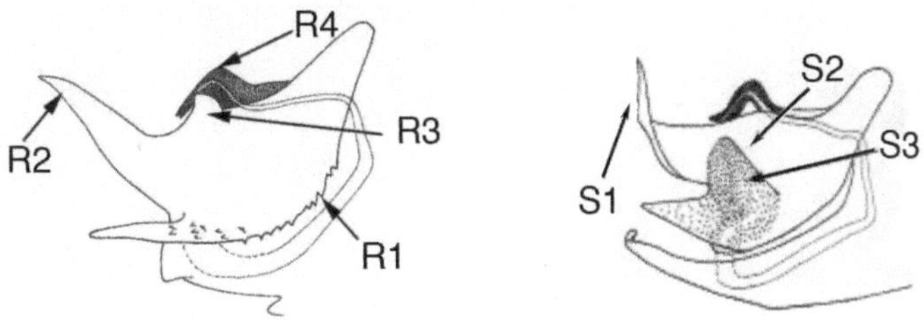

Holmelgonia afromontana sp. nov.
urn:lsid:zoobank.org:act:F482911E-4987-4F7E-BC74-D4101DE9EE29
Figs 4-5, 8, 11-16, 26-29, 39

Diagnosis

The male is characterized by the dorsal tibial apophysis, which is smoothly rounded at the extremity and has a ventral concavity provided with a short, sharp, tooth-shaped excrescence inside. Proximal and distal arms of the paracymbium with short setae, distal arm strongly sclerotized and curved at the tip. The female is recognized by details of the epigyne of which the central depression and the kidney-shaped spermathecae are the most characteristic.

Etymology

The specific name '*afromontana*' refers to the high altitude habitat in which the species was collected.

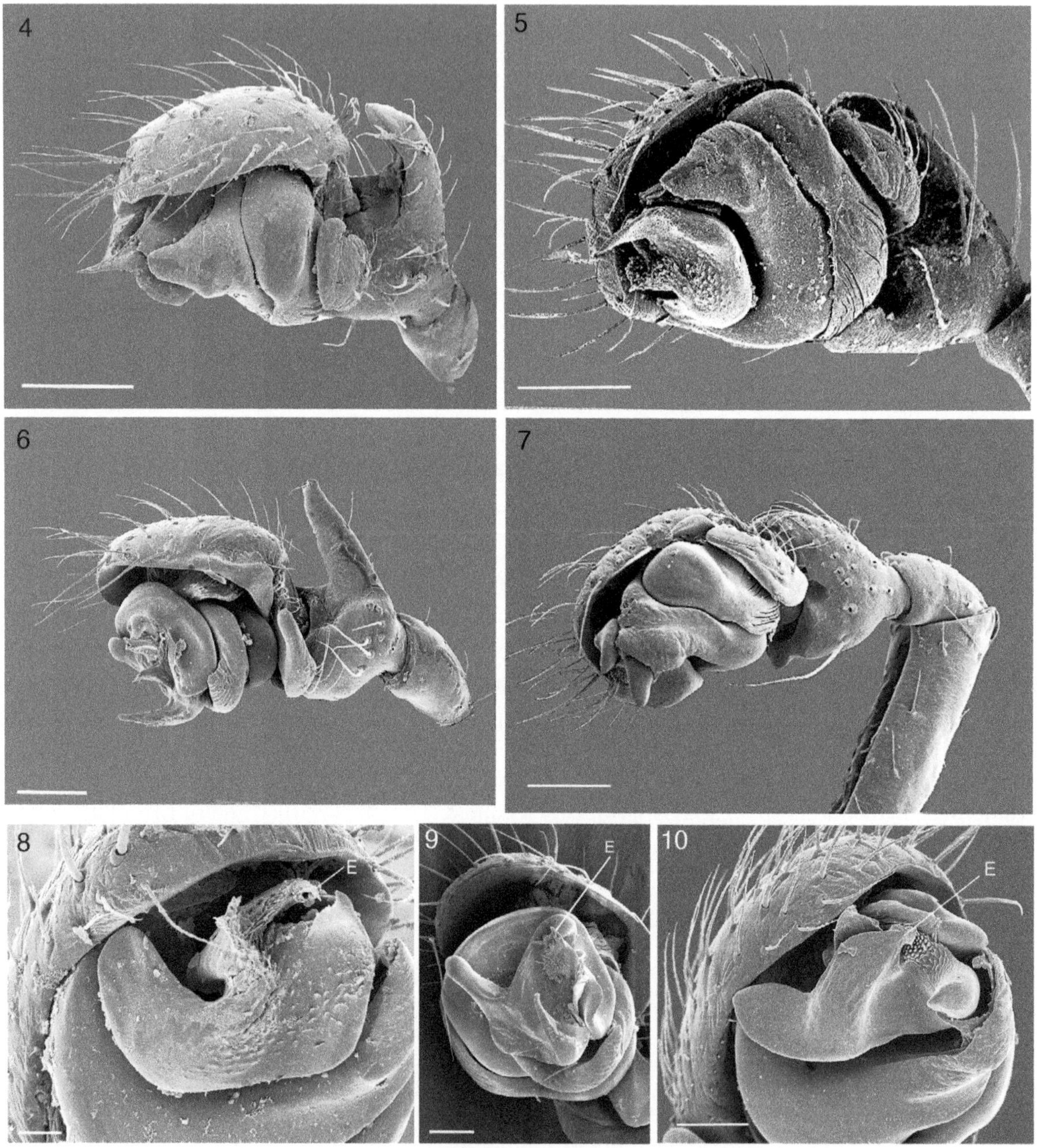

Figs 4-10. SEM micrographs. — **4-5, 8**. *Holmelgonia afromontana* sp. nov. (♂, MRAC 227153). **4**. Palp, retrolateral view. **5**. As previous, ventral view. **8**. Embolic division, frontal view. — **6, 9**. *Holmelgonia bosnasutus* sp. nov. (♂, MRAC 227145). **6**. Palp, retrolateral view. **9**. Embolic division, ventral view. — **7, 10**. *Holmelgonia disconveniens* sp. nov. (♂, MRAC 241448). **7**. Palp, retrolateral view. **10**. Embolic division, frontal view. Scale bars 4-7 = 100 μm; 9-10 = 50 μm; 8 = 20 μm. E: embolus.

Material examined

Holotype

BURUNDI: ♂, Parc National de la Kibira, Rwegura, Mt. Musumba, 2.86445° S, 29.49640° E, site 5, forest with *Carapa grandiflora* and *Polyscias fulva,* 25 Jun. 2008, 2252 m, pitfalls, B. Nzigidahera coll. (MRAC 227154).

Paratypes

BURUNDI: All from Parc National de la Kibira, Rwegura, Mt Musumba, pitfalls, B. Nzigidahera coll. – 1 ♀, together with holotype (MRAC 227154). – 2.86435° S, 29.49546° E, site 6, forest with *Carapa grandiflora* and *Polyscias fulva,* 2150 m: 1 ♂, 2 ♀♀, 10 Mar. 2008 (MRAC 227159); 1 ♂, 25 Aug. 2008 (INECN); 3 ♂♂, 10 Nov. 2008 (INECN); 1 ♂, 10 Nov. 2008 (INECN); 2 ♂♂, 25 Jul. 2008 (INECN); 1 ♂, 1 juv., 10 Jul. 2008 (MRAC 227157); 1 ♀, 25 Mar. 2008 (MRAC 227161). – 2.86445° S, 29.49640° E, site 5, forest with *Carapa grandiflora* and *Polyscias fulva,* 2252 m: 1 ♂, 10 Apr. 2008 (MRAC 227148); 1 ♂, 10 May 2008 (MRAC 227156); 1 ♂, 1 ♀, 25 Aug. 2008 (MRAC 227151); 1 ♂, 10 Aug. 2008 (INECN); 1 ♂, 10 Aug. 2008 (MRAC 227158); 2 ♂♂, 10 Jun. 2008 (INECN); 5 ♂♂, 25 Feb. 2008 (INECN); 1 ♂, 25 Nov. 2008 (INECN); 1 ♂, 1 ♀, 10 Feb. 2008 (INECN); 2 ♂♂, 1 juv., 25 Sep. 2008 (INECN). – 2.86455° S, 29.50427° E, site 4, forest with *Macaranga neomildbraediana* and *Polyscias fulva,* 2352 m: 1 ♀, 25 Jun. 2008 (MRAC 227152); 3 ♂♂, 10 Feb. 2008 (MRAC 227153); 1♂, 25 Feb. 2008 (INECN); 1 ♂, 10 Oct. 2008 (INECN); 1 ♂, 1 juv., 10 Sep. 2008 (INECN); 1♂, 25 Feb. 2008 (INECN). – 2.87149° S, 29.49641° E, site 3, 2444 m: 2 ♂♂, 26 Jan. 2008 (INECN). – 2.87696°S, 29.49709° E, site 1, 2650 m: 3 ♀♀, 25 Oct. 2008 (INECN).

Other material

BURUNDI: Parc National de la Kibira, Forêt de Rwegura, pitfalls, B. Nzigidahera coll., 2.90249°E, 29.51345°S, forest with *Macaranga neomildbreadiana* and *Xymalos monospora*, 2223 m, 1 ♂, 23 Nov. 2003 (INECN).

Description

Male (holotype, Figs 4-5, 8, 11-13, 26-27, 29)

Measurements. Total length 1.70, carapace 0.70 long, 0.60 wide.

Cephalothorax. Carapace medium brown, with black striae radiating from foveal area. Chelicerae yellow, mottled with grey. Retromargin of chelicerae with 5 teeth; promargin with 4 small teeth. Clypeus 2 times ALE, grey. Sternum dark brown mottled with black, shield-shaped, 0.48 long and 0.44 wide.

Eyes. Ocular area black. Posterior eye row strongly procurved, anterior eye row recurved; AME 0.04, ALE 0.08, PME 0.08, PLE 0.08.

Legs. Pale yellow; tibia I with two spines. Femora without dorsal spines, with ventral spines in two rows. Metatarsi without dorsal spines.

Leg measurements

Legs	Fe	P	T	Mt	t	Total
I	0.72	0.22	0.66	0.60	0.52	2.72
II	0.70	0.22	0.62	0.54	0.50	2.58
III	0.62	0.20	0.56	0.54	0.40	2.32
IV	0.82	0.20	0.80	0.72	0.44	2.98

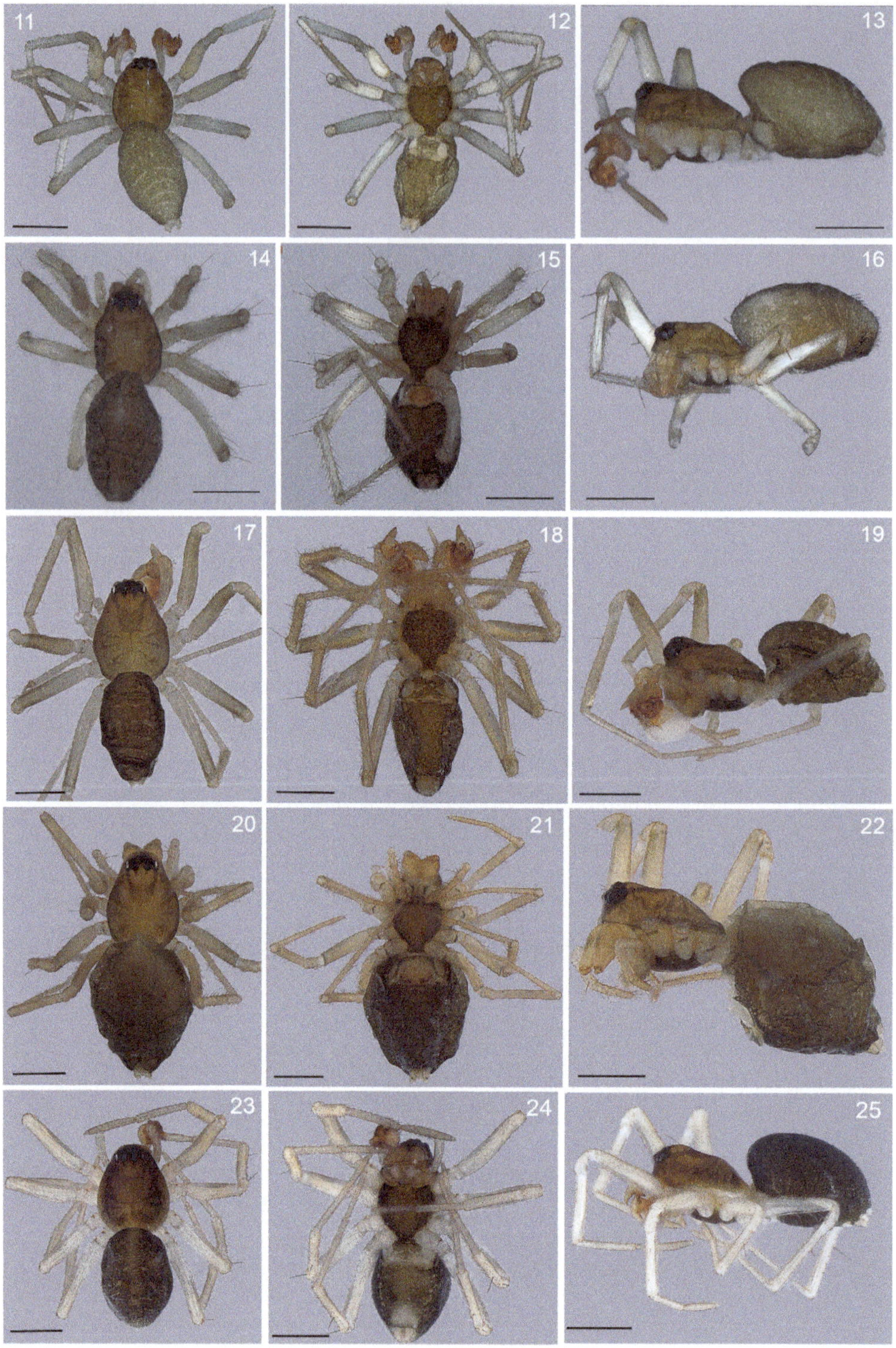

Figs 11-25. — **11-16**. *Holmelgonia afromontana* sp. nov. (♂ holotype, ♀ paratype 227154). **11**. Habitus, ♂, dorsal view. 12. Habitus, ♂, ventral view. **13**. Habitus, ♂, lateral view. **14**. Habitus, ♀, dorsal view. **15**. Habitus, ♀, ventral view. **16**. Habitus, ♀, lateral view. — **17-22**. *Holmelgonia bosnasutus* sp. nov. (♂ holotype, female paratype 227507). **17**. Habitus, ♂, dorsal view. **18**. Habitus, ♂, ventral view. **19**. Habitus, ♂, lateral view. **20**. Habitus, ♀, dorsal view. **21**. Habitus, ♀, ventral view. **22**. Habitus, ♀, lateral view. — **23-25**. *Holmelgonia disconveniens* sp. nov. (♂ holotype). **23**. Habitus, dorsal view. **24**. Habitus, ventral view. **25**. Habitus, lateral view. Scale bars = 0.5 mm.

LEG CHAETOTAXY

Legs	TM	Tibial spines			
		Proximal		Distal	
		Position	Length/DT	Position	Length/DT
I	0.40	0.16	1.80	0.50	1.60
II	0.20	0.14	2.20	0.42	1.40
III	0.20	0.12	3.25	-	-
IV	0.30	0.12	3.50	-	-

ABDOMEN. Oval, pale grey, dorsally with pale spots and four faint horizontal chevrons becoming smaller towards the rear end. Spinnerets pale yellow, conical.

PALP (Figs 4-5, 8, 26-27, 29). Dorsal tibial apophysis ventrally concave, with tooth-shaped excrescence inside concavity; dorsally with smoothly rounded tip. Paracymbium: proximal arm tapered, with row of short hairs; distal arm with two lobes at extremity, caudal one with some hairs. Prolateral part of subtegulum with oblique grooves. Embolic division (Fig. 29) fairly simple; embolus long, frontal appendage short, truncated, margin with many tiny teeth; dorsal appendage short, rounded.

Female (paratype, MRAC 227154, Figs 14-16, 28, 39)

Habitus similar to male.

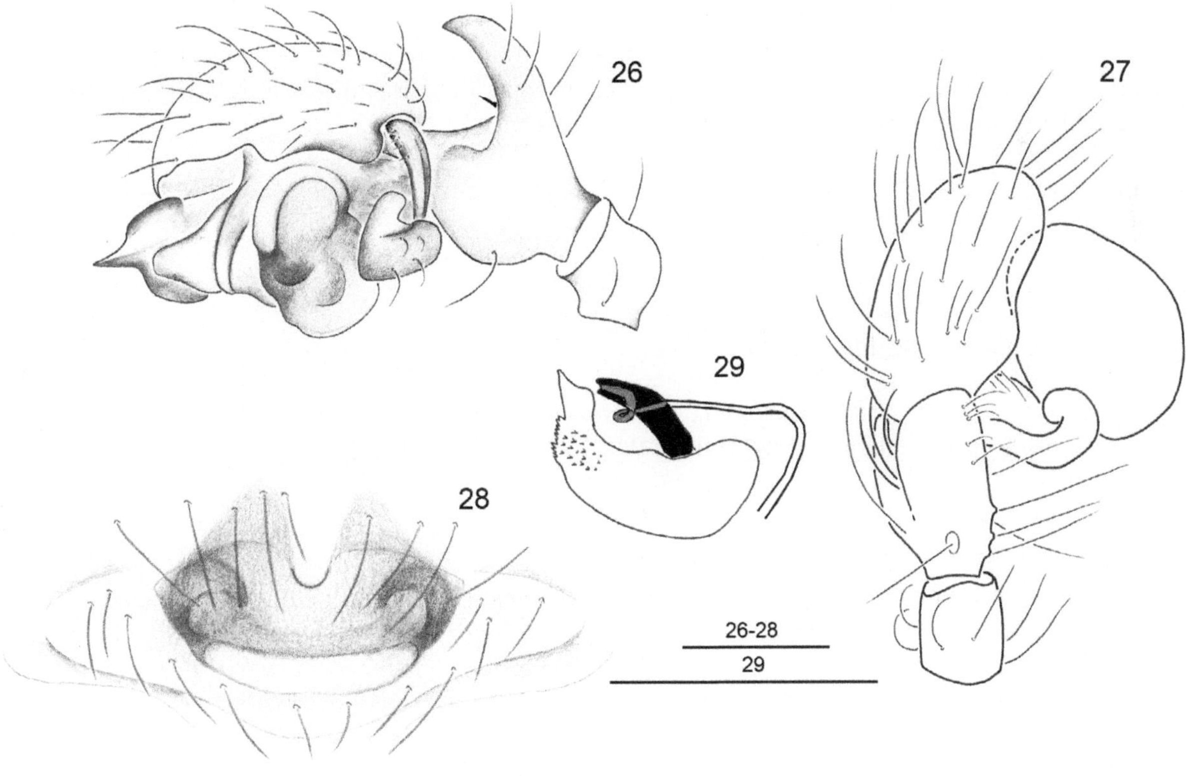

Figs 26-29. *Holmelgonia afromontana* sp. nov. (♂ holotype, ♀ paratype 227154). **26**. Palp, ♂, retrolateral view. **27**. Palp, ♂, dorsal view. **28**. Epigyne, ventral view. **29**. Embolic division of ♂ palp, ventrolateral view. Scale bars = 0.1 mm.

MEASUREMENTS. Total length 1.64. Carapace 0.70 long, 0.60 wide.

CEPHALOTHORAX. Clypeus 1.66 times ALE. Sternum 0.48 long, 0.46 wide.

EYES. AME 0.04, ALE 0.08, PME 0.08, PLE 0.08.

LEG MEASUREMENTS

Legs	Fe	P	T	Mt	t	Total
I	0.70	0.22	0.58	0.52	0.40	2.42
II	0.74	0.22	0.50	0.50	0.44	2.40
III	0.64	0.20	0.44	0.44	0.36	2.08
IV	0.80	0.20	0.66	0.66	0.46	2.78

LEG CHAETOTAXY

Legs	TM	Tibial spines			
		Proximal		Distal	
		Position	Length/DT	Position	Length/DT
I	0.33	0.22	2.00	0.42	1.80
II	0.36	0.10	2.00	0.38	2.00
III	0.40	0.12	3.25	-	-
IV	0.54	0.16	4.00	-	-

EPIGYNE (Fig. 30). Fairly large ventral plate broadly indented in front, posterior margin slightly procurved. Copulatory ducts originate from centre near posterior margin, entering spermathecae near lateral margin; spermathecae large, kidney-shaped, with transverse long axis.

Variation

Carapace colour may vary from medium brown to yellowish grey.

Distribution

Known only from the type locality (Fig. 1).

Holmelgonia bosnasutus sp. nov.
urn:lsid:zoobank.org:act:098B9E4B-00C0-437E-A36E-EEE46A2D9826
Figs 6, 9, 17-22, 30-34, 40

Diagnosis

The male is characterized by the sharply pointed dorsal tibial apophysis with undulated margin, each undulation provided with a hair, combined with the broad prolateral apophysis, the broad proximal arm and smoothly rounded tip of the distal arm of the paracymbium. Females are recognized by the shape of the epigyne, provided with two holes clearly visible and opening in the posterior margin.

Etymology

The specific name '*bosnasutus*' refers to the shape of the epigyne which is reminscent of a cow's nose.

Material examined

Holotype

BURUNDI: ♂, Parc National de la Kibira, Mt. Musumba, Rwegura, 2°52' S, 29°30' E, 28 Dec. 2004, site 2, forest with *Hagenia abyssinica*, B. Nzigidahera coll. (MRAC 227508).

Paratypes

BURUNDI: All from Parc National de la Kibira, forêt de Rwegura, pitfalls, B. Nzigidahera coll. – 2.89565° S, 29.50345° E, site 2, forest with *Polyscias fulva* and *Macaranga neomildbraediana,* 2120 m: 1 ♂, 29 Jun. 2005 (INECN); 1 ♂, 28 Dec. 2004 (MRAC 227567). – 2.88562°S, 29.50221°E, site 3, forest with *Polyscias fulva* and *Hagenia abyssinica,* 2160 m: 1 ♀, 27 May 2005 (MRAC 227507); 1 ♂, 15 Mar. 2005 (INECN). – 2.89435° S, 29.50078°E, site 4, 2400 m: 2 ♂♂, 28 Feb. 2005 (INECN); 1 ♂, 28 Dec. 2004 (INECN).

Other material

BURUNDI: Parc National de la Kibira, Rwegura, Mt Musumba, B. Nzigidahera coll., 2.86435°S, 29.49540°E, site 7, tea plantation, 2100 m: 1 ♂, 25 Aug. 2008 (MRAC 227146); 1 ♂, 10 Jul. 2008 (MRAC 227144); 2 ♂♂, 1J, 10 Oct. 2008 (MRAC 227145); 1 ♂, 10 Nov. 2008 (INECN); 1 ♂, 25 Mar. 2008 (INECN). – 2.86435°S, 29.49546E, site 6, forest with *Carapa grandiflora* and *Polyscias*

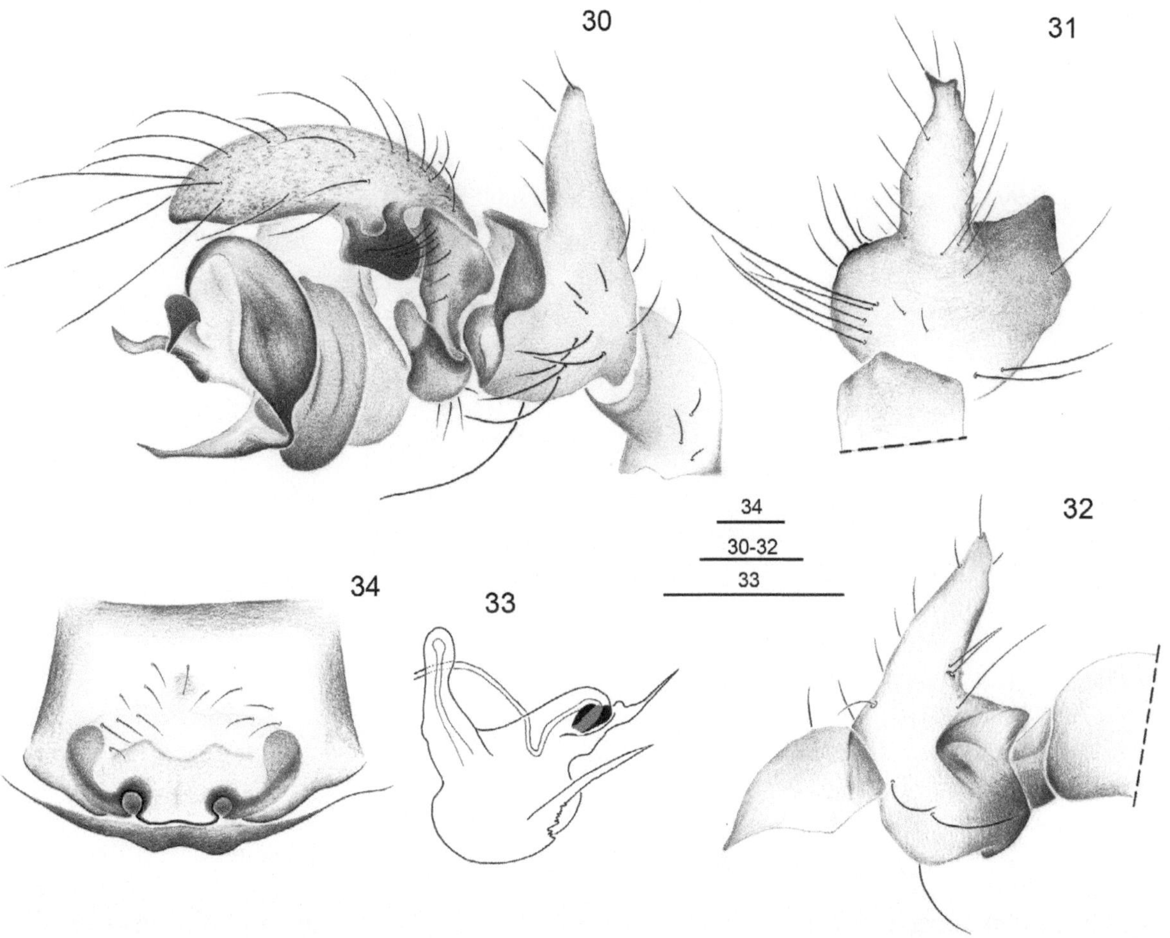

Figs 30-34. *Holmelgonia bosnasutus* sp. nov. (♂ holotype, ♀ paratype 227507). **30**. Palp, ♂, retrolateral view. **31**. Palpal tibia, ♂, dorsal view. **32**. Palpal tibia, ♂, prolateral view. **33**. Embolic division of ♂ palp, ventrolateral view. **34**. Epigyne, ventral view. Scale bars = 0.1 mm.

fulva, 2150 m: 1 ♂, 10 Sep. 2008 (INECN). – 2.86455°S, 29.50427°E, site 4, forest with *Macaranga neomildbraediana* and *Polyscias fulva,* 2352 m: 1 ♀, 10 Jul. 2008 (MRAC 227137); 1 ♂, 25 May 2008 (MRAC 227194); 2♂♂, 25 Dec. 2008 (INECN); 1 ♂, 25 Jan. 2008 (INECN). – 2.87060° S, 29.49669° E, site 2, forest with *Hagenia abyssinica*, 2548 m: 1 ♂, 10 Jul. 2008 (MRAC 227195); 1 ♂, 25 Apr. 2008 (MRAC 227196); 1 ♂, 25 Jul. 2008 (INECN); 1 ♂, 10 Sep. 2008 (INECN). – 2.87696°S, 29.49709°E, site 1, *Philippia benguelensis*, 2650 m: 1 ♂, 25 Nov. 2008 (INECN).

Description

Male (holotype, Figs 6, 9, 17-19, 30-33)

Measurements. Total length 2.00, carapace 0.88 long, 0.70 wide.

Cephalothorax. Carapace greyish brown with black striae radiating from fovea; with pale area behind posterior eyes, followed by rectangular spot from which two striae run towards the PLE. Chelicerae coloured as carapace, retromargin with 5 teeth; promargin with 5 small teeth. Clypeus 2.80 times ALE. Sternum 0.54 long, 0.44 wide, dark greyish brown, shield-shaped.

Eyes. Ocular area black. Anterior eye row straight, posterior eye row strongly procurved. AME 0.03, ALE 0.08, PME 0.08, PLE 0.08.

Legs. Yellow to greyish yellow; femora without dorsal spines, but with ventral spines in two rows. Metatarsi with dorsal spines.

Leg measurements

Legs	Fe	P	T	Mt	t	Total
I	0.90	0.20	0.84	0.84	0.60	3.44
II	0.64	0.18	0.58	0.74	0.62	2.76
III	0.60	0.16	0.60	0.64	0.54	2.54
IV	0.72	0.14	0.58	0.94	0.60	2.98

Leg chaetotaxy

Legs	TM	Tibial spines			
		Proximal		Distal	
		Position	Length/DT	Position	Length/DT
I	0.38	0.18	2.75	0.75	1.75
II	0.42	0.20	3.25	0.70	2.00
III	0.54	0.20	3.25	-	-
IV	0.38	0.30	3.00	-	-

Abdomen. Oval, dark greyish brown, with nine thin, pale transverse lines. Venter dark brown. Spinnerets pale greyish, short, conical.

Palp (Figs 6, 9, 30-33). Retrolateral tibial apophysis with sharp tip and undulated lateral margins, each undulation provided with a hair; laterally with short, broad prolateral apophysis with blunt tip. Paracymbium with broad proximal arm and distal arm with smoothly rounded tip. Proximal part of tegulum with ventrolateral, oblique grooves. Embolic division with short curved embolus, long sharp ventral appendage, sinuous, frontal appendage with sharp tip and long, broad, rounded dorsal appendage.

Female (paratype, MRAC 227507, Figs 20-22, 34, 40)

Very similar to male.

MEASUREMENTS. Total length 2.13. Carapace 0.90 long, 0.72 wide.

CEPHALOTHORAX. Clypeus 3.0 times ALE. Sternum 0.49 long, 0.49 wide.

EYES. As in male: AME 0.03, ALE 0.08, PME 0.08, PLE 0.08.

LEGS. TM: I: 0.41, II: 0.40, III: 0.50.

LEG MEASUREMENTS

Legs	Fe	P	T	Mt	t	Total
I	0.90	0.20	0.80	0.74	0.60	3.24
II	0.66	0.12	0.54	0.68	0.60	2.60
III	0.70	0.14	0.42	0.60	0.50	2.36
IV	0.78	0.14	0.34	0.82	0.42	2.50

EPIGYNE (Figs 34, 40). With two holes at posterior margin, resembling the shape of a cow's nose. Copulatory ducts short, curved, originating from triangular atrium near posterior margin; entering spermathecae near lateral margin, spermathecae oval, longest axis longitudinal.

Distribution

Known only from the type locality, Mt Musumba in Kibira National Park (Fig. 1).

Holmelgonia disconveniens sp. nov.
urn:lsid:zoobank.org:act:2787FB55-2335-4C64-BFB8-ABC0825A7F56
Figs 2-3, 7, 10, 23-25, 35-38

Diagnosis

The male is characterized by the triangular dorsal tibial apophysis, with a tooth at the tip and sinuous prolateral margin, by the fairly large, sharp, sclerotized prolateral apophysis, and by the paracymbium without hairs, its distal arm with a well delimited knob.

Etymology

The specific name '*disconveniens*' refers to the contrasting tinge of dark body and pale legs.

Material examined

Holotype

BURUNDI: ♂, Parc National de la Kibira, Rwegura, Mt Musumba, 2.86435° S, 29.49540° E, site 1, tea plantation, 10 Sep. 2008, 2100 m, pitfalls, B. Nzigidahera coll. (MRAC 229928).

Paratypes

BURUNDI: 1 ♂, 10 Sep. 2008, same data as for holotype (MRAC 229929). – 2.86435° S, 29.49546° E, site 6, forest with *Carapa grandiflora* and *Polyscias fulva,* 2150 m: 1 ♂, 10 Aug. 2008 (MRAC 229741); 1 ♂, 10 Sep. 2008 (MRAC 229740). – 2.86435°S, 29.49540°E, site 7, tea plantation, 2100 m, site 7: 1 ♂, 10 Sep. 2008 (MRAC 241448); 1 ♂, 10 Sep. 2008 (MRAC 241449). – 2.86455° S, 29.50427° E, site 4, forest with *Macaranga neomildbraediana* and *Polyscias fulva,* 2352 m: 2 ♂♂, 25 Nov. 2008 (INECN); 1 ♂, 10 Jan. 2009 (INECN); 4 ♂♂, 25 Aug. 2008 (MRAC 227149); 1 ♂, 25 Jun. 2008 (MRAC 227150).

Description

Male (holotype, Figs 7, 10, 23-25, 35-38)

MEASUREMENTS. Total length 1.8, carapace 0.80 long 0.70 wide.

CEPHALOTHORAX. Carapace dark brown, with faint dark striae radiating from fovea; striae larger towards the margin; a faint pale semicircular area behind eyes formed by two brown bands originating behind PLE, connected in front of fovea. Chelicerae yellowish brown. Retromargin with 4 teeth; promargin with 5 small teeth. Clypeus grey, 2.33 times ALE. Sternum 0.56 long, 0.50 wide, dark brown, shield-shaped.

EYES. Ocular area black. Anterior eye row recurved, posterior eye row strongly procurved; AME 0.05, ALE 0.08, PME 0.07, PLE 0.07.

LEGS. Pale yellow, tarsi and metatarsi pale grey, femora without dorsal spines, with ventral spines in two rows. Metatarsi without dorsal spines.

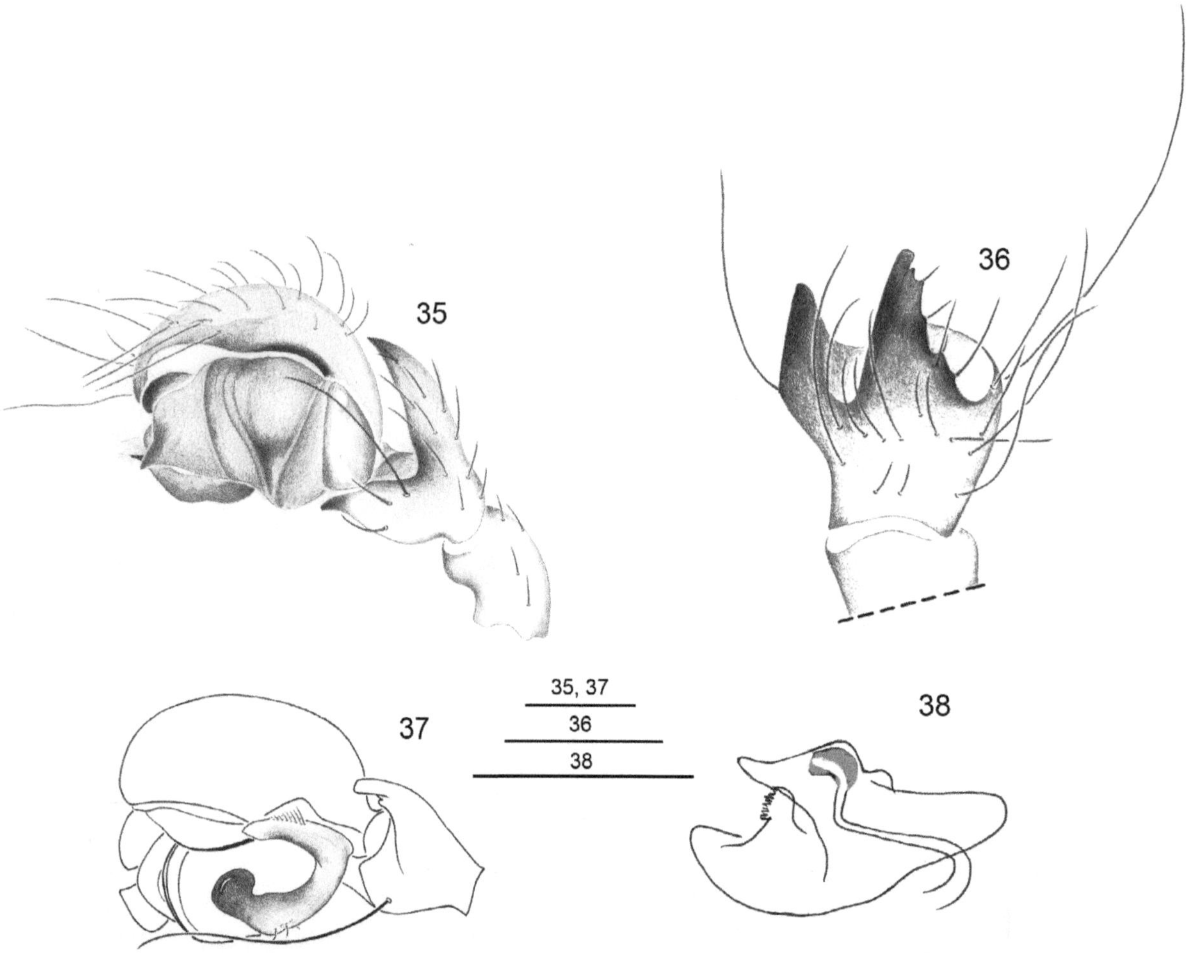

Figs 35-38. *Holmelgonia disconveniens* sp. nov. (♂ holotype). **35**. Palp, retrolateral view. **36**. Palpal tibia, dorsal view. **37**. Palpal tibia, dorsolateral view. **38**. Embolic division of palp, ventrolateral view. Scale bars = 0.1 mm.

Leg measurements

Legs	Fe	P	T	Mt	t	Total
I	0.70	0.22	0.58	0.56	0.48	2.54
II	0.66	0.22	0.56	0.56	0.42	2.42
III	0.60	0.20	0.50	0.50	0.52	2.32
IV	0.76	0.22	0.64	0.70	0.56	2.88

Leg chaetotaxy

Legs	TM	Tibial spines			
		Proximal		Distal	
		Position	Length/DT	Position	Length/DT
I	0.55	0.12	0.60	0.40	1.00
II	0.42	0.18	0.60	0.44	1.00
III	0.40	0.12	3.00	-	-
IV	0.45	0.12	3.25	-	-

Abdomen. Oval, greyish black, with faint central dark brown band and small pale points, with six transverse pale chevrons in posterior half. Spinnerets yellowish grey, conical.

Palp (Figs 7, 10, 35-38). Dorsal tibial apophysis triangular, with two teeth near tip; retrolateral tibial apophysis almost as long as dorsal one, ending in sharp, slightly curved tip. Paracymbium with proximal arm without hairs; distal arm ending in well delimited, rounded knob. Subtegulum with transverse ventral grooves. Embolic division with short, curved embolus, posterior appendage very short and smoothly rounded; dorsal appendage triangular; frontal appendage short, club-shaped, with frontal margin with many small teeth.

Female

Unknown.

Distribution

Known only from the type locality (Fig. 1).

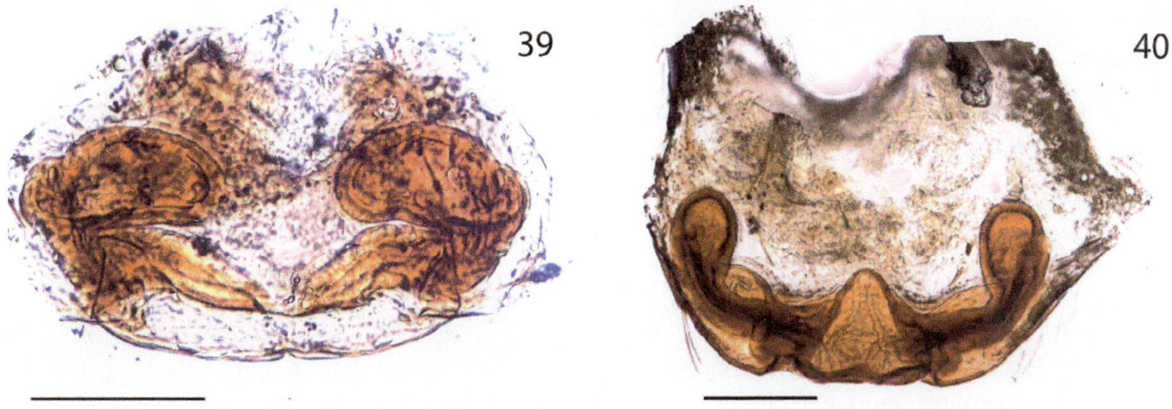

Figs 39-40. — **39**. *Holmelgonia afromontana* sp. nov. (♀, 227154). Epigyne, cleared, dorsal view. — **40**. *Holmelgonia bosnasutus* sp. nov. (♀, 227507). Epigyne, cleared, dorsal view. Scale bars = 0.1 mm.

Discussion

Linyphiidae are a striking example of a family in which genera have been erected based on secondary sexual characters that appear to be species specific. This phenomenon was discussed in Jocqué *et al.* (2013), who argued that genera in some families are particularly difficult to define when they contain so-called 'templates' (as defined by Jocqué & Bosselaers 2011), lacking clear somatic autapomorphies. The definition of *Holmelgonia* is no exception to this: apart from the absence of stridulating files on the chelicerae, no somatic characters could be found to define it. Even the chaetotaxy, and more specifically the position of the metatarsal trichobothria, varies to an extent that is unusual in erigonine genera. The genus has therefore to be defined on the base of features of the secondary genitalia. The palp has a tibia with a dorsal process and the ED is rather complex, with several protuberances. Yet, in two of the species described here the ED is fairly simple. The latter feature, however, is not restricted to *Holmelgonia* and can be found in many genera in the Erigoninae. It thus remains questionable whether the species attributed to *Holmelgonia* really do constitute a natural group.

This study is a result of the intense sampling of ground spiders carried out in the forest ecosystems of western Burundi during several years (2003, 2005 and 2008). The three species described here were only found in the mountain forest of Kibira National Park. *Holmelgonia bosnasutus* sp. nov. occupies a wide altitudinal gradient, ranging from 2100 to 2650 m. The species *H. afromontana* sp. nov. and *H. disconveniens* sp. nov. are confined to a narrower altitudinal zone between 2100 and 2352 m.

This study shows again that the Albertine Rift in Central Africa has only been superficially explored with regard to its spider fauna. This was already clear with the discovery of species swarms in the Gnaphosidae, in which the genera *Zelotibia* Russell-Smith & Murphy, 2005 (20 species) and *Zelowan* Murphy & Russell-Smith, 2010 (18 species) were found to have radiated in that mountain range (Russell-Smith & Murphy 2005; Murphy & Russell-Smith 2010; Nzigidahera & Jocqué 2009). Further studies in preparation show that similar radiations can be expected in other genera of the Linyphiidae and in other families such as the Tetragnathidae, particularly in ground-living genera that have lost the web-building habit. There is a considerable urgency to describe the fauna in this region, where human populations are expanding rapidly and deforestation is consequently very intense.

Acknowledgements

ABIC (African Biodiversity Information Centre) of the Royal Museum for Central Africa and GBIF are thanked for several travel grants to BN. We are indebted to the Royal Museum for Central Africa for accommodation for BN during his studies in Belgium. We thank Alain Reygel, who made the drawings with his usual skill. We thank two anonymous reviewers for improvements to the first version.

References

Frick H. & Scharff N. 2013. Phantoms of Gondwana?—phylogeny of the spider subfamily Mynogleninae (Araneae: Linyphiidae). *Cladistics* 29: 1-40.

Holm Å. 1962. The spider fauna of the East African mountains. Part I: Fam. Erigonidae. *Zoologiska Bidrag från Uppsala* 35: 19-204.

Jocqué R., Alderweireldt M. & Dippenaar-Schoeman A. 2013. Biodiversity, an African perspective. *In*: Penney D. (ed.) *Spider Research in the 21st Century*: 18-57. Siri Scientific Press. Rochdale, U.K.

Jocqué R. & Bosselaers J. 2011. Revision of *Pseudocorinna* Simon and a new related genus (Araneae: Corinnidae): two more examples of spider templates with a large range of complexity in the genitalia. *Zoological Journal of the Linnean Society* 162: 271-350.

Jocqué R. & Scharff N. 1986. Spiders of the family Linyphiidae (Araneae) from the Tanzanian mountain areas Usambara, Uluguru and Rungwe. *Annales du Musée royal de l'Afrique Centrale, Sciences Zoologiques* 248: 1-61.

Jocqué R. & Scharff N. 2007. *Holmelgonia,* a new name for the genus *Elgonia* Holm, 1989 (Araneae, Linyphiidae). *Journal of Afrotropical Zoology* 3: 161.

Murphy J.A. & Russell-Smith A. 2010. *Zelowan,* a new genus of African zelotine ground spiders (Araneae: Gnaphosidae). *Journal of Afrotropical Zoology* 6: 59-82.

Nzigidahera B. & Jocqué R. 2009. An update of *Zelotibia* (Araneae, Gnaphosidae), a spider genus with a species swarm in the Albertine Rift. *ZooKeys* 13: 1-28.

Platnick N.I. 2013. *The World Spider Catalog.* Version 13.5. American Museum of Natural History. Available from http://research.amnh.org/iz/spiders/catalog/ (accessed 20 January 2013).

Russell-Smith A. & Murphy J.A. 2005. *Zelotibia,* a new zelotine spider genus from central Africa (Araneae, Gnaphosidae). *Journal of Afrotropical Zoology* 2: 103-122.

Scharff N. 1990. Spiders of the family Linyphiidae from the Uzungwa Mountains, Tanzania (Araneae). *Entomologica Scandinavica Supplements* 36: 1-95.

Scharff N. 1992. The linyphiid spider fauna of eastern Africa (Araneae: Linyphiidae)—distribution patterns, diversity and endemism. *Biological Journal of the Linnean Society* 45: 117-54.

Scharff N. 1993. The Linyphiid spider fauna (Araneae: Linyphiidae) of mountain forests in the Eastern Arc mountains. *In*: Lovett J.C. & Wasser S.K. (eds) *Biogeography and Ecology of the Rain Forests of Eastern Africa:* 115-132. Cambridge University Press.

Seyfulina R.R. & Jocqué R. 2007. An interactive key to the Afrotropical genera of Linyphiidae.

Seyfulina R.R. & Jocqué R. 2009. *Venia kakamega* gen. n., sp. n., a new, canopy-dwelling Afrotropical erigonine spider (Araneae, Linyphiidae). *Journal of Afrotropical Zoology* 5: 3-13.

13

Two new *Thiratoscirtus* species from Gabon (Araneae, Salticidae, Thiratoscirtinae)

Michael SEITER[1] & Wanda WESOŁOWSKA[2]

[1] Group of Arthropod Ecology and Behavior, Division of Plant Protection, Department of Crop Sciences, University of Natural Resources and Life Sciences, Peter Jordan Straße 82, 1190 Vienna, Austria.

[2] Department of Biodiversity and Evolutionary Taxonomy, Wrocław University, Przybyszewskiego 63/77, 51-148 Wrocław, Poland.

Abstract. Two new species of jumping spiders, *Thiratoscirtus oberleuthneri* (♂) and *Th. lamboji* (♀), are described from Gabon, one of the least explored areas of the Afrotropics. Both species live in rainforest, at the forest floor. They are members of a very poorly known subfamily of salticids, the Thiratoscirtinae.

Key words. arachnology, taxonomy, Afrotropical region

Introduction

Thiratoscirtinae is a subfamily of jumping spiders that has been established by Bodner & Maddison (2012) on molecular grounds. It includes a few genera distributed in Western and Central Africa. The majority of the thiratoscirtines are forest-dwellers. They have a high cephalothorax, relatively large anterior median eyes and an eye field that is wider at the first row of eyes than at the posterior row. Both the species affiliations and the phylogenetic relationships within this subfamily are poorly understood and require further studies.

The males of *Thiratoscirtus* Simon, 1886, the type genus of the subfamily, have a characteristic dense tuft of long bristles on the prolateral surface of the palpal tibia. The tarsus of the female pedipalp is armed with 1-3 spines. The diagnostic features of the genus have been summarized in Wesołowska & Russell-Smith (2011) and Wiśniewski & Wesołowska (2013). Currently, 21 species of *Thiratoscirtus* are known (World Spider Catalog 2014). We here provide the descriptions of two new species of *Thiratoscirtus* from Gabon.

Unfortunately, for each of the newly described species only one specimen of one sex was found. Because the features of these species (especially the structure of their genitalia) are unique, their identification will not be problematic in the future. We think, their descriptions are important due to our inadequate knowledge of the Western African salticids (Gabon is practically understudied) and our poor understanding of the thiratoscirtines. Knowledge of the subfamily will probably be essential to understanding the phylogenetic relations among jumping spiders of this part of Africa.

Material and methods

The specimens were collected by direct searching in different parts of Gabon during a field trip in February 2011. For detailed information see the specific section below. All individuals have been stored in ethanol (70% solution). Specimens were studied and measured under a stereo microscope (Nikon SMZ1000), equipped with a camera and an ocular micrometer scale. A male pedipalp and the epigyne were dissected for a more detailed study. The epigyne was macerated in 5% hot KOH for a few minutes, dehydrated with 100% ethanol, cleared in xylene and put in eugenol in temporary mounts. After examination, the genitalia were placed in micro-vials with ethanol and stored in the vials containing the specimens from which they had been removed. Terminology is standard for Araneae. Species were measured as in Metzner (1999), dimensions are given in millimeters. Descriptions of a colour pattern are based on the alcohol-preserved specimens.

The type material is kept in the Natural History Museum of Vienna (NHMW).

Results

Class Arachnida Cuvier, 1812
Order Araneae Clerck, 1757
Family Salticidae Blackwall, 1841
Genus ***Thiratoscirtus*** Simon, 1886

Thiratoscirtus oberleuthneri sp. nov.
urn:lsid:zoobank.org:act:69137100-27F6-42ED-8FD0-65BCAA296BA3
Figs 1–3

Differential diagnosis

The male is distinctive in having a long anterior lobe of the bulb and in the presence of large terminal apophysis. An anterior lobe of the bulb is also present in *T. perspicuus* Wiśniewski & Wesołowska, 2013, but in the latter species it is clearly smaller and the embolus of that species is more delicate than in *T. oberleuthneri* sp. nov. The habitus of the newly described species is similar to that of *T. torquatus* Simon, 1903.

Etymology

The species name is a patronym dedicated to Anton Oberleuthner, who assisted during the field trip in Gabon.

Type material

Holotype

GABON: ♂, close to Midzic, 0°49.781' N, 11°29.395' E, 329 m a.s.l., handcatch, 10–25 Feb. 2011, leg. M. Seiter (NHMW 21938).

Description

Dimensions

Cephalothorax: length 3.3, width 2.6, height 1.8. Abdomen: length 3.4, width 1.8. Eye field: length 1.5, anterior width 2.3, posterior width 2.1.

Male

General appearance as in Fig. 1A–B. Carapace oval, moderately high with gently sloping posterior part, fovea sulciform, clearly visible. Eye field occupies almost half of carapace's length, distance between anterior lateral eyes slightly larger than between posterior laterals. Ocular area dark brown, almost black near eyes, thoracic part light brown anteriorly, darkening to black at margins, striae radiating from fovea. Carapace covered with fine light hairs, denser and longer on anterior part of eye field, a few long brown bristles near anterior row of eyes and laterally from eye field. Clypeus low, brown. Chelicerae large, with two teeth on promargin and single larger tooth on retromargin. Mouth parts brown. Sternum light brown, darker at edges. Abdomen slightly elongated, narrower than carapace, blackish with mosaic of small light dots, along dorsum wide serrated yellowish white band, tinged with brown anteriorly, sigilla visible. Venter dark grey, with four lines formed by light dots. Dorsum of abdomen covered with dense short brown hairs, with admixture of white in anterior part, a few brown bristles on anterior edge. Spinnerets black. Legs generally brown with yellowish transverse stripes: coxae and trochanters pale yellow, femora dark brown with two light rings, patellae light at base but dark distally, tibiae brown with median light ring, metatarsi as patellae, tarsi light. Hairs on legs brown. Spination of leg I: femur 0-1-1 dorsally, patella with one spine on pro- and retrolateral side, tibia 1-1 retrolaterally and 2-1-2 ventrally, metatarsus 1-1 pro- and retrolaterally and 2-2 ventrally. Pedipalp dark brown, clothed in dense long brown hairs, especially on prolateral side of tibia and prolaterally at tip of cymbium (Fig. 1C). Bulb with very long anterior membranous protuberance curtaining embolus (Figs 1C, 2A), long terminal apophysis prolaterally from embolus (Fig. 2A–B), tibial apophysis long (Fig. 2B–D), cymbium narrow (Fig. 2D).

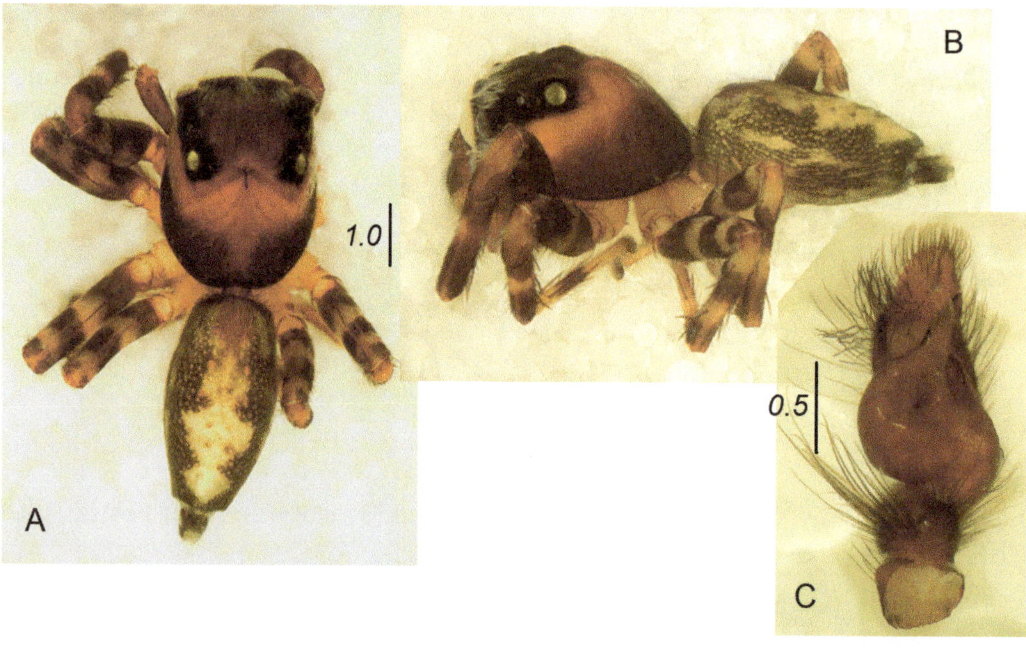

Fig. 1. *Thiratoscirtus oberleuthneri* sp. nov., holotype, male. **A**. General appearance, dorsal view. **B**. General appearance, lateral view. **C**. Palp, ventral view. Scale bars: A–B = 1 mm, C = 0.5 mm.

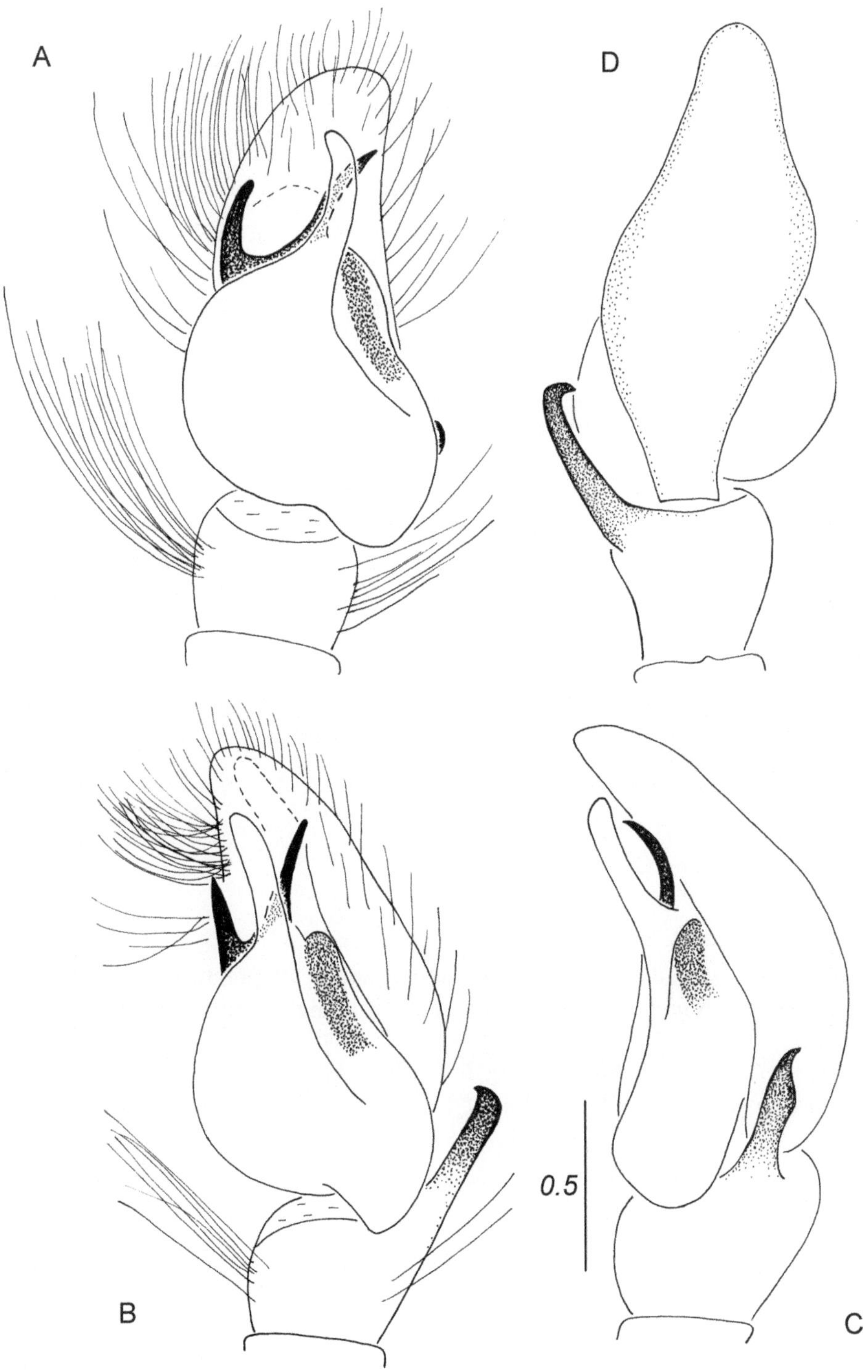

Fig. 2. *Thiratoscirtus oberleuthneri* sp. nov., holotype, palpal organ. **A**. Ventral view. **B**. Ventrolateral view. **C**. Lateral view. **D**. Dorsal view. Scale bar for all figures = 0.5 mm.

Fig. 3. Natural habitat of *Thiratoscirtus oberleuthneri* sp. nov. in Midzic, Gabon (photo by A. Lamboj).

Female

Unknown.

Distribution

This species is known only from the type locality in Gabon.

Habitat

The habitat is typical of a lowland rainforest in Gabon: sunlit, temperature around 33°C and high humidity. The specimen was found on the underside of a leaf in the vicinity of a small river (Fig. 3).

Thiratoscirtus lamboji sp. nov.
urn:lsid:zoobank.org:act:D14E2B1B-CFB0-4CA6-A70A-E96CB92BBB53
Figs 4–6

Differential diagnosis

The female can be recognized by the structure of the epigyne, which is similar to that in *T. atakpa* Wesołowska & Edwards, 2012, but differs by having clearly longer, loop-forming, seminal ducts.

Etymology

The species name is a patronym dedicated to Anton Lamboj, who assisted during the field trip in Gabon and contributed by taking the habitat pictures.

Material examined

Holotype

GABON: ♀, 122 km south of Makokou, 0°17.213' N, 13°14.743' E, 500 m a.s.l., handcatch, 10–25 Feb. 2011, leg. M. Seiter (NHMW 21939).

Description

Dimensions

Cephalothorax: length 2.6, width 1.8, height 1.4. Abdomen: length 2.6, width 2.2. Eye field: length 1.1, anterior width 1.7, posterior width 1.5.

Male

Unknown.

Female

General appearance as in Fig. 4A–B. Carapace oval, high, sloping abruptly behind eye field, fovea sulciform. Distance between anterior lateral eyes slightly larger than between posterior laterals, eye field black, iridescent. Thoracic part greyish brown, foveal area yellowish, yellow streak extending to posterior edge of carapace. Carapace clothed in short light hairs, denser in anterior part of eye field, hairs form small patch between anterior median eyes. Reddish fawn hairs surround anterior eyes from above and form patches between median and lateral anterior eyes. Large red patches laterally from eyes of the last row made by the same hairs. Clypeus moderately high, dark. Chelicerae with two teeth on promargin and single larger tooth on retromargin, dorsal surface of chelicerae blackish. Mouth parts brown, sternum yellow. Abdomen oval, slightly swollen, greyish brown with broad serrated median streak, abdominal sides yellow marked with grey, venter whitish yellow. Dorsal surface covered with

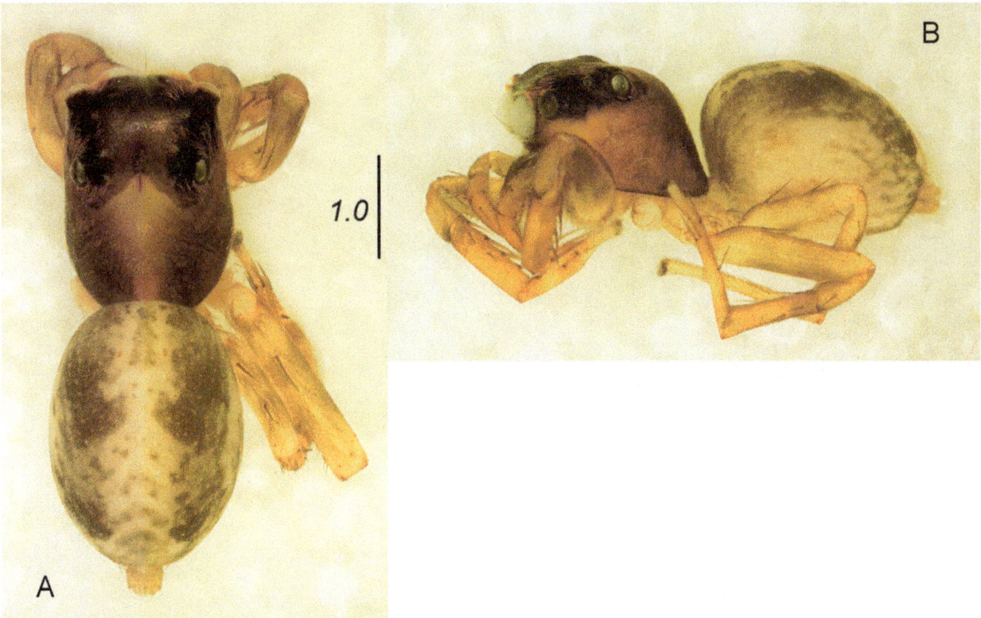

Fig. 4. *Thiratoscirtus lamboji* sp. nov, holotype, female. **A**. General appearance, dorsal view. **B**. General appearance, lateral view. Scale bar for all figures = 1 mm.

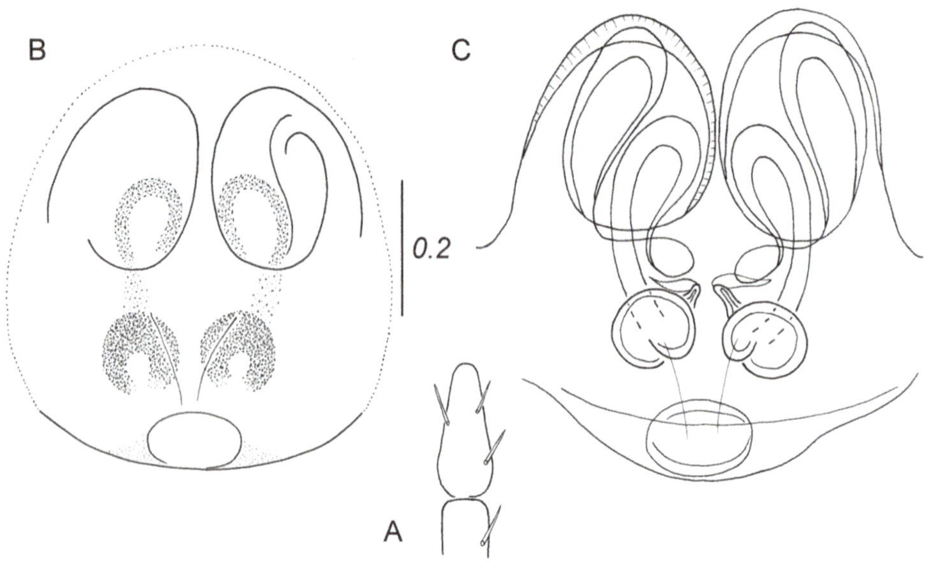

Fig. 5. *Thiratoscirtus lamboji* sp. nov., holotype. **A**. Female palp. **B**. Epigyne. **C**. Internal structure of epigyne. Scale bar for all figures = 0.2 mm.

Fig. 6. Natural habitat of *Thiratoscirtus lamboji* sp. nov., 122 km south of Makokou, Gabon (photo by A. Lamboj).

short brown hairs. Spinnerets light. Legs yellow, first pair slightly darker, femora and tibiae tinged with grey. Spination of leg I: femur with 1-1 dorsally and 0-1 prolaterally, tibia 4 pairs ventrally and 1-1-0 prolaterally, metatarsus 2 pairs ventrally (ventral spines on tibia and metatarsus long). Pedipalps yellow, with single prolateral spine and two retrolateral spines on tarsus (Fig. 5A). Epigyne large with two oval anterior depressions (Fig. 5B), partially plugged with waxy secretion, lips of the depressions strongly sclerotized. Posterior epigynal edge strongly sclerotized, with oval convex process. Seminal ducts long, forming loop, spermathecae small (Fig. 5C).

Distribution

This species is known only from the type locality in Gabon.

Habitat

The specimen was found in a sandy habitat (Fig. 6) in the vicinity of a secondary lowland rainforest. It was exposed to very high temperatures.

Discussion

Gabon belongs to the least explored areas of the Afrotropics, and the salticid fauna of this country is poorly known. Prószyński (2013) lists only 19 species from Gabon, ten of them as species *inquirendae*. These figures rather show the degree of our ignorance, than the biological reality. Taking into account the considerable size of Gabon, its tropical localization and the fact that it remains one of the most extensively forested countries in Africa (Le Houérou 2009), the number of salticid species, that might be found there surely is much higher. The large forested areas of Gabon remain of great importance for the study of the biodiversity. Therefore, any new data on the spiders of this area are valuable.

Our paper contributes to the knowledge of thiratoscirtines of Central Africa. Despite the fact that the presented descriptions of the two species are based only on a single sex, we consider them important for further studies, and hope that data on the other sex will be supplied in the future.

There is no doubt that many more species of *Thiratoscirtus* will be found in Gabon since the molecular study of Bodner & Maddison (2012) mentioned no less than seven undescribed species from the country.

Acknowledgements

The first author is greatly indebted to Anton Lamboj (Austria) and Anton Oberleuthner (Austria) for their assistance in the field and their contribution to jumping spider fauna research. This study would not be possible without the help of Mario Freudenschuss (Austria). We would also like to thank the anonymous reviewers for their help and comments.

References

Bodner M.R. & Maddison W.P. 2012. The biogeography and age of salticid spider radiations (Araneae: Salticidae). *Molecular Phylogenetics and Evolution* 65: 213–240.

Le Houérou H.N. 2009. *Bioclimatology and Biogeography of Africa.* Springer-Verlag, Berlin & Heidelberg.

Metzner H. 1999. Die Springspinnen (Arachnida, Araneae, Salticidae) Griechenlands. *Andrias* 14: 1–279.

Prószyński J. 2013. *Monograph of Salticidae (Araneae) of the World 1995-2013*

Wesołowska W. & Russell-Smith A. 2011. Jumping spiders (Araneae: Salticidae) from Southern Nigeria. *Annales Zoologici* 61 (3): 553–619.

Wiśniewski K. & Wesołowska W. 2013. New *Thiratoscirtus* species from equatorial Africa (Araneae: Salticidae). *Genus* 24 (2): 251–260.

World Spider Catalog. 2014. *World Spider Catalog*. Natural History Museum Bern

Permissions

All chapters in this book were first published in EJT, by Consortium of European natural history institutions; hereby published with permission under the Creative Commons Attribution License or equivalent. Every chapter published in this book has been scrutinized by our experts. Their significance has been extensively debated. The topics covered herein carry significant findings which will fuel the growth of the discipline. They may even be implemented as practical applications or may be referred to as a beginning point for another development.

The contributors of this book come from diverse backgrounds, making this book a truly international effort. This book will bring forth new frontiers with its revolutionizing research information and detailed analysis of the nascent developments around the world.

We would like to thank all the contributing authors for lending their expertise to make the book truly unique. They have played a crucial role in the development of this book. Without their invaluable contributions this book wouldn't have been possible. They have made vital efforts to compile up to date information on the varied aspects of this subject to make this book a valuable addition to the collection of many professionals and students.

This book was conceptualized with the vision of imparting up-to-date information and advanced data in this field. To ensure the same, a matchless editorial board was set up. Every individual on the board went through rigorous rounds of assessment to prove their worth. After which they invested a large part of their time researching and compiling the most relevant data for our readers.

The editorial board has been involved in producing this book since its inception. They have spent rigorous hours researching and exploring the diverse topics which have resulted in the successful publishing of this book. They have passed on their knowledge of decades through this book. To expedite this challenging task, the publisher supported the team at every step. A small team of assistant editors was also appointed to further simplify the editing procedure and attain best results for the readers.

Apart from the editorial board, the designing team has also invested a significant amount of their time in understanding the subject and creating the most relevant covers. They scrutinized every image to scout for the most suitable representation of the subject and create an appropriate cover for the book.

The publishing team has been an ardent support to the editorial, designing and production team. Their endless efforts to recruit the best for this project, has resulted in the accomplishment of this book. They are a veteran in the field of academics and their pool of knowledge is as vast as their experience in printing. Their expertise and guidance has proved useful at every step. Their uncompromising quality standards have made this book an exceptional effort. Their encouragement from time to time has been an inspiration for everyone.

The publisher and the editorial board hope that this book will prove to be a valuable piece of knowledge for researchers, students, practitioners and scholars across the globe.

List of Contributors

Christopher H. DIETRICH
Illinois Natural History Survey, Prairie Research Institute, University of Illinois, 1816 S. Oak St., Champaign, IL 61820

Ana Clara GONÇALVES
Illinois Natural History Survey, Prairie Research Institute, University of Illinois, 1816 S. Oak St., Champaign, IL 61820

Arnaud HENRARD
Section Invertebrates non-insects, Royal Museum for Central Africa, Leuvense Steenweg 13, B-3080 Tervuren, Belgium
Earth and Life Instititute, Biodiversity Research Center, Université Catholique de Louvain, Place Croix du Sud 1-4, B-1348 Louvain la Neuve, Belgium

Rudy JOCQUÉ
Section Invertebrates non-insects, Royal Museum for Central Africa, Leuvense Steenweg 13, B-3080 Tervuren, Belgium

Barbara C. BAEHR
Queensland Museum, PO Box 3300, South Brisbane, QLD 4101, Australia
CSER, School of Environmental and Life Sciences, University of Newcastle, Callaghan, NSW 2308, Australia

Jérôme CONSTANT
Royal Belgian Institute of Natural Sciences, O.D. Phylogeny and Taxonomy, Entomology, Vautier street 29, B-1000 Brussels, Belgium

Alexander FEDOSOV
A.N. Severtsov Institute of Ecology and Evolution, Russian Academy of Sciences, 119071, Leninsky prospect 33, Moscow, Russia

Baldomero M. OLIVERA
Department of Biology, University of Utah, Salt Lake City, UT 84112, USA

Maren WATKINS
Department of Biology, University of Utah, Salt Lake City, UT 84112, USA

Varvara BARKALOVA
Moscow State University, Faculty of Biology, 1-12, Leninskie Gory, Moscow, 119991 Russia

Daniele POLOTOW
Instituto Butantan, Laboratório Especial de Coleções Zoológicas, Av. Vital Brasil, 1500, Butantã, São Paulo, CEP 05503-000, Brazil

Rudy JOCQUÉ
Royal Museum for Central Africa, Tervuren, Belgium

Sven BOSTRÖM
Department of Zoology, Swedish Museum of Natural History, Box 50007, SE-104 05 Stockholm, Sweden

Oleksandr HOLOVACHOV
Department of Zoology, Swedish Museum of Natural History, Box 50007, SE-104 05 Stockholm, Sweden

Joachim BRESSEEL
Royal Belgian Institute of Natural Sciences, O.D. Phylogeny and Taxonomy, Entomology, Vautier street 29, B-1000 Brussels, Belgium

Jérôme CONSTANT
Royal Belgian Institute of Natural Sciences, O.D. Phylogeny and Taxonomy, Entomology, Vautier street 29, B-1000 Brussels, Belgium

Rudy JOCQUÉ
Royal Museum for Central Africa, B-3080 Tervuren, Belgium

Arnaud HENRARD
Royal Museum for Central Africa, B-3080 Tervuren, Belgium
Earth and Life Institute, Biodiversity Research Center, UCL-17.07.04, Bâtiment Carnoy, Croix du Sud, 5, B-1348, Louvain-la-Neuve, Belgium

Kai HELLER
Erwin-Salomon-Str. 25, D-25451 Quickborn, Germany

Heikki HIPPA
Gribbylunds allé 2, SE-183 65 Täby, Sweden

Pekka VILKAMAA
Finnish Museum of Natural History, Zoology Unit, P.O. Box 17, FI-00014 University of Helsinki, Finland

Daniel LEDUC
National Institute of Water and Atmospheric Research, Private Bag 14-901, Wellington, New Zealand

Dominick VERSCHELDE
Ghent University - Zoology Museum, K.L. Ledeganckstraat 35, B-9000 Ghent, Belgium

Robert Mesibov
Queen Victoria Museum and Art Gallery, Launceston, Tasmania, Australia 7250

Benoît NZIGIDAHERA
Institut National pour l'Environnement et la Conservation de la Nature (INECN), B.P. 2757 Bujumbura, Burundi

Rudy JOCQUÉ
Institut National pour l'Environnement et la Conservation de la Nature (INECN), B.P. 2757 Bujumbura, Burundi

Michael Seiter
Group of Arthropod Ecology and Behavior, Division of Plant Protection, Department of Crop Sciences, University of Natural Resources and Life Sciences, Peter Jordan Straße 82, 1190 Vienna, Austria

Wanda Wesołowska
Department of Biodiversity and Evolutionary Taxonomy, Wrocław University, Przybyszewskiego 63/77, 51-148 Wrocław, Poland